Advances in Fermented Foods and Beverages

Advances in Fermented Foods and Beverages

Contributors

G Spano, P Russo et al.

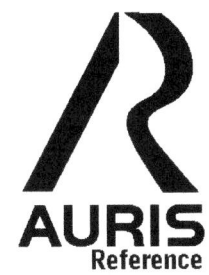

AURIS
Reference

www.aurisreference.com

Advances in Fermented Foods and Beverages

Contributors: G Spano, P Russo et al.

Published by Auris Reference Limited
www.aurisreference.com

United Kingdom

Advances in Fermented Foods and Beverages

ISBN: 978-1-78154-808-0

British Library Cataloguing in Publication Data
A CIP record for this book is available from the British Library

Printed in the United Kingdom

Exclusively distributed by CBS Publishers & Distributors Pvt. Ltd.

Sales & Distribution Rights only for India, Pakistan, Bangladesh, Sri Lanka, Nepal and Bhutan.This book is not to be sold outside these territories.

Contents

vi

List of Abbreviations

AAB	Acetic acid bacteria
ACE	Angiotensin I-converting enzyme
AF	Acceptability factor
AGEs	Advanced glycation end-products
ATP	Adenosine 5-triphosphate
BA	Biogenic amines
CFU	Colony-forming units
CRF	Chronic renal failure
CVD	Cardiovascular disease
DAD	Diode array detector
DAD	Diode array detector
EC	Ethylcarbamate
EFSA	European Food Safety Agency
FOSHU	Foods for Special Health Use
GABA	γ-aminobutyric acid
GC–MS	Gas chromatography mass spectrometry
GK	Goto-Kakizaki
Glu-AGEs	Glucose-derived AGEs
GOLD	GO-lysine dimers
GRAS	Generally Recognized As Safe
HPAEC	High performance anion exchange chromatography
HPLC	High-performance liquid chromatography
HS-SPME	Headspace solid-phase micro extraction
IARC	International Agency for Research on Cancer
IBD	Inflammatory bowel diseases
LAB	Lactic acid bacteria
LLE	Liquid-liquid extraction
LOD	Limit of detection
LOQ	Limit of quantification
LP	Lactoperoxidase
MAOI	Monoamine oxidase inhibitor
MC	Methylcarbamate
MRS	Man-Rogosa-Sharpe
NCR	Nitrogen catabolite repression
NSLAB	Non-starter lactic acid bacteria
OD	Optical density
QPS	Qualified Presumption of Safety'
ROS	Reactive oxygen species
RRC	Rayeb red-carrot juice
RYC	Rayeb-yellow carrot juice
SPE	Solid phase extraction

TCA	Tricarboxylic acid
UC	Ulcerative colitis
UTIs	Urinary tract infections
VRBG	Violet red bile glucose

List of Contributors

G Spano
Università degli Studi di Foggia, Department of Food Science, Foggia, Italy

P Russo
Università degli Studi di Foggia, Department of Food Science, Foggia, Italy

A Lonvaud-Funel
Universite Victor Segalen Bordeaux II, Bordeaux, France

P Lucas
[2]Universite Victor Segalen Bordeaux II, Bordeaux, France

H Alexandre
Institut Universitaire de la Vigne et du Vin, Jules Guyot, Universite de Bourgogne, Dijon, France

C Grandvalet
Institut Universitaire de la Vigne et du Vin, Jules Guyot, Universite de Bourgogne, Dijon, France

E Coton
ADRIA Normandie, Food Safety and Technology Department, Villers-Bocage, France

M Coton
ADRIA Normandie, Food Safety and Technology Department, Villers-Bocage, France

L Barnavon
Inter Rhône, Inter Rhòne, Technical Department, Orange, France

B Bach
Inter Rhône, Inter Rhòne, Technical Department, Orange, France

F Rattray
Chr. Hansen A/S, Hoersholm, Denmark

A Bunte
Chr. Hansen A/S, Hoersholm, Denmark

C Magni
Instituto de Biologia Molecular y Celular de Rosario, FCByF, University of Rosario, Rosario, Santa Fe, Argentina

V Ladero
IPLA-CSIC, Villaviciosa, Asturias, Spain

M Alvarez
IPLA-CSIC, Villaviciosa, Asturias, Spain

M Fernández
IPLA-CSIC, Villaviciosa, Asturias, Spain

P Lopez
Centro de Investigaciones Biológicas (CIB), Department of Molecular Microbiology and Infection Biology, Madrid, Spain

P F de Palencia
Centro de Investigaciones Biológicas (CIB), Department of Molecular Microbiology and Infection Biology, Madrid, Spain

A Corbi
Centro de Investigaciones Biológicas (CIB), Department of Molecular Microbiology and Infection Biology, Madrid, Spain

H Trip
Groningen Biomolecular Sciences and Biotechnology Institute, University of Groningen, Department of Molecular Microbiology, Groningen, The Netherlands

J S Lolkema
Groningen Biomolecular Sciences and Biotechnology Institute, University of Groningen, Department of Molecular Microbiology, Groningen, The Netherlands

Emiliane Andrade Araújo
Universidade Federal de Viçosa, Campus Rio Paranaíba, Rio Paranaíba, MG, Brazil

Ana Clarissa dos Santos Pires
Departamento de Tecnologia de Alimentos, Universidade Federal de Viçosa, Viçosa, MG, Brazil

Antônio Fernandes de Carvalho
Departamento de Tecnologia de Alimentos, Universidade Federal de Viçosa, Viçosa, MG, Brazil

Maximiliano Soares Pinto
Instituto de Ciências Agrárias, Universidade Federal de Minas Gerais, Montes Claros, MG, Brazil

Gwénaël Jan
INRA, UMR1253 Science et Technologie du Lait et de l'Œuf, Rennes, France
Masayoshi Takeuchi
Department of Advanced Medicine, Medical Research Institute, Kanazawa Medical University, Uchinadamachi, Ishikawa, Japan,

Jun-ichi Takino
Laboratory of Biochemistry, Faculty of Pharmaceutical Sciences, Hiroshima International University, Kure, Hiroshima, Japan,

Satomi Furuno
Department of Pathophysiological Science, Faculty of Pharmaceutical Science, Hokuriku University, Kanazawa, Ishikawa, Japan,

Hikari Shirai
Department of Pathophysiological Science, Faculty of Pharmaceutical Science, Hokuriku University, Kanazawa, Ishikawa, Japan,

Mihoko Kawakami
Department of Pathophysiological Science, Faculty of Pharmaceutical Science, Hokuriku University, Kanazawa, Ishikawa, Japan,

Michiru Muramatsu
Department of Pathophysiological Science, Faculty of Pharmaceutical Science, Hokuriku University, Kanazawa, Ishikawa, Japan,

Yuka Kobayashi
Department of Pathophysiological Science, Faculty of Pharmaceutical Science, Hokuriku University, Kanazawa, Ishikawa, Japan,

Sho-ichi Yamagishi
Department of Pathophysiology and Therapeutics of Diabetic Vascular Complications, Kurume University School of Medicine, Kurume, Fukuoka, Japan

Tendekayi H. Gadaga
Department of Environmental Health, University of Swaziland,

Molupe Lehohla
Department of Pharmacy, National University of Lesotho

VictorNtuli
Department of Biology, National University of Lesotho

Fábio Faria-Oliveira
Laboratório de Biologia Celular e Molecular, Núcleo de Pesquisas em Ciências Biológicas, Universidade Federal de Ouro Preto, Brazil

Raphael H.S. Diniz
Laboratório de Biologia Celular e Molecular, Núcleo de Pesquisas em Ciências Biológicas, Universidade Federal de Ouro Preto, Brazil

Fernanda Godoy-Santos
Laboratório de Biologia Celular e Molecular, Núcleo de Pesquisas em Ciências Biológicas, Universidade Federal de Ouro Preto, Brazil

Fernanda B. Piló
Laboratório de Biologia Celular e Molecular, Núcleo de Pesquisas em Ciências Biológicas, Universidade Federal de Ouro Preto, Brazil

Hygor Mezadri
Laboratório de Biologia Celular e Molecular, Núcleo de Pesquisas em Ciências Biológicas, Universidade Federal de Ouro Preto, Brazil

Ieso M. Castro
Laboratório de Biologia Celular e Molecular, Núcleo de Pesquisas em Ciências Biológicas, Universidade Federal de Ouro Preto, Brazil

Rogelio L. Brandão

Laboratório de Biologia Celular e Molecular, Núcleo de Pesquisas em Ciências Biológicas, Universidade Federal de Ouro Preto, Brazil

Anne Pihlanto
MTT Agrifood Finland, Biotechnology and Food Research, Jokioinen, Finland

Ho-Sang Shin
Department of Environmental Education, Kongju 314-701, Republic of Korea.

Eun-Young Yang
Department of Environmental Science, Kongju National University, Kongju 314-701, Republic of Korea.

Ingrid Torres-Rodríguez
Departamento de Ingeniería Celular y Biocatálisis, Instituto de Biotecnología, Universidad Nacional Autónoma de México (UNAM)

María Elena Rodríguez-Alegría
Departamento de Ingeniería Celular y Biocatálisis, Instituto de Biotecnología, Universidad Nacional Autónoma de México (UNAM)

Alfonso Miranda-Molina
Departamento de Ingeniería Celular y Biocatálisis, Instituto de Biotecnología, Universidad Nacional Autónoma de México (UNAM)

Martha Giles-Gómez
Departamento de Ingeniería Celular y Biocatálisis, Instituto de Biotecnología, Universidad Nacional Autónoma de México (UNAM)

Rodrigo Conca Morales
Departamento de Ingeniería Celular y Biocatálisis, Instituto de Biotecnología, Universidad Nacional Autónoma de México (UNAM)

Agustín López-Munguía
Departamento de Ingeniería Celular y Biocatálisis, Instituto de Biotecnología, Universidad Nacional Autónoma de México (UNAM)

Francisco Bolívar
Departamento de Ingeniería Celular y Biocatálisis, Instituto de Biotecnología, Universidad Nacional Autónoma de México (UNAM)

Adelfo Escalante
Departamento de Ingeniería Celular y Biocatálisis, Instituto de Biotecnología, Universidad Nacional Autónoma de México (UNAM)

Hideki Okada
Department of Food and Nutrition Sciences, Graduate School of Dairy Science Research, Rakuno Gakuen University

Eri Fukushi
Department of Food and Nutrition Sciences, Graduate School of Dairy Science Research, Rakuno Gakuen University

Akira Yamamori
Department of Food and Nutrition Sciences, Graduate School of Dairy Science Research, Rakuno Gakuen University

Naoki Kawazoe
Department of Food and Nutrition Sciences, Graduate School of Dairy Science Research, Rakuno Gakuen University

Shuichi Onodera
Department of Food and Nutrition Sciences, Graduate School of Dairy Science Research, Rakuno Gakuen University

Jun Kawabata
Department of Food and Nutrition Sciences, Graduate School of Dairy Science Research, Rakuno Gakuen University

Norio Shiomi
Department of Food and Nutrition Sciences, Graduate School of Dairy Science Research, Rakuno Gakuen University

Amany E. El-Abasy
Food Science and Technology Department, Faculty of Agriculture, Alexandria University, Alexandria, Egypt.

Hany A. Abou-Gharbia
Food Science and Technology Department, Faculty of Agriculture, Alexandria University, Alexandria, Egypt.

Hamida M. Mousa
Food Science and Technology Department, Faculty of Agriculture, Alexandria University, Alexandria, Egypt.

Mohammed M. Youssef
Food Science and Technology Department, Faculty of Agriculture, Alexandria University, Alexandria, Egypt.

Freek Spitaels
Laboratory of Microbiology, Faculty of Sciences, Ghent University, Ghent, Belgium,

Anneleen D. Wieme
Laboratory of Microbiology, Faculty of Sciences, Ghent University, Ghent, Belgium, Laboratory of Biochemistry and Brewing, Faculty of Bioscience Engineering, Ghent University, Ghent, Belgium,

Maarten Janssens
Research Group of Industrial Microbiology and Food Biotechnology (IMDO), Faculty of Sciences and Bioengineering Sciences, Vrije Universiteit Brussel, Brussels, Belgium,

Maarten Aerts
Laboratory of Microbiology, Faculty of Sciences, Ghent University, Ghent, Belgium,

Heide-Marie Daniel
Mycothe`que de l'Universite´ catholique de Louvain (MUCL), Belgian Coordinated Collection of Microorganisms (BCCM), Earth and Life Institute, Applied Microbiology, Mycology, Universite´ catholique de Louvain, Louvain-la-Neuve, Belgium

Anita Van Landschoot
Laboratory of Biochemistry and Brewing, Faculty of Bioscience Engineering, Ghent University, Ghent, Belgium,

Luc De Vuyst
Research Group of Industrial Microbiology and Food Biotechnology (IMDO), Faculty of Sciences and Bioengineering Sciences, Vrije Universiteit Brussel, Brussels, Belgium,

Peter Vandamme
Laboratory of Microbiology, Faculty of Sciences, Ghent University, Ghent, Belgium,

Albert Mas
Facultad de Enología, Universitat Rovira i Virgili, Marcel·lí Domingo s/n, 43003 Tarragona, Spain

Jose Manuel Guillamon
Facultad de Enología, Universitat Rovira i Virgili, Marcel·lí Domingo s/n, 43003 Tarragona, Spain
Departamento de Biotecnologia de Alimentos, Instituto de Agroquímica y Tecnología de los Alimentos (CSIC), Agustín Escardino, 7, 46980 Valencia, Spain

Maria Jesus Torija
Facultad de Enología, Universitat Rovira i Virgili, Marcel·lí Domingo s/n, 43003 Tarragona, Spain

Gemma Beltran
Facultad de Enología, Universitat Rovira i Virgili, Marcel·lí Domingo s/n, 43003 Tarragona, Spain

Ana B. Cerezo
Facultad de Farmacia, Universidad de Sevilla, Profesor García González, 2, 41012 Sevilla, Spain

Ana M. Troncoso
Facultad de Farmacia, Universidad de Sevilla, Profesor García González, 2, 41012 Sevilla, Spain

M. Carmen Garcia-Parrilla
Facultad de Farmacia, Universidad de Sevilla, Profesor García González, 2, 41012 Sevilla, Spain

Isabela Ferrari Pereira Lima
Bioprocess Engineering and Biotechnology Division, Chemical Engineering Department, Federal University of Paraná, Curitiba, PR 81531-991, Brazil

Juliano De Dea Lindner
Bioprocess Engineering and Biotechnology Division, Chemical Engineering Department, Federal University of Paraná, Curitiba, PR 81531-991, Brazil
Research and Development Department, Incorpore Foods, Camboriú, SC 88340-000, Brazil

Vanete Thomaz Soccol
Post-Graduation Program, Positivo University, Curitiba, PR 81280-330, Brazil

José Luiz Parada
Post-Graduation Program, Positivo University, Curitiba, PR 81280-330, Brazil

Carlos Ricardo Soccol
Bioprocess Engineering and Biotechnology Division, Chemical Engineering Department, Federal University of Paraná, Curitiba, PR 81531-991, Brazil

Dirk W. Lachenmeier
Chemisches und Veterinäruntersuchungsamt (CVUA) Karlsruhe, Weissenburger Strasse 3, 76187 Karlsruhe, Germany

Fotis Kanteres
Centre for Addiction and Mental Health (CAMH), 33 Russell Street, Toronto, ON, M5S 2S1, Canada

Thomas Kuballa
Chemisches und Veterinäruntersuchungsamt (CVUA) Karlsruhe, Weissenburger Strasse 3, 76187 Karlsruhe, Germany

Mercedes G. López
Unidad de Biotecnología e Ingeniería Genética de Plantas, Centro de Investigación y Estudios Avanzados del IPN, 36500 Irapuato, Gto., Mexico

Jürgen Rehm
Centre for Addiction and Mental Health (CAMH), 33 Russell Street, Toronto, ON, M5S 2S1, Canada

Dalla Lana School of Public Health, University of Toronto, 55 College Street, To-
ronto, ON, M5T3M7, Canada
Institute for Clinical Psychology and Psychotherapy, TU Dresden, Chemnitzer
Strasse 46, 01187 Dresden, Germany

Preface

Fermentation in food processing is the conversion of carbohydrates to alcohols and carbon dioxide or organic acids using yeasts, bacteria, or a combination thereof, under anaerobic conditions. Fermentation usually implies that the action of microorganisms is desirable. Fermentation is one of the oldest forms of food preservation in the world. This book, Advances in Fermented Foods and Beverages, reviews the health benefits of fermented foods and beverages, the microbiology of fermentation, and key aspects of fermented food production. In first chapter, we describe the physiological role and toxic effects of BA, their presence in fermented food products, their production by microorganisms, environmental conditions, some methods available for detecting the presence of BA or BA-producing microorganisms and, finally, methods to reduce BA content in fermented food. In second chapter, we discuss the application of probiotic microorganisms in fermented dairy products, particularly cheeses. In addition, we also discuss the benefits of probiotic fermented foods on human health. Third chapter indicates that some lactic acid bacteria beverages, carbonated drinks, sugar-sweetened fruit drinks, sports drinks, mixed fruit juices, confectionery (snacks), dried fruits, cakes, cereals, and prepared foods contain markedly higher Glu-AGE levels than other classes of beverages and foods. Fourth chapter describes the traditional methods of preparing fermented foods and beverages of Lesotho. Fifth chapter aims to contribute to a comprehensible analysis of the role of yeast and LAB on the production of fermented beverages from South America. The microbiological diversity associated with the fermentation of a wide diversity of raw materials, from sugarcane to cassava, as well as new potential biotechnological applications will be addressed. Sixth chapter reviews on liberation during fermentation, of bioactive peptides with properties relevant to cardiovascular health including the effects on blood pressure and oxidative stress. The focus is mainly to those peptides with in vivo blood pressure lowering effects. Moreover, bioavailability of peptides and aspects of necessary further information is given. Screening and characterization of extracellular polysaccharides produced by leuconostoc kimchii isolated from traditional fermented pulque beverage has been described in eighth chapter and ninth chapter presents structural analysis of three novel trisaccharides isolated from the fermented beverage of plant extracts. Tenth chapter aims to formulate and evaluate yoghurt and Rayeb mixes with red and yellow carrot juices. Eleventh chapter presents a study determined the microbiota involved in the fermentation of lambic beers by sampling two fermentation batches during two years in the most traditional lambic brewery of Belgium, using culture-dependent and culture-independent methods. The aim of thirteenth chapter was to develop an innovative, non-dairy, functional, probiotic, fermented beverage using herbal mate extract as a natural ingredient which would also be hypocholesterolemic and hepatoprotective.

Chapter 1

BIOGENIC AMINES IN FERMENTED FOODS

G Spano[1], P Russo[1], A Lonvaud-Funel[2], P Lucas[2], H Alexandre[3], C Grandvalet[3], E Coton[4], M Coton[4], L Barnavon[5], B Bach[5], F Rattray[6], A Bunte[6], C Magni[7], V Ladero[8], M Alvarez[8], M Fernández[8], P Lopez[9], P F de Palencia[9], A Corbi[9], H Trip[10] and J S Lolkema[10]

[1]Università degli Studi di Foggia, Department of Food Science, Foggia, Italy

[2]Universite Victor Segalen Bordeaux II, Bordeaux, France

[3]Institut Universitaire de la Vigne et du Vin, Jules Guyot, Universite de Bourgogne, Dijon, France

[4]ADRIA Normandie, Food Safety and Technology Department, Villers-Bocage, France

[5]Inter Rhône, Inter Rhòne, Technical Department, Orange, France

[6]Chr. Hansen A/S, Hoersholm, Denmark

[7]Instituto de Biologia Molecular y Celular de Rosario, FCByF, University of Rosario, Rosario, Santa Fe, Argentina

[8]IPLA-CSIC, Villaviciosa, Asturias, Spain

[9]Centro de Investigaciones Biológicas (CIB), Department of Molecular Microbiology and Infection Biology, Madrid, Spain

[10]Groningen Biomolecular Sciences and Biotechnology Institute, University of Groningen, Department of Molecular Microbiology, Groningen, The Netherlands

ABSTRACT

Food-fermenting lactic acid bacteria (LAB) are generally considered to be non-toxic and non-pathogenic. Some species of LAB, however, can produce biogenic amines (BAs). BAs are organic, basic, nitrogenous compounds, mainly formed through decarboxylation of amino acids. BAs are present in a wide range of foods, including dairy products, and can occasionally accumulate in high concentrations. The consumption of food containing large amounts of these amines can have toxicological consequences. Although there is no specific legislation regarding BA content in many fermented products, it is generally assumed that they should not be allowed to accumulate. The ability of microorganisms to decarboxylate amino acids is highly variable, often being strain specific, and therefore the detection of bacteria possessing amino acid decarboxylase activity is important to estimate the likelihood that foods contain BA and to prevent their accumulation in food products. Moreover, improved knowledge of the factors involved in the synthesis and accumulation of BA should lead to a reduction in their incidence in foods.

INTRODUCTION

Lactic acid bacteria (LAB) can produce metabolic energy and/or increase their acid resistance by using catabolic pathways that convert amino acids into amine-containing compounds referred to as biogenic amines (BA) (Griswold et al., 2006). Usually, the consumption of foods containing large amounts of these amines can have toxicological consequences (Shalaby, 1996). These problems are more severe in consumers with less efficient detoxification systems because of their genetic constitution or their medical treatments (Bodmer et al., 1999). Foods likely to contain high levels of BA include fish, fish products and fermented foodstuffs (for example, meat, dairy products and vegetables) and beverages (for example, wine, cider and beer). The most important BAs—both qualitatively and quantitatively—in foods and beverages are histamine, tyramine, putrescine, cadaverine and β-phenylethylamine, products of the decarboxylation of histidine, tyrosine, ornithine, lysine and β-phenylalanine, respectively. According to their chemical structure, they can be classified as aliphatic (putrescine, cadaverine, spermine and spermidine), aromatic (tyramine and phenylethylamine) or heterocyclic (histamine and tryptamine). According to their number of amine groups, they can be divided into monoamines (tyramine and phenylethylamine), diamines (putrescine and cadaverine) or polyamines (spermine and spermidine). Recently, the genes of diverse pathways producing BA were identified in LAB (for a review, see Linares et al., 2010). Interestingly, the pathways seem to be strain dependent rather than species specific, suggesting that horizontal gene transfer may account for their dissemination in LAB (Lucas et al., 2005; Marcobal et al., 2006a; Coton and Coton, 2009). In addition, the enzymes of pathways involved in BA production can be encoded by unstable plasmids (Lucas et al., 2005; Satomiet al., 2008) and only strains harbouring BA-related plasmids are able to produce BA (Lucas et al., 2005).

The presence of BA in foods has traditionally been used as an indicator of undesired microbial activity. Relatively high levels of certain BAs have also been reported to indicate the deterioration of food products and/or their defective manufacture. Their toxicity has led to the general agreement that they should not be allowed to accumulate in food.

BA accumulation in foods requires the availability of precursors (that is, amino acids), the presence of microorganisms with amino acid decarboxylases and favourable conditions for their growth and decarboxylating activity (Fernández et al., 2007; Arena et al., 2008). Therefore, BA production by LAB may be controlled at various levels during food fermentation, including food fermentation practices and factors involved in food fermentation processes.

In this article we describe the physiological role and toxic effects of BA, their presence in fermented food products, their production by microorganisms, environmental conditions, some methods available for detecting the presence of BA or BA-producing microorganisms and, finally, methods to reduce BA content in fermented food.

PHYSIOLOGICAL ROLE AND TOXICOLOGICAL EFFECTS OF BAS

In eukaryotic cells, BA biosynthesis is essential, as these compounds function as precursors for the synthesis of hormones, alkaloids, nucleic acids and proteins (Premont et al., 2001). Some BAs have an important role as neurotransmitters, whereas others, such as putrescine and spermidine, are needed for critical biological functions (Igarashi et al., 2001). In prokaryotic cells, the physiological role of BA synthesis mainly seems to be related to defence mechanisms used by bacteria to withstand acidic environments (Rhee et al., 2002; Lee et al., 2007). Decarboxylation increases survival under acidic stress conditions (Rhee et al., 2002) through the consumption of protons and the excretion of amines and CO_2, helping to restore the internal pH (van de Guchte et al., 2002). BA production may also offer a way of obtaining energy: electrogenic amino acid/amine antiport can lead to generation of proton motive force (Molenaar et al., 1993). This function would be particularly important for microorganisms lacking a respiratory chain for generating high yields of adenotriphosphate (Vido et al., 2004).

Some studies suggest new and interesting hypotheses on the physiological role of amines in microorganisms (Tkachenko et al., 2001). In Escherichia coli, the expression of oxyR, the gene that protects E. coli against oxidative stress, was enhanced by physiological concentrations of the BA putrescine. Moreover, putrescine was shown to produce a protective effect if the DNA is damaged by reactive oxygen species (Tkachenko et al., 2001). Cells of E. coli grown in M9 minimal medium and subjected to a hyperosmotic shock by addition of 0.5 M NaCl immediately started to excrete putrescine, suggesting that putrescine may be involved in osmotic stress tolerance in E. coli. Therefore, bacteria that possess amino acid decarboxylase activity could overcome or reduce the effects of factors that induce stress responses in the cell, such as oxygen and NaCl, with the production of BA.

Although BA is required for many critical biological functions, the consumption of foods containing large amounts of BA can have toxicological consequences. After food consumption, small quantities of BA are commonly metabolised in the human gut to physiologically less active forms through the action of amine oxidases (monoamine oxidases (MAOs) and diamine oxidase). Histamine can also be detoxified by methylation (through the action of methyl transferases) or acetylation (Lehane and Olley, 2000). However, the intake of foods with high BA loads, or inadequate detoxification, either for genetic reasons (Caston *et al.*, 2002) or because of the inhibitory effects of some medicines or alcohol (Bodmer*et al.*, 1999), can lead to BA entering the systemic circulation and causing the release of adrenaline and noradrenaline, provoking gastric acid secretion, increased cardiac output, migraine, tachycardia, increased blood sugar levels and higher blood pressure (Shalaby, 1996). BA levels are also higher in patients with Parkinson's disease, schizophrenia and depression (Premont *et al.*, 2001).

The establishment of what constitutes a toxic level of BA is difficult, as this depends on the characteristics of different individuals. Human sensitivity varies according to the individual detoxifying activities of some enzymes involved in BA metabolism, such as histamine methyltransferase or others less specific, such as MAO and diamine oxidase. These enzymes are inhibited by several types of drugs, such as the neuromuscular blocking drugs d-tubocurarine, pancuronium and alcuronium, and ethanol (Sattler *et al.*, 1985) or antidepressant drugs (Livingston and Livingston, 1996). As a consequence of this synergistic action, the simultaneous consumption of fermented foods and beverages may cause disorders, including life-threatening serotonin syndrome, even if each separate product might not be considered as hazardous (Lonvaud-Funel, 2001). Because of the wide range of possible monoamine oxidase inhibitor (MAOI) drug and tyramine-rich food interactions, the use of MAOIs has been limited, despite their clinical benefits (Livingston & Livingston, 1996). This risk has also prompted clinicians to propose the so-called 'MAOI diet', in which the tyramine intake is controlled by restricting known tyramine-rich food stuffs corresponding mainly to fermented products (aged cheese; aged or cured meats; sauerkraut; soy sauce and tap beer) (Gardner *et al.*, 1996). Secondary amines, such as putrescine and cadaverine, can also react with nitrite to form carcinogenic nitrosamines (ten Brink *et al.*, 1990), and the adherence to intestinal mucosa of some enteropathogens, such as *E. coli* O157:H7, is increased in the presence of tyramine (Lyte, 2004). It has been suggested that BAs have been the causative agents behind a number of food poisoning episodes, the most notorious being caused by histamine. Histamine poisoning is also known as 'scombroid poisoning' owing to the association of this illness with the consumption of scombroid fish (Taylor, 1983). With respect to cheese, BA food poisoning can be caused by high levels of tyramine, especially in combination with the use of MAOIs as antidepressants. This effect is known as the 'cheese reaction' (Silla Santos, 1996).

There is little specific legislation with regard to BA content in foods. Although for fish products there are clear limits for histamine (Commission Regulation (EC) 2073/2005), upper limits for BA in other foods have only been recommended or suggested (for example, 100 mg of histamine per kg of food or 2 mg of histamine per litre of alcoholic beverage). Generally, in alcoholic beverages, the toxic dose is considered to be between 8 and 20 mg/l for histamine, 25 and 40 mg/l for tyramine, whereas as little as 3 mg/l of phenylethylamine can cause negative physiological effects (Soufleros et al., 1998).

In addition to toxicological effects, BA in wine can also have consequences for wine retailers trying to export wines, as some countries have established maximum limits for histamine content in wine (Martín-Álvarez et al., 2006).

BAS IN WINE AND DAIRY PRODUCTS

In wine, more than 20 amines have been identified and their total concentration has been reported to range from a few mg/l to about 50 mg/l, depending on the quality of the wine (Lonvaud-Funel, 2001; Landete et al., 2005). Similar BA contents have also been described in ciders (Garai et al., 2006, 2007). The variability of the amine contents in wine could be explained on the basis of differences in the wine-making process, time and storage conditions, raw material quality and possible microbial contamination during winery operations (Lonvaud-Funel, 2001). BA in wine may have two different sources: raw materials and fermentation processes. Some amines are already found in grapes, namely, histamine and tyramine, as well as several volatile amines and polyamines. Histamine, tyramine and putrescine are the BAs found in higher concentrations in wine, but cadaverine, phenylethylamine and isoamylamine are also present in smaller amounts. Putrescine and cadaverine are normally associated with poor sanitary conditions of grapes (Leitão et al., 2005).

Extensive research has been conducted to correlate BA production in wine with species of LAB involved in the wine-making process. It is widely known that*Pediococcus, Lactobacillus, Leuconostoc* and *Oenococcus* spp. are implicated in BA production in wine (Landete et al., 2007). Different strains of *Lactobacillus hilgardii, Lactobacillus buchneri, Lactobacillus brevis* and

*Lactobacillus mali*produce a variety of BA in wine (Moreno-Arribas and Lonvaud-Funel, 1999;Moreno-Arribas *et al.*, 2000, 2003; Martín *et al.*, 2005; Constantini *et al.*, 2006;Landete *et al.*, 2007). *Leuconostoc mesenteroides* has a high potential to produce tyramine or histamine in wine (Moreno-Arribas *et al.*, 2003; Landete *et al.*, 2007). *Oenococcus oeni* is able to significantly contribute to the overall BA content of wines, mainly producing histamine. The ability of *O. oeni* to produce BA varies among strains (Coton *et al.*, 1998; Guerrini *et al.*, 2002). Moreover, LAB strains have the ability to simultaneously produce different amines (Coton *et al.*, 1998; Moreno-Arribas *et al.*, 2000; Guerrini *et al.*, 2002), suggesting that some strains might possess more than one amino acid decarboxylase activity under specific culture conditions.

Together with wine, dairy products (in particular cheese) can accumulate high levels of BA. In the raw material (milk), polyamines are the most abundant. However, in the final product, tyramine, histamine, putrescine, cadaverine and, at lower concentrations, β-phenylethylamine and tryptamine, are all detected. The BA content of different types of cheese varies; indeed, it can also vary within the same type of cheese and even between different sections of the same cheese (Novella-Rodríguez *et al.*, 2003a).

The main BA producers in cheese are Gram-positive bacteria, with LAB being the main histamine and tyramine producers. The genera *Enterococcus*, *Lactobacillus,Leuconostoc* and *Streptococcus* include some strains that have been described as BA producers. These can be present in milk microbiota or introduced through contamination before, during or after the processing of dairy products. BA$^+$-LAB may even form part of the starters or adjunct cultures. Several authors have reported the presence of tyrosine and histamine decarboxylase activity in strains from various starter cultures (Linares *et al.*, 2010). It is therefore important to include the inability to produce BA as an indispensable condition of strains intended to be used as starters.

Regarding the safety of starters, the European Food Safety Agency (EFSA) has recently introduced a system for a pre-market safety assessment of selected taxonomic groups of microorganisms leading to a 'Qualified Presumption of Safety' (QPS) European equivalent of the Generally Recognized As Safe (GRAS) status (EFSA, 2007). *Lactobacillus* associated with food, including *L. buchneri*, *L. brevis*, *L. hilgardii*, have obtained a QPS status (EFSA, 2007), although some strains of these species have been described as BA producers (Lucas *et al.*, 2005; Martín *et al.*, 2005; Coton and Coton, 2009). This could raise the question of the addition of 'absence of BA production and BA production-associated genes' as qualification criteria in the QPS context. In addition, because of a recent increase in BA content in fermented food, as recently reported at the EFSA Network meeting on Microbiological Risk

Assessment held in early June 2009, risks associated with BA in foods have been discussed at the 51st plenary meeting of the scientific panel on biological hazards held in Parma (Italy) from 9 to 10 September 2009.

DETECTION OF BAS IN FERMENTED FOOD

Early detection of BA-producing bacteria is essential in the food industry to avoid the risk of amine formation. Several methods to detect the production of BA through microorganisms have been developed, from simple methods such as paper chromatography or spectrofluorimetric determination to more sophisticated techniques such as automated systems for the detection of microbial metabolic activities or automated conductance measurements (Marcobal et al., 2006b; Önal, 2007; Linares et al., 2010). With respect to the detection of BA-producing microorganisms, screening methods were initially based on the use of differential media containing a pH indicator to identify BA-producing strains (Maijala, 1993; Bover-Cid and Holzapfel, 1999). However, several studies describing the loss of ability to produce BA in LAB after prolonged storage or cultivation of isolated strains in synthetic media have been reported. For instance, the instability of histidine decarboxylase cells of *L. hilgardii* is easily explained by the loss of histidine decarboxylase plasmid, which depends greatly on bacterial culture conditions (Lucas et al., 2005).

The improvement of fast, reliable and culture-independent molecular tools, usually based on PCR approaches, has recently allowed a fast and accurate detection of BA-producing bacteria in fermented beverages. In fact, using several target genes, it has become possible to identify and/or quantify all of the LAB involved in BA production in a given sample (Lucas and Lonvaud-Funel, 2002; Coton and Coton, 2005; Marcobal et al., 2006b; Ladero et al., 2008; Nannelli et al., 2008). A relationship between the presence of the gene encoding the decarboxylase and the capacity to synthesise BA has been reported by several authors (Fernández et al., 2004; Landete et al., 2005; Lucas et al., 2005). PCR has successfully been used with milk curd and cheese samples in this respect (Fernández et al., 2004), as well as for the detection of tyramine-producing bacteria during cheese manufacture (Fernández et al., 2006) or wine fermentation (Nannelli et al., 2008). A multiplex PCR method for the simultaneous detection of histamine-, tyramine- and putrescine-producing LAB has recently been proposed in order to identify BA-producing strains in wine and cider (Marcobal et al., 2005). In addition to end-point PCR analysis, a real-time quantitative PCR has been developed for detecting histamine-producing LAB in cheese and wine (Fernández et al., 2006; Lucas et al., 2008) or successfully used in the different steps of cheese manufacture and wine fermentation (Ladero et al., 2008; Nannelli et al., 2008).

METHODS TO REDUCE BAS CONTENT IN FERMENTED FOOD

Many LAB strains are used as starter cultures in several fermented foods and beverages. In general, the choice of starter cultures is fundamental to guarantee the quality of the final products. For this reason, the inability to form BA should be an important criterion in the selection of starter cultures for the production of fermented food and beverages. Inoculation with starter cultures that are unable to produce BA is a viable option for the control of these compounds in wine (Martín-Álvarez *et al.*, 2006). It seems that co-inoculation of *O. oeni* starter cultures, together with alcoholic fermentation, has the potential to curb BA formation even more than conventional inoculation for malolactic fermentation after the completion of alcoholic fermentation. BAs may also be oxidised by the action of amino oxidase. The potential role of microorganisms involved in food fermentations with amino oxidase activity has been investigated with the aim of preventing or reducing the accumulation of BA in foods (Leuschner *et al.*, 1998). Unfortunately, at this stage, amine degradation seems to be restricted to aerobic microorganisms that are of limited use in fermented foods such as wine, which characteristically constitutes an anaerobic environment.

In addition to amino oxidase activity and the use of microbial starters unable to produce BA, several food-processing parameters may affect the final content of BA in food. For example, the relationship between the treatment of milk and BA content has been studied by several authors. The highest degradation rate of histamine is usually observed at 37 °C, although degradation is still considerable at 22 and 15 °C (Dapkevicius *et al.*, 2000). The analysis of the BA content of different types of cheese showed that these compounds were more common in those made from raw milk (Fernández *et al.*, 2007). Other milk treatments, such as the use of pressure, have also been investigated as a means of reducing BA contents. However, no differences in BA profiles were observed between cheese made with sterilised or pasteurised milk, suggesting that the control of BA-producing microorganisms by adequate treatment of milk is one of the most important factors for reducing BA accumulation in dairy products (Novella-Rodríguez *et al.*, 2003b). In the case of wine, the agricultural and oenological practices may control the accumulation of BA. Indeed, viticulture region and grape varieties seem to influence the amounts of BA, as wines of some regions present higher contents of amines than do wines from other regions (Marques *et al.*, 2008).

CONCLUSIONS

The demand for safer foods has promoted more research into BA over the past few years, but some questions still remain unanswered. Fermented food such as wine and cheese can accumulate large quantities of BA. The synthesis and accumulation of BA in foods require the presence of bacteria with decarboxylase deaminase activity, environmental conditions that allow for their growth and for enzyme activity, and the presence of the appropriate amino acid substrates. The influence of processing parameters such as grape composition, milk quality and the treatment of milk and wine has been analysed, and there is general agreement on the importance of these factors in reducing the presence of BA in dairy products and fermented beverages. Furthermore, there is no doubt regarding the importance of selecting starter strains unable to synthesise BA. Knowledge of the metabolic pathways involved in BA production and the factors affecting BA accumulation in food may also be useful in suggesting possible means of reducing BA contents. In addition, because many LABs are normal inhabitants of the intestinal microbiota, it may be worthwhile tracing BA-producing LAB, following consumption by human subjects, to analyse their survival in the gastrointestinal tract and their contribution to BA production in the human body.

Finally, although BAs are present in many different foods and beverages and their concentrations vary widely between and within food types, a shared regulation limiting the amounts of BA in foods is still lacking (except for histamine in fish). Information regarding their presence in foods is also important for the food trade sector (in particular import and export) because recommended upper levels of BA content vary between countries. Therefore, even though information on BA is currently not included in food composition databases, information on their existence, distribution and concentration in fermented foods is crucial and may be useful for the food industry, health professionals and consumers.

ACKNOWLEDGEMENTS

This work was funded by the EU commission in the framework of the BIAMFOOD project (Controlling Biogenic Amines in Traditional Food Fermentations in Regional Europe—project No. 211441).

CONFLICT OF INTEREST

The authors declare no conflict of interest.

REFERENCES

1.   Arena ME, Landete JM, Manca de Nadra MC, Pardo I, Ferrer S (2008). Factors affecting the production of putrescine from agmatine

by*Lactobacillus hilgardii* X1B isolated from wine. *J Appl Microbiol* **105**, 158–165.

2. Bodmer S, Imark C, Kneubühl M (1999). Biogenic amines in foods: histamine and food processing. *Inflam Res* **48**, 296–300.

3. Bover-Cid S, Holzapfel WH (1999). Improved screening procedure for biogenic amine production by lactic acid bacteria. *Int J Food Microbiol* **53**, 33–41.

4. Caston JC, Eaton CL, Gheorghui BP, Ware LL (2002). Tyramine induced hypertensive episodes, panic attacks in hereditary deficient monoamine oxidase patients: case reports. *JSC Med Assoc* **98**, 187–192.

5. Constantini A, Cersosimo M, Del Prete V, Garcia-Moruno E (2006). Production of biogenic amines by lactic acid bacteria: screening by PCR, thin layer chromatography, and HPLC of strains isolated from wine and must. *J Food Protect* **69**, 391–396.

6. Coton E, Coton M (2005). Multiplex PCR for colony direct detection of Gram positive histamine- and tyramine-producing bacteria. *J Microbiol Methods***63**, 296–304.

7. Coton E, Coton M (2009). Evidence of horizontal transfer as origin of strain to strain variation of the tyramine production trait in *Lactobacillus brevis.Food Microbiol* **26**, 52–57.

8. Coton E, Rollan G, Bertrand A, Lonvaud-Funel A (1998). Histamine-producing lactic acid bacteria in wines: early detection, frequency and distribution. *American J Enol Viticul* **49**, 199–204.

9. Dapkevicius MLNE, Nout MJR, Rombouts FM, Houben JH, Wymenga W (2000). Biogenic amine formation and degradation by potential fish silage starter microorganisms. *Int J Food Microbiol* **57**, 107–114.

10. EFSA (2007). Opinion of the Scientific Committee on a request from EFSA on the introduction of a Qualified Presumption of Safety (QPS) approach for assessment of selected microorganisms referred to EFSA. *The EFSA J* **587**, 1–16.

11. Fernández M, del Río B, Linares DM, Martín MC, Alvarez MA (2006). Real time polymerase chain reaction for quantitative detection of histamine-producing bacteria: use in cheese production. *J Dairy Sci* **89**, 3763–3769.

12. Fernández M, Linares DM, Alvarez MA (2004). Sequencing of the tyrosine decarboxylase cluster of *Lactococcus lactis* IPLA 655 and the development of a PCR method for detecting tyrosine decarboxylating lactic acid bacteria.*J Food Prot* **67**, 2521–2529.

13. Fernández M, Linares DM, Rodríguez A, Alvarez MA (2007). Factors affecting tyramine production in *Enterococcus durans* IPLA 655. *Appl Microbiol Biotechnol* **73**, 1400–1406.

14. Garai G, Dueñas MT, Irastorza A, Martín-Álvarez PJ, Moreno-Arribas MV (2006). Biogenic amines in natural ciders. *J Food Prot* **69**, 3006–3012.

15. Garai G, Dueñas MT, Irastorza A, Moreno-Arribas MV (2007). Biogenic amine production by lactic acid bacteria isolated from cider. *Lett Appl Microbiol* **45**, 473–478.

16. Gardner DM, Shulman KI, Walker SE, Tailor SA (1996). The making of a user friendly MAOI diet. *J Clin Psychiatry* **57**, 99–104.

17. Griswold AR, Jameson-Lee M, Burne RA (2006). Regulation and physiologic significance of the agmatine deiminase system of *Streptococcus mutans*UA159. *J Bacteriol* **188**, 834–841.

18. Guerrini S, Mangani S, Granchi L, Vincenzini M (2002). Biogenic amine production by *Oenococcus oeni*. *Current Microbiol* **44**, 374–378.

19. Igarashi K, Ito K, Kashiwagi K (2001). Polyamine uptake systems in*Escherichia coli*. *Res Microbiol* **152**, 271–278.

20. Ladero VM, Linares DM, Fernández M, Alvarez MA (2008). Real time quantitative PCR detection of histamine-producing lactic acid bacteria in cheese: relation with histamine content. *Food Res Int* **41**, 1015–1019.

21. Landete JM, Ferrer S, Pardo I (2007). Biogenic amine production by lactic acid bacteria, acetic bacteria and yeast isolated from wine. *Food Control*18, 1569–1574.

22. Landete JM, Ferrer S, Polo L, Pardo I (2005). Biogenic amines in wines from three Spanish regions. *J Agricul Food Chem* **53**, 1119–1124.

23. Lee YH, Kim BH, Kim JH, Yoon WS, Bang SH, Park YK (2007). CadC has a global translational effect during acid adaptation in *Salmonella enterica*serovar *typhimurium*. *J Bacteriol* **189**, 2417–2425.

24. Lehane L, Olley J (2000). Histamine fish poisoning revisited. *Int J Food Microbiol* **58**, 1–37

25. Leitão MC, Marques AP, San Romão MV (2005). A survey of biogenic amines in commercial Portugese wines. *Food Control* **16**, 199–204.

26. Leuschner RG, Heidel M, Hammes WP (1998). Histamine and tyramine degradation by food fermenting microorganisms. *Int J Food Microbiol* **39**, 1–10.

27. Linares DM, Cruz Martín M, Ladero V, Alvarez MA, Fernández M (2010). Biogenic amines in dairy products. *Critical Rev Food Sci Nutr*

(in press).

28. Livingston MG, Livingston HM (1996). Monoamine oxidase inhibitors. An update on drug interactions. *Drug Saf* **14**, 219–227.

29. Lonvaud-Funel A (2001). Biogenic amines in wines: role of lactic acid bacteria. *FEMS Microbiol Lett* **199**, 9–13.

30. Lucas P, Lonvaud-Funel A (2002). Purification and partial gene sequence of the tyrosine decarboxylase of *Lactobacillus brevis* IOEB 9809. *FEMS Microbiol Lett* **211**, 85–89.

31. Lucas PM, Claisse O, Lonvaud-Funel A (2008). High frequency of histamine producing bacteria in the enological environment and instability of the histidine decarboxylase production phenotype. *Appl Environ Microbiol* **74**, 811–817.

32. Lucas PM, Wolken WAM, Claisse O, Lolkema JS, Lonvaud-Funel A (2005). Histamine-producing pathway encoded on an unstable plasmid in*Lactobacillushilgardii* 0006. *Appl Environ Microbiol* **71**, 1417–1424.

33. Lyte M (2004). The biogenic amine tyramine modulates the adherence of*Escherichia coli* O157:H7 to intestinal mucosa. *J Food Prot* **67**, 878–883.

34. Maijala RL (1993). Formation of histamine and tyramine by some lactic acid bacteria in MRS-broth and modified decarboxylation agar. *Lett Appl Microbiol* **17**, 40–43.

35. Marcobal A, de las Rivas B, Moreno-Arribas MV, Muñoz R (2005). Multiplex PCR method for the simultaneous detection of histamine-, tyramine-, and putrescine producing lactic acid bacteria in foods. *J Food Prot* **68**, 874–878.

36. Marcobal Á, de las Rivas B, Moreno-Arribas MV, Muñoz R (2006a). Evidence for horizontal gene transfer as origin of putrescine production in*Oenococcusoeni* RM83. *App Environ Microbiol* **72**, 7954–7958.

37. Marcobal A, de las Rivas B, Muñoz R (2006b). Methods for the detection of bacteria producing biogenic amines on foods: a survey. *J Con Prot Food Safety* **1**, 187–196.

38. Marques AP, Leitão MC, San Romão MV (2008). Biogenic amines in wines: influence of oenological factors. *Food Chem* **107**, 853–860.

39. Martín MC, Fernández M, Linares DM, Alvarez MA (2005). Sequencing, characterization and transcriptional analysis of the histidine decarboxylase operon of *Lactobacillus buchneri*. *Microbiol* **151**, 1219–1228.

40. Martín-Álvarez PJ, Marcobal Á, Polo C, Moreno-Arribas MV (2006). Influence of technological practices on biogenic amine contents in red

wines. *Eur Food Res Technol* **222**, 420–424.

41. Molenaar D, Bosscher JS, Ten Brink B, Driessen AJM, Konings WN (1993). Generation of a proton motive force by histidine decarboxylation and electrogenic histidine/histamine antiport in *Lactobacillus buchneri*. *J Bacteriol* **175**, 2864–2870

42. Moreno-Arribas V, Lonvaud-Funel A (1999). Tyrosine decarboxylase activity of *Lactobacillus brevis* IOEB 9809 isolated from wine and *L. brevis* ATCC 367. *FEMS Microbiol Lett* **180**, 55–60.

43. Moreno-Arribas V, Polo MC, Jorganes F, Muñoz R (2003). Screening of biogenic amine production by lactic acid bacteria isolated from grape must and wine. *Int J Food Microbiol* **84**, 117–123

44. Moreno-Arribas V, Torlois S, Joyex A, Bertrand A, Lonvaud-Funel A (2000). Isolation, properties and behaviour of tyramine-producing lactic acid bacteria from wine. *J Appl Microbiol* **88**, 584–593.

45. Nannelli F, Claisse O, Gindreau E, de Revel G, Lonvaud-Funel A, Lucas PM (2008). Determination of lactic acid bacteria producing biogenic amines in wine by quantitative PCR methods. *Lett Appl Microbiol* **47**, 594–599.

46. Novella-Rodríguez S, Veciana-Nogués MT, Izquierdo-Pulido M, Vidal-Carou MC (2003a). Distribution of biogenic amines and polyamines in cheese. *J Food Sci* **68**, 750–755.

47. Novella-Rodríguez S, Veciana-Nogués MT, Trujillo-Mesa AJ, Vidal-Carou MC (2003b). Profile of biogenic amines in goat cheese made from pasteurized and pressurized milks. *J Food Sci* **67**, 2940–2944.

48. Önal A (2007). A review: current analytical methods for the determination of biogenic amines in food. *Food Chem* **103**, 1475–1486.

49. Premont RT, Gainetdinov RR, Caron MG (2001). Following the trace of elusive amines. *Proc Natl Acad Sci USA* **98**, 9474–9475

50. Rhee JE, Rhee JH, Ryu PY, Choi SH (2002). Identification of the *cadBA*operon from *Vibrio vulnificus* and its influence on survival to acid stress.*FEMS Microbiol Lett* **208**, 245–251.

51. Satomi M, Furushita M, Oikawa H, Yoshikawa-Takahashi M, Yano Y (2008). Analysis of a 30 kbp plasmid encoding histidine decarboxylase gene in*Tetragenococcus halophilus* isolated from fish sauce. *Int J Food Microbiol***126**, 202–209.

52. Sattler J, Hesterberg R, Lorenz W, Schmidt U, Crombach M, Stahlknecht CD (1985). Inhibition of human and canine diamine oxidase by drugs used inan intensive care unit: relevance for clinical side effects? *Agents*

Actions **16**, 91–94.

53. Shalaby AR (1996). Significance of biogenic amines to food safety and human health. *Food Res Int* **29**, 675–690.

54. Silla Santos MH (1996). Biogenic amines: their importance in foods. *Int J Food Microbiol* **29**, 213–231.

55. Soufleros E, Barrios ML, Bertrand A (1998). Correlation between the content of biogenic amines and other wine compounds. *Am J Enol Viticul***49**, 266–269.

56. Taylor SL (1983). Monograph on histamine poisoning. In: Codex Alimentarius Commission, FAO/WHO, Rome. 19th session of the Codex Committee on Food Hygiene, Washington, DC, 26–30 September 1983.

57. Ten Brink B, Damink C, Joosten HMLJ, Huisint-Veld JHJ (1990). Occurrence and formation of biologically amines in food. *Int J Food Microbiol* **11**, 73–84.

58. Tkachenko A, Nesterova L, Pshenichnov M (2001). The role of the natural polyamine putrescine in defense against oxidative stress in *Escherichia coli*. *Arch Microbiol* **176**, 155–157.

59. Van de Guchte M, Serror P, Chervaux C, Smokvina T, Ehrlich SD, Maguin E (2002). Stress responses in lactic acid bacteria. *Antonie van Leeuwenhoek***82**, 187–216.

60. Vido K, Le Bars D, Mistou MY, Anglade P, Gruss A, Gaudu P (2004). Proteome analyses of heme-dependent respiration in *Lactococcus lactis*: involvement of the proteolytic system. *J Bacteriol* **186**, 1648–1657.

Chapter 2

PROBIOTICS IN DAIRY FERMENTED PRODUCTS

Emiliane Andrade Araújo[1], Ana Clarissa dos Santos Pires [2], Antônio Fernandes de Carvalho[2], Maximiliano Soares Pinto[3] and Gwénaël Jan[4]

[1]Universidade Federal de Viçosa, Campus Rio Paranaíba, Rio Paranaíba, MG, Brazil

[2]Departamento de Tecnologia de Alimentos, Universidade Federal de Viçosa, Viçosa, MG, Brazil

[3]Instituto de Ciências Agrárias, Universidade Federal de Minas Gerais, Montes Claros, MG, Brazil

[4]INRA, UMR1253 Science et Technologie du Lait et de l'Œuf, Rennes, France

INTRODUCTION

Since ancient times, food has been considered essentialand indispensable to human life. Numerous studies clearlyshow that an individual's quality of life is linked to daily diet and lifestyle (Moura, 2005).

Interest in the role of probiotics for human health began as early as 1908 when Metchnikoff associated the intake of fermented milk with prolonged life (Lourens-Hattingh and Vilijoen, 2001b). However, the relationship between intestinal microbiota and good health and nutrition has only recently been investigated. Therefore, it was not until the 1960's that health benefit claims began appearing on foods labels. In recent years,there has been an increasing interest in probiotic foods, which has stimulated innovation and fueled the development of new products around the world. Probiotic bacteria have increasinglybeen incorporated into foods in order to improve gut health by maintaining the microbial gastrointestinal balance. The most popular probioticfoodsare produced in the dairy industry because fermented dairy products have been shown to be the most efficient delivery vehicle for live probiotics to date. In this chapter, we will discuss the application of probiotic microorganisms in fermented dairy products, particularly cheeses. In addition, we will also discuss the benefits of probiotic fermented foods on human health.

PROBIOTIC CONCEPTS

The word "probiotic" comes from Greek and means "for life" (Fuller, 1989). Over the years, the term "probiotic" has been given several definitions. "Probiotic" is used to refer to cultures of live microorganisms which, when administered to humans or animals, improve properties of indigenous microbiota (Margoles and Garcia, 2003). In the food industry, the term is described as "live microbial food ingredients that are beneficial to health" (Clancy, 2003).

It is important to mention that for a microorganism to be considered probiotic, (Figure 1), it must survive passage through the stomach and maintain its viability and metabolic activity in the intestine (Hyun and Shin, 1998). Native inhabitants of the human or animal gastrointestinal tract, such as lactobacilli and bifidobacteria, are considered to be probiotic, but often display low stress tolerance, which reduces their viability in probiotic applications. Microorganisms traditionally grown in fermented foods, such as lactic acid bacteria, propionibacteria and yeasts, are also considered for these applications..

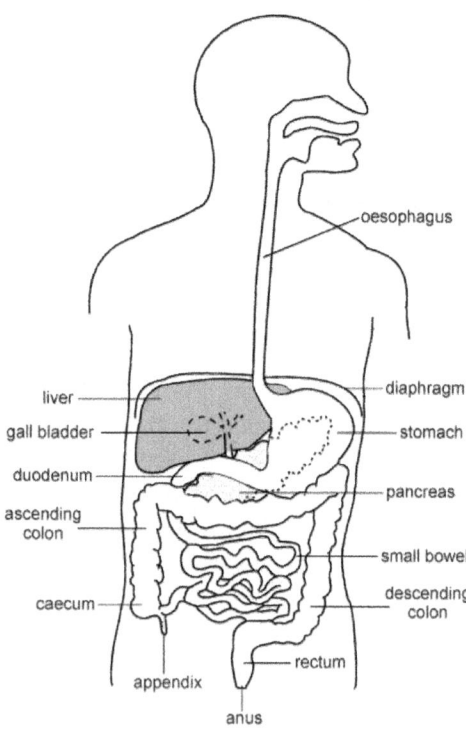

Figure 1: Schematic representation of gastrointestinal tract

It is essential that commercialized probiotic products which make health claims meet the minimum criterion of one millionviable probiotic cells per milliliter of product at the expiration date. Accordingly, the minimumdosage of probiotic cells per day for any beneficial effecton the consumer is considered to be 10^8–10^9 probiotic CFU ml^{-1}or CFU g^{-1}, which corresponds to an intake of 100g product containing10^6–10^7CFU ml^{-1}or CFU g^{-1} per day (Lorens-Hattingh and Viljoen, 2001a).

Selection of Probiotic Microorganisms

The human intestinal tract constitutes a complex ecosystem of microorganisms. The bacterial population in the large intestine is very high and can reach maximum counts of 10^{12} CFU g^{-1}. In the small intestine, the bacterial content is considerably lower at only 10^4–10^8 CFU g^{-1}. In the stomach only 10^1-10^2 CFU g^{-1} are found due to the low pH of the environment (Lorens-Hattingh and Viljoen, 2001b).

It is known that microbiota in the human intestine changes during human development. The intestine of newborn babies is fully sterile, however immediately after birth, colonization of many kinds of bacteria begins. On the first and second days after birth, coliforms, enterococci, clostridia and lactobacilli have been shown to be present present in infants' feces. Within three to four days, bifidobacteria begins colonization and becomes predominant around the fifth day. Simultaneously, coliform counts decrease. Breast-fed babies show 1 log-count more of bifidobacteria in feces than bottle-fed babies. Enterobacteriaceae,streptococci, and other putrefactive bacteria counts are higher in bottle-fed babies, suggesting that breast-fed babies are more resistant to gastrointestinal infections than the bottle-fed infants (Lorens-Hattingh and Viljoen, 2001b).

In addition to the microbiota changes that occur during human aging, the microbiota in the gastrointestinal system can also change because of the food and health conditions of an individual. For example, use of antibiotics can damage the equilibrium of intestinal microbiota, reducing counts of bifidobacteria and lactobacilli and increasing clostridia. The ensuing imbalance can cause diarrhea in elderly and immunocompromised people.

To help improve the balance of intestinal microbiota, probiotic microorganisms can be added to the human diet in order to stimulate thegrowth of preferred microorganisms, crowd outpotentially harmful bacteria, and reinforce thebody's natural defense mechanisms.

The selection of probioticmicroorganismsis based on safety, functionaland technological aspects, as reported by (Saarela et al., 2000). These are summarized in Figure 2.

Certain probiotic bacteria have been extensively studied and are already on the market, as shown in Table 1.

Before probiotic strains can be delivered toconsumers, they must first be able to be manufacturedunder industrial conditions. They must then surviveand retain their functionality during storageas frozen or freeze-dried cultures, as well as in thefood products into which they are finally formulated. Moreover, they must be able to be incorporated into foods without producing off-flavorsor textures (Saarela et al., 2000).

Functional food requirements musttake into consideration the following aspects in relation to the probiotics: The preparationshould remain viable for large-scale production; itshould remain stable and viable during storage and use; it should be able to survive in the intestinalecosystem (Prado et al., 2008).

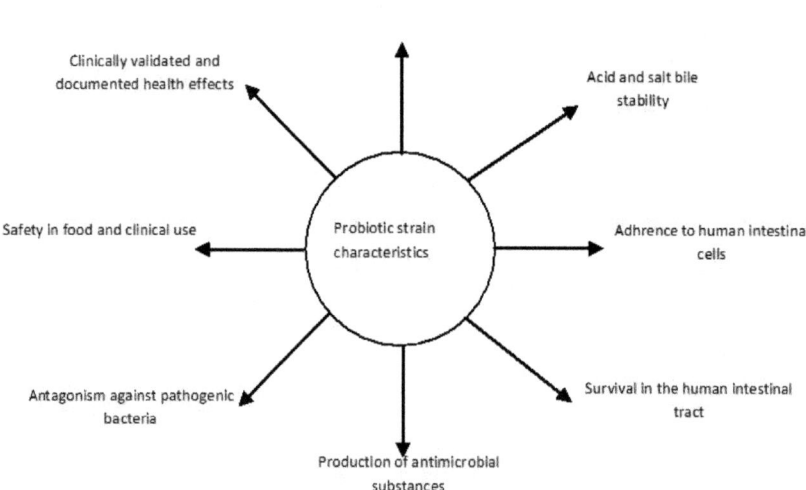

Figure 2: Theoretical basis for selection of probiotic microorganism selection (adapted from Saarela et al., 2000).

Table 1: Probiotic bacteria marketed worldwide

Strains	Origin
Lactobacillus casei Shirota	Yakult, Japan
Lactobacillus reuteri MM53	BioGaia, Sweden
Bifidobacterium lactis HN019	Danisco, France

Lactobacillus rhamnosus GG	Valio, Finland
*Lactobacillus acidophilus*NCFM	Nestle, Switzerland
Lactobacillus casei DN-173 010	Danone, France
Lactobacillus casei CRl-431	Chr. Hansen, USA
*Bifidobacterium animalis*BB12	Chr. Hansen, Denmark
*Bifidobacterium animalis*DN173010	Danone, France

BENEFICIAL EFFECTS OF PROBIOTICS

The role of balanced nutrition for health maintenance has attracted the attention of the scientific community, which in turn has produced numerous studies in order to prove the performance of certain foods in reducing the risk of Some diseases. There has also been considerable growing interest in encouraging research into new natural components (Thamer and Penna, 2006).

In a healthy host, a balance exists among members of the gut microbiota, such that potential pathogenic and non-pathogenic organisms can be found in apparent harmony. In the case ofbacterial infection, this balance can become disturbed, leading to often dramatic changes in the composition.

For most bacterial infections, nonspecific antibiotics are used, killing both non-pathogenic members of gut microbiota as well as pathogenic members. This can lead to a substantial delay in the restoration of healthy gut microbiota (Reid et al, 2011). The restoration of the gut microbiota balance is believed to be important because maintaining a healthy and balanced gut microbiota throughout life is thought to help preserve health and favor longevity.

The most comprehensive analysis of human microbiota to date examined 27 distinct sites in the body and revealed the presence of 22 bacterial phyla, with most sequences (92.3%) related to just four phyla: Actinobacteria (36.6%), Firmicutes (34.3%), Proteobacteria (11.9%) and Bacteroidetes (9.5%) (Costelo, 2008).

The metabolic capacity of gut bacteria is extremely diverse. This diversity is influenced by the large number of bacterial genera and species. Lactic acid species are present, as well as peptide-degrading bacteria, amino acids, and other methanogenic bacteria components of the gut microbiota which grow with the intermediate products of fermentation such as hydrogen, lactate, succinate and ethanol (Topping and Clifton, 2001).

In host's diet residue (matter undigested by its digestive systemincluding resistant starch, fibers, proteins and peptides) substrates for primary fermentation can be found. Other important available substrates derive from mucin glycoproteins, exfoliated epithelial cells and pancreatic Secretions (MacFarlane et al., 1992).

Hydrolysis and carbohydrate metabolism in the large intestine is influenced by a variety of physical, chemical, biological and environmental parameters. Probably the nature and quantity of available substrate that has greater meaning, making the diet easier and the main mechanism by which to influence the profile of fermentation. Other factors affecting the colonization and growth of bacteria in the intestine are intestinal pH, which inhibits the production of metabolites (acids and peroxides) and specific inhibitory substances (bacteriocins), bile salts and molecules and cells which constitute the immune system (Rastall et al., 2000).

Knowledge of intestinal gut microbiota and their interactions led to the development of food strategies aimed at the stimulation and maintenance of normal bacteria present in the gut (Gibson and Fuller, 2000).

According to Wohlgemuth (2010), strategies for studying mechanisms of probiotic action involve in-vitro models, orconventional or gnotobiotic animal models, plus development of a simplified human intestinal gut microbiota. Wo hlgemuth'sarticleproposescertain requirements that a model should ideally fulfill:

- Selected bacterial species should represent numerically dominant organisms of the human gut microbiota.
- By and large, the metabolic activity of this community should mimic that of normal human gut microbiota.
- The genome sequence of all members of the microbial community should be known.
- The members of this consortium should form a stable community in rodents.It should be possible to maintain this community under gnotobiotic conditions from generation to generation.
- The composition of the microbial community should be modifiable when required.

It is possible to increase the number of health-promoting microorganisms in gut microbiota through the introduction of probiotics in the diet. The probiotics will selectively modify the composition of the gut microbiota, providing the probiotic microorganisms demonstrate a competitive advantage over other bacteria in the ecosystem (Crittenden, 1999). Probiotic therapeutic properties are listed in Table 2.

Table 2: Therapeutic Properties of Probiotics

Probiotic therapeutic properties
Influence on host gut microbiota and pathogenic bacteria
Improvement of specific enzymatic activities
Production of antibacterial substances
Competitive exclusion of pathogenic bacteria
Induction of defensin production
Improvement of intestinal barrier function
Modulation of host immune functions
Modulation of intestinal carcinogenesis
Modulation of cholesterol uptake

There is a growing body of evidence that ingested beneficial bacteria, called probiotics, can beneficially modulate chronic intestinal inflammation, diarrhea, constipation, vaginitis, irritable bowel syndrome, atopic dermatis, food allergies and liver disease (Wallace et al., 2011, Nutrition reviews).

Probably the most promising area is the alleviation of symptoms linked to inflammatory bowel diseases (IBD), a growing health concern. As an example, the probiotic preparation VSL#3 induced remission in children (n=18) with mild to moderate ulcerative colitis (UC) (Huynh et al., 2009, Inflamm. Bowel Dis.) Accordingly, VSL#3 was tested in a 1-year, placebo-controlled, double-blind clinical study on UC children (n=29). Remission was achieved in 36.4% of children receiving IBD therapy and placebo, but in 92.8% of children receiving IBD therapy and VSL#3 (Milele et al., 2009, Am J Gastroeterol.) Similar promising results were obtained with the probiotic *Escherichia coli* Nissle 1917 strain (Kruis et al., 2004, Gut ; Do et al., Ann Pharmacother, 2010). However, a review of available data indicates that more clinical studies are needed to confirm the beneficial effects of these products in UC and in inactive pouch patients (Jonkers et al., 2012, Drugs). This review also states that there is no evidence to support the use of probiotics in Crohn's disease.

Other studies confirm these findings. Miele et al. (2009) reported that all of 29 patients studied responded to inflammatory bowel disease therapy. Remission was achieved in 92.8% of patients treated with mixed probiotics and 36.4% of patients treated with placebo. Overall, 21.4 % patients treated with a mix of probiotics and 73.3 % patients treated with placebo relapsed within 1 year of follow-up. Urinary tract infections (UTIs) are a common and frequently recurrent infection among women. Depletion of vaginal lactobacilli is associated with UTI risk, which suggests that repletion of the bacteria may be beneficial. Young women with a history of recurrent UTI were randomized to receive either a probiotic or placebo daily. Recurrent UTI occurred in 15% of women receiving probiotic compared with 27% of women receiving placebo (Stapleton et al., 2011).

Probiotics have considerable potential for preventive and therapeutic applications in gastrointestinal disorders. However, it is important to note that many probiotic health claims have not yet been substantiated through experimental evidence. In addition, the efficacy demonstrated for a single given bacterial strain cannot be extrapolated to other probiotic organisms. Moreover, the mechanisms underlying probiotic action have not yet been fully elucidated. A better understanding of these mechanisms will be able to shed light on the disparate clinical data and provide new tools to help the prevention or treatment of health disorders (Wohlgemuth et al., 2010; Yan et al., 2011).

APPLICATION OF PROBIOTIC BACTERIA IN DAIRY FOODS

There is evidence that food matrices play an importantrole in the beneficial health effects of probioticson the host (Espirito Santo et al., 2011).

Fermented foods, particularly dairy foods,are commonly used as probiotic carriers. Fermented beverages provide an important contribution to the human diet in many countries because fermentationis an inexpensive technology which preserves food, improves its nutritional value and enhances its sensoryproperties (Gadaga et al., 1999). However, the increasing demand for new probiotic products has encouraged the development of other matrices to deliver probiotics, such as ice cream, infant milk power and fruit juice.

Davidson et al. (2000) evaluated the viability of probiotic strains in low-fat ice cream. They used cultures containing *Streptococcus salivarius* ssp. *thermophilus* and *Lactobacillusdelbrueckii* ssp.*Bulgaricus*, *Bifidobacterium longum* and *Lactobacillusacidophilus*, and verified that culture bacteria did not decrease inthe yogurt during frozen storage. Also, the presence of probiotic bacteria did not alter the sensory characteristics of the ice cream. The icecream matrix may offer a good vehicle for probiotic culturesdue to its composition, which includes milk proteins, fat andlactose, as well as other compounds. Moreover, its frozen state contributes to its efficiency. However, a probiotic ice cream product shouldhave relatively high pH values –5.5 to 6.5, in order to favor an increased survival of lactic cultures during storage.The lower acidity also results in increased consumer acceptance, especially among consumers who prefer milder Products. (Cruz et al., 2009b). Growth of a probiotic yeast, *Saccharomyces boulardii*, in associationwith the bio-yogurt microflora, which is done by incorporating the yeast into commercial bio-yogurt, has been suggested as a way to stimulate growth of probiotic organisms and to assure their survival during storage. Lorens-Hattingh and Viljoen(2001a) studied the ability of probiotic yeast to grow and survive in dairy products, namely bio-yogurt, UHT yogurt and UHT milk. *S. boulardii* was incorporatedinto these dairy products and stored at 4 °Cover a 4-week period. It was observed that the probiotic yeast species, *S. boulardii*, had the ability to grow inbio-yogurt and reach maximum counts exceeding $10^7 CFU\ g^{-1}$. The number of yeast populations was substantially higher in the fruit-based yogurt, mainly due to the presence of sucrose and fructose derived from the fruit. Despite the inability of *S. boulardii* to utilize lactose, the yeast species utilized available organic acids, galactose and glucose derived from bacterial metabolismof the milk sugar lactose present in the dairy products.

The viability of strains of *L. acidophilus* and *Bifidobacterium animalis* ssp. *lactis* in stirred yoghurts with fruit preparationsof mango, mixed berry, passion fruit and strawberrywas evaluated during shelf-life (Godward et al., 2000;Kailasapathy et al., 2008). The authors observed that regardless of concentrations, the additionof any of the fruit preparations had no effect on the counts of the two probiotics tested.

Fermented milks supplemented with lemon and orangefibers increased the counts of *L. acidophilus*and*L. casei* during cold storage compared to the control set. This was not the case for *B. bifidum*, possibly owing to the well-knownsensitivity of bifidobacteria species to an acidicenvironment (Sendra et al., 2008).

PROBIOTIC CHEESES

Probiotic foods are currently primarily found in fermented milk drinks and yogurt, both of which have limited shelf life compared to cheeses. Incorporation of probiotic cultures in cheeses offers the potential not only to improve health but also product quality. It also opens the way to increasing the range of probiotic products on the market. The manufacture of most cheeses involves combining four ingredients: milk, rennet, microorganisms and salt These are processed using a number of common steps such as gel formation, whey expulsion, acid production and salt addition. Variations in ingredient blends and subsequent processing have led to the evolution of all cheese varieties.

Cheeses are dairy products which have a strong potential for delivering probiotic microorganisms into the human intestine, due to their specific chemical and physical characteristics. Cheeses have higher pH levels, lower titratable acidity, higher buffering capacity, more solid consistency, relatively higher fat content, higher nutrient availability and lower oxygen content than yogurts. These qualities protect probiotic bacteria during storage and passage through the gastrointestinal tract (Karimi et al., 2011; Ong et al., 2006).

As mentioned above, the physicochemical properties of food influence probiotic bacteria survival in the digestive tract, due to the low pH in the stomach, typically between 2.5 and 3.5 (Holzapfel et al., 1998), and the antimicrobial activity of pepsin that serve as effective barriers against the entrance of bacteria into the intestinal tract. Values of pH between 1 and 5 are commonly employed in determining the *in vitro* acid tolerance of *Lactobacillus* and *Bifidobacterium* spp. (Charteris et al., 1998). Bile salt concentrations between 0.15% and 0.3% have been recommended as appropriate for selection of probiotic bacteria for human consumption (Yang and Adams, 2004).

A variety of microorganisms, typically food-grade lactic acid bacteria (LAB), have been evaluated for their probiotic potential and have been applied as adjunct cultures in various food products or therapeutic preparations (Rodgers, 2008). *Lactobacillus* and *Bifidobacterium*species may be found in many foods; some are frequently regarded as probiotics due to their capacity to improve certain biological functions in the host. Complex interactions occur among resident microbiota, epithelial and immune cells and probiotics. These interactions play a major role in the development and maintenance of the beneficial activities for healthy humans (Medici el al., 2004).

According to Karimi et al. (2012), recommendations for the minimum viable counts of each probiotic strain in gram or millilitre of probiotic products vary when it comes to providing health benefits related to probiotic organisms. For example, the minimum viable levels of 10^5 cfu g^{-1}have been recommended (Shah, 1995); while 10^6 cfu g^{-1} (Karimi and Amiri-Rigi, 2010; Talwalkar and Kailasapathy, 2004) and 10^7 cfu g^{-1} (Samona and Robinson, 1994) have been suggested for probiotics in different products. However, populations of 10^6-10^7 CFU/g in the final product have been shown to be more acceptable asefficientlevels of probiotic cultures in processed foods (Talwalkar, Miller, Kailasapathy and Nguyen, 2004), with numbers attaining 10^8 - 10^9 CFU when provided by a daily consumption of 100 g or 100 mL of probiotic food, and hence benefiting human health (Jayamanne & Adams, 2006). It is important to emphasize that the incorporation of probiotic cultures into cheeses would produce functional foods only if the cultures remained viable in recommended numbers during maturation and shelf life of the products.

One of the preconditions for a bacterial strain to be called probiotic is the strain's ability to survive in the gastrointestinal environment, although the importance of viability for the beneficial effects of probiotics has not been well defined since inactivated and dead cells can also have immunological and health-promoting effects (Ghadimi et al., 2008; Lopez et al., 2008).

Moreover, there are significant technological challenges associated with the introduction and maintenance of high numbers of probiotic microorganisms in foods that depend on the form of the probiotic inoculant, and with the viability and maintenance of probiotic characteristics in the food product up to the time of consumption. Spray drying has been used as a preservation method for microbial cultures. Gardiner et al. (2002) produced spray-dried probiotic milk powder containing the probiotic *Lactobacillus paracasei* NFBC 338. The powder contained 1 x 10^9 CFU.g^{-1} *L. paracasei* which was used as adjunct inoculums during probiotic Cheddar cheese manufacture. After three months of ripening, the count was 7.7 x 10^7 CFU.g^{-1}, without any adverse effects on the cheese. The researchers' data shows that probiotic spray-dried powder may be a useful means for adding probiotic strains to dairy products.

In order to use probiotic bacteria in the manufacture of cheese products, the process may have to be modified and adapted to the requirements of the strains employed. Overall, probiotic strains should be technologically compatible with the food manufacturing process involved. With regard to the development of probiotic cheeses, this means that such strains should be cultivable to high cell density for inoculation into the cheese vat, or that the strains are capable of proliferating during the manufacturing and/or ripening process (Ross et al., 2002). In general, a probiotic cheese should have the same attributes as a conventional cheese: the incorporation of probiotic bacteria should not imply a loss of quality of the product. In this context, the level of proteolysis and lipolysis must be the same or even better than for cheese which does not have functional food appeal (Cruz et al., 2009a).

Proteolysis plays a critical role in determining typical sensory characteristics and represents a significant quality indicator for certain cheeses. Proteolysis is caused by enzymes found in milk (plasmin), rennet (pepsin and chymosin) and microbial enzymes released by starter cultures. The activities of these enzymes hydrolyze the fractions of caseins, which leads to the formation of peptides. These peptides may be further hydrolyzed with proteolytic enzymes originating from microbiota such as starter bacteria, non-starter lactic acid bacteria (NSLAB) and probiotic adjuncts to the cheeses, into smaller peptides and free amino acids, which are important for flavor development in some cheeses (Ong et al., 2007; Cliffe et al., 1993; Lynch et al., 1999).

Three batches of Cheddar cheeses (Batch 1, with only starter lactococci; Batch 2, with lactococci and *Lactobacillus acidophilus* 4962, *Lb. casei* 279, *Bifidobacterium longum* 1941; Batch 3, with lactococci and *Lb. acidophilus* LAFTIs L10, *Lb. paracasei* LAFTI L26, *B. lactis* LAFTI B94) were manufactured in triplicate to study the survival and influence of probiotic bacteria on proteolytic patterns and production of organic acid during a ripening period of 6 months at 4 °C. All probiotic adjuncts survived the manufacturing process and maintained their viability of 7.5 log10 cfu g^{-1} at the end of the ripening term. The number of lactococci decreased by one to two log cycles, but their counts were not significantly different (P> 0.05) in either the control or the probiotic cheeses. No significant differences were observed in composition (fat, protein, moisture, salt content), although acetic acid concentration was higher in the probiotic cheeses. Proteolysis assessment during ripening showed no significant differences (P> 0.05) in the level of water-soluble nitrogen (primary proteolysis), but the levels of secondary proteolysis indicated by the concentration of free amino acids were significantly higher (P> 0.05) in probiotic cheeses. These data thus suggested that Cheddar cheese is an effective vehicle for the delivery of probiotic organisms (Ong et al., 2006).

Phillips et al. (2006) have also studied probiotic Cheddar cheese. Theymanufactured six batches of Cheddar cheese containing different combinations of commercially-available probiotic cultures. Duplicate cheeses contained organisms from each supplier, *Bifidobacterium* spp., *Lactobacillus acidophilus* and either *Lactobacillus casei*, *Lactobacillus paracasei*, or *Lactobacillus rhamnosus*. Using selective media, the different strains were assessed for viability during Cheddar cheese maturation over 32 weeks. *Bifidobacterium* sp. remained at high numbers with the three strains present in cheese at 4×10^7, 1.4×10^8, and 5×10^8 CFU/g respectively after 32 weeks. Similarly, the *L. casei* (2×10^7 CFU/g),*L. paracasei* (1.6×10^7 CFU/g), and *L. rhamnosus* (9×10^8 CFU/g) strains survivedwell.However, the *L. acidophilus* strains performed poorly. Both decreased in a similar manner and were recorded at 3.6×10^3CFU/g and 4.9×10^3 CFU/g after 32 weeks.

Numerous scientific papers have been published on the development of fresh cheeses containing recognized and potentially probiotic cultures. Theyhave described suitable viable counts as well as a positive influence on texture and sensorial properties of the cheeses. Cottage cheese in particular shows an adequate profile for the incorporation of probiotic cells and/or prebiotic substances. In addition, cottage cheese is a healthy alternative to many other cheeses by virtue of its low fat content.

Araújo et al. (2010) developed a symbiotic cottage cheese containing *Lactobacillus delbrueckii* UFV H2b20 and inulin, and evaluated the survival of this bacterium when thecheese was exposed to conditions simulating those found in the gastro-intestinal tract. Throughout the entire storage period of the cheese, the probiotic cell counts were higher than recommended levels for probiotic products. The probiotic bacterium exhibited satisfactory resistance to low pH values and to high concentrations of bile salts. The addition of probiotic cells and inulin generated no alterations in the physicochemical characteristics of cheese. By allowing the viable microorganism has characteristics desirable for incorporation of a probiotic strain. Probiotic cells could be added to the dressing, creamy liquid that surrounds the granules of cheese because after this step there is not exposition at high temperature.

Although cottage cheese is well adapted to the health requirements of modern populations, its consumption has been in decline over the past few years. By developing new production processes, cottage cheese, apart from carrying the nutritional qualities of milk, may also furnish consumers with a source of lactic acid bacteria, probiotic microorganisms and prebiotics. The lactic acid bacteria perform more critical functions in cottage cheese than just producing lactic acid.They also aid the manufacture process and increase the final rheological and sensorial qualities of the cheese. Controlling of the fermentation process with lactic acid bacteria allows for the enhancement of the sensorial quality of the cheese and could hence play a crucial role in increasing consumption of cottage cheese.

Souza, et al. (2008) and Souza and Saad (2009) studied the manufacture of Minas fresh cheese supplemented solely with the probiotic strain of *L. acidophilus* La-5. Cheeses manufactured solely with La-5 presented populations above 1×10^6 CFU/g, reaching 1×10^7 CFU/g on the 14th day of storage. The Argentinean fresh cheese is a soft rindless cheese with a ripening period of 12 days at 5 °C before its commercial distribution. This cheese presents the following physicochemical characteristics: pH 5.29, moisture 58% (w/w), fat 12% (w/w), proteins 23% (w/w), salt 0.9% (w/w), ashes 3.4% (w/w), dry matter 40.8% (w/w) and calcium 0.6% (w/w). This product has proven to be an adequate vehicle for probiotic bacteria during storage and until consumption.It offers offer a certain degree of protection of the viability of bacteria during the *in vitro* simulation of gastric transit (Vinderola et al., 2000).

Kasimoglu et al. (2004) have shown that *L. acidophilus* strain can be used for the manufacture of probiotic Turkish white cheese. The final numbers of *L. acidophilus* were greater than the minimum (10^7 cfu g^{-1}) required to make health benefits claims. Furthermore, *L. acidophilus* can be used to enhance flavor, texture, and a produce a high level of proteolysis. Moreover, probiotic cheese which was vacuum packed following salting was shown to be more acceptable than the corresponding cheese stored in brine following salting. Therefore, vacuum packaging is the preferred means for storing probiotic Turkish white cheeses.

CONCLUDING REMARKS AND FUTURE TRENDS

In conclusion, probiotic microorganisms, including bacteria and yeasts, are attracting a growing interest due to their promising physiological effects as well asthevalue they add to probiotic-containing food products. There is a growing body of evidence that probiotics may play a beneficial role in human health (Ouwehand et al., 2002; Collado et al., 2009).Established effects in humans include alleviation of symptoms linked to lactose intolerance or to irritable bowel syndrome. They also include reduced diarrhea associated with antibiotic treatment, rotavirus or traveler's diseases. It should be emphasized that the beneficial properties of probiotic microorganisms are highly dependent on the strains, which means that each strain or product requires demonstration of the specific effects *in vivo*. The possibility of using certain probiotics to modulate the immune system, particularly at the mucosal level (O'Flaherty et al., 2010) is the most promising application. In this respect, promising healing effects were obtained using the probiotic mixture VSL#3 on ulcerative colitis patients (Miele et al., 2009; Huynh et al., 2009;Ng et al., 2010). These clinical studies, which still need to be confirmed by larger studies, strongly suggest that selected strains of probiotics may help in treating the bowel diseases which constitute a growing health concern in developing countries. Clearly, animal studies suggest other promising probiotic effects incuding inflammatory diseases, allergies and associated asthma, and colorectal cancer. These applications open exciting avenues that must be investigated at both molecular and clinical levels.

Understanding the impact of ingested bacteria on health, as well as the impact of gut microbiota perturbation (dysbiosis) on emerging diseases, including immune disorders and cancer remains a great challenge. In developed countries, gut microbiota have evolved with a reduced diversity of bacterial species (Yatsunenko et al., 2012). This is particularly true in Crohn's disease patients (Manichanh et al., 2006), who lack immunomodulatory anti-inflammatory bacteria, including *Faecalibacterium prausnitzii*(Sokol et al., 2008).A similar reduced diversity was also described in the case of colorectal cancer, (Chen et al., 2012) confirming the involvement of dysbiosis in digestive cancers (Azcarate-Peril et al., 2011). The composition of gut microbiota is linked to long term dietary patterns (Wu et al., 2011).This suggests that ingested bacteria can participate in the prevention and/or treatment of emerging diseases. This hypothesis has been reinforced by recent epidemiological studies which show that raw milk prevents the onset of allergy and asthma in children (Loss et al., 2011; Waser et al., 2007; Braun-Fahrlander et al., 2011). The authors suggested a protective immunomodulatory role of raw milk bacteria (Braun-Fahrlander et al., 2011).

Most interestingly, bacterial species used as dairy starters display promising properties in this field. For example, immunomodulatory anti-inflammatory properties were described in certain strains of*Propionibacterium freudenreichii*

(Foligné et al., 2010; Deutsch et al., 2012), *Streptococcus thermophilus* (Ogita et al., 2011), *Lactobacillus delbrueckii* subsp. *bulgaricus* and subsp. *lactis* (Santos-Rocha et al., 2012), as well as *Lactobacillus helveticus* (Guglielmetti et al., 2010*)*. Modulation of colon cancer cell growth was also reported in vitro and/or in animal models for *P. freudenreichii*(Cousin et al., 2010; Lan et al., 2008), when the cells were exposed to yogurt containing *S. thermophilus and L. bulgaricus*(Narushima et al., 2010; Perdigon et al., 2002) and *L. helveticus*(de Moreno et al., 2010).Future trends may thus include the development of specific fermented dairy products designed for specific population. These could usebacteria strains and employ both technological capabilities and probiotic potential to affect immune system modulation, gut physiology and cancer cells.

ACKNOWLEDGEMENT

We would like to thank to Mary Margaret Chappell for reading and contributing. The authors are supported by grants from the FAPEMIG, CAPES and CNPq.

REFERENCES

1.   E. A. Araujo, A. F. Carvalho, E. S. Leandro, M. M. Furtado, C. A. Moraes, 2010Probiotics in Dairy Fermented ProductsJournal of Functional Foods28589

2.   M. A. Azcarate-Peril, M. Sikes, J. M. Bruno-Barcena, 2011Probiotics in Dairy Fermented ProductsAm. J. Physiol Gastrointest. Liver Physiol,, 301:G401G424.

3.   D. Benton, C. Williams, A. Brown, 2007Probiotics in Dairy Fermented ProductsEuropean Journal of Clinical Nutrition61355361

4.   C. Braun-Fahrlander, M. E. Von, 2011Can farm milk consumption prevent allergic diseases? Clin. Exp. Allergy, 412935

5.   W. P. Charteris, P. M. Kelly, L. Morelli, J. K. Collins, 1998Development and application of in vitro methodology to determine the transit tolerance of potentially probiotic Lactobacillus and Bifidobacteriumspecies in the upper human gastrointestinal tract. Journal of Applied Microbiology, 84759768

6.   W. J. L. Chen, J. W. Anderson, D. Jennings, 1984Probiotics in Dairy Fermented ProductsProc. Soc. Exp. Biol. Med,175215218

7.   W. Chen, F. Liu, Z. Ling, X. Tong, C. Xiang, 2012Human intestinal lumen and mucosa-associated microbiota in patients with colorectal cancer.PLoS. ONE., 7:e39743.

8.   R. Clancy, 2003Probiotics in Dairy Fermented ProductsFEMS Immunology and Medical Microbiology38912

9.   A. J. Cliffe, J. D. Marks, F. Mulholland, 1993Isolation and characterization of non-volatile flavors from cheese: Peptide profile of flavor fractions from Cheddar cheese, determined by reverse-phasehigh performance liquid chromatography. International Dairy Journal, 3379387

10.   M. C. Collado, E. Isolauri, S. Salminen, Y. Sanz, 2009The impact of probiotic on gut health.Curr. Drug Metab, 106878

11.   E. K. Costello, C. L. Lauber, M. Hamady, N. Fierer, J. I. Gordon, R. Knight, 2009Bacterial community variation in human body habitats across space and time. Science, 32616941697

12.   F. J. Cousin, S. Jouan-Lanhouet, M. T. Dimanche-Boitrel, L. Corcos, G. Jan, 2012Milk Fermented by Propionibacterium freudenreichii Induces Apoptosis of HGT-1 Human Gastric Cancer Cells.PLoS. ONE., 7:e31892.

13.   R. G. Crittenden, In. Prebiotics, G. W. . Tannock, Ed., 1999Probiotics: a critical review. Norfolk: Horizon Scientific Press, 141156

14.   A. G. Cruz, F. C. A. Buriti, C. H. B. Souza, J. A. F. Faria, S. M. I. Saad, 2009aProbiotics in Dairy Fermented ProductsTrends in Food Science & Technology20344354

15.   A. G. Cruz, A. E. C. Antunes, A. L. O. P. Sousa, J. A. F. Faria, S. M. I. Saad, 2009bProbiotics in Dairy Fermented ProductsFood Research International4212331239

16.   A. G. Cruz, J. A. F. Faria, A. G. F. Van Dender, 2007Probiotics in Dairy Fermented ProductsFood Research International40951956

17.   R. H. Davidson, S. E. Duncan, C. R. Hackney, W. N. Eigel, J. W. Boling, 2000Probiotics in Dairy Fermented ProductsJournal of Dairy Science83666673

18.   L. A. de Moreno, G. Perdigon, 2010Probiotics in Dairy Fermented ProductsProc. Nutr. Soc., 69421428

19.   S. M. Deutsch, S. Parayre, A. Bouchoux, F. Guyomarc'h, J. Dewulf, M. Dols-Lafargue, F. Baglinière, F. J. Cousin, H. Falentin, G. Jan, B. Foligné, 2012Probiotics in Dairy Fermented ProductsAppl. Environ. Microbiol., 7817651775

20.   Santo. A. P. Espirito, P. Perego, A. Converti, M. N. Oliveira, 2011Influence of foodmatrices on probiotic viability: A reviewfocusing on the fruitybases. Trends in Food Science & Technology, 22377385

21.   B. Foligné, S. M. Deutsch, J. Breton, F. J. Cousin, J. Dewulf, M. Samson,

B. Pot, G. Jan, 2010Probiotics in Dairy Fermented ProductsAppl. Environ. Microbiol., 7682598264

22. R. Fuller, 1989Probiotics in Dairy Fermented ProductsJournal of Applied Bacteriology, 66365378

23. T. H. Gadaga, A. N. Mutukumira, J. A. Narvhus, S. B. Feresu, 1999A review of traditional fermented foods and beverages ofZimbabwe. International Journal of Food Microbiology, 53111

24. G. E. Gardiner, P. Bouchier, E. O'sullivan, J. Kelly, K. Collins, G. Fitzgerald, R. P. Ross, C. Stanton, 2002A spray-dried culture for porbiotic Cheddar chesse manufacture. International Dairy Journal, 12749756

25. D. Ghadimi, R. Folster-Holst, M. De Vrese, P. Winkler, K. J. Heller, J. Schrezenmeir, 2008Effects of probiotic bacteria and their genomic DNA on TH1/TH2-cytokineproduction by peripheral blood mononuclear cells (PBMCs) of healthy and allergicsubjects. Immunobiology, 213677692

26. G. R. Gibson, R. Fuller, 2000Probiotics in Dairy Fermented ProductsJournal Nutrition, 130391395

27. G. Godward, K. Sultana, K. Kailasapathy, P. Peiris, R. Arumugaswamy, N. Reynolds, 2000The importance ofstrain selection on the viability and survival of probiotic bacteria indairy foods. Milchwissenschaft, 55441445

28. S. Guglielmetti, V. Taverniti, M. Minuzzo, S. Arioli, I. Zanoni, M. Stuknyte, F. Granucci, M. Karp, D. Mora, 2010Probiotics in Dairy Fermented ProductsInfect. Immun., 7847344743

29. W. H. Holzapfel, P. Haberer, J. Snel, U. Schillinger, In't. Huis, J. H. J. Velt, 1998Overview of gut flora and probiotics. International Journal of Food Microbiology, 4185101

30. H. Q. Huynh, J. Debruyn, L. Guan, H. Diaz, M. Li, S. Girgis, J. Turner, R. Fedorak, K. Madsen, 2009Probiotics in Dairy Fermented ProductsDis., 15760768

31. C. Hyun, H. Shin, 1998Utilization of bovine plasma obtained from aslaughterhouse for economic production of probiotics. Journal of Fermentation and Bioenginering, 863437

32. V. S. Jayamanne, M. R. Adams, 2006Determination of survival,identity, and stress resistance of probiotic bifidobacteriain bio-yoghurts. Letters in Applied Microbiology189 EOF194 EOF

33. K. Kailasapathy, I. Harmstorf, M. Phillips, 2008Survival ofLactobacillus acidophilus and Bifidobacterium animalis ssp lactisin stirred fruit

yogurts. LWT-Food Science and Technology, 4113171322

34.   R. Karimi, A. Amiri-Rigi, 2010Probiotics in Dairy Products. Marz Danesh Publication,Tehran.

35.   R. Karimi, A. M. Mortazavian, A. Amiri-Rigi, 2012Probiotics in Dairy Fermented ProductsFood Microbiology2919

36.   R. Karimi, A. M. Mortazavian, Cruz. A. G. Da, 2011Probiotics in Dairy Fermented ProductsDairy Science and Technology91283308

37.   A. Kasimoglu, M. Goncuoglu, S. Akgun, 2004Probiotic White cheese withLactobacillus acidophilus. International Dairy Journal, 1410671073

38.   A. Lan, A. Bruneau, M. Bensaada, C. Philippe, P. Bellaud, S. Rabot, G. Jan, 2008Probiotics in Dairy Fermented ProductsBr. J Nutr., 10012511259

39.   M. Lopez, N. Li, J. Kataria, M. Russell, J. Neu, 2008Probiotics in Dairy Fermented ProductsJournal of Nutrition13822642268

40.   G. Loss, S. Apprich, M. Waser, W. Kneifel, J. Genuneit, G. Buchele, J. Weber, B. Sozanska, H. Danielewicz, E. Horak, R. J. Van Neerven, D. Heederik, P. C. Lorenzen, M. E. Von, C. Braun-Fahrlander, 2011Probiotics in Dairy Fermented ProductsJ. Allergy Clin. Immunol., 128766773

41.   A. Lourens-Hattingh, B. C. Viljoen, 2001aProbiotics in Dairy Fermented ProductsFood Research International34791796

42.   A. Lourens-Hattingh, B. C. Viljoen, 2001bProbiotics in Dairy Fermented ProductsInternational Dairy Journal, 11117

43.   C. M. Lynch, D. D. Muir, J. M. Banks, P. L. H. Mcsweeney, P. F. Fox, (1999). Influence of adjunct cultures of Lactobacillus paracasei ssp. paracasei or Lactobacillus plantarum on Cheddar cheese ripening. Journal of Dairy Science, 82 16181628 .

44.   G. T. Macfarlane, G. R. Gibson, J. H. Cummings, 1992Probiotics in Dairy Fermented ProductsJournal Applied Bacteriology, 72(3), 57 EOF64 EOF

45.   C. Manichanh, L. Rigottier-Gois, E. Bonnaud, K. Gloux, E. Pelletier, L. Frangeul, R. Nalin, C. Jarrin, P. Chardon, P. Marteau, J. Roca, J. Dore, 2006Reduced diversity of faecal microbiota in Crohn's disease revealed by a metagenomic approach.Gut, 55205211

46.   A. Margoles, L. Garcia, 2003Characterisation of a bifidobacteriumstrain wish acquired resistance to cholate: A preliminary study.International Journal of Food Microbiology, 80191198

47.   M. Medici, C. G. Vinderola, G. Perdigon, 2004Gut mucosal

immunomodulation by probiotic fresh chesse. International Dairy Journal, 14611618

48.   E. Miele, F. Pascarella, E. Giannetti, L. Quaglietta, N. Robert, R. N. Baldassano, Staiano. A. Annamaria, 2009Effect of a Probiotic Preparation (VSL#3) on Induction and Maintenance of Remission in Children With Ulcerative Colitis. American Journal of Gastroenterology, 104437443

49.   E. Miele, F. Pascarella, E. Giannetti, L. Quaglietta, R. N. Baldassano, A. Staiano, 2009Effect of a probiotic preparation (VSL#3) on induction and maintenance of remission in children with ulcerative colitis. Am.J. Gastroenterol., 104437443

50.   M. R. L. Moura, 2005Alimentos Funcionais: seus benefícios e a legislação: Avaiable in: http://acd.ufrj.br/consumo/leituras/ ld.htm#leituras.

51.   S. Narushima, T. Sakata, K. Hioki, T. Itoh, T. Nomura, K. Itoh, 2010Probiotics in Dairy Fermented ProductsExp. Anim., 59487494

52.   S. C. Ng, S. Plamondon, M. A. Kamm, A. L. Hart, H. O. Al-Hassi, T. Guenther, A. J. Stagg, S. C. Knight, 2010Immunosuppressive effects via human intestinal dendritic cells of probiotic bacteria and steroids in the treatment of acute ulcerative colitis.Inflamm. Bowel. Dis., 1612861298

53.   S. O'flaherty, D. M. Saulnier, B. Pot, J. Versalovic, 2010How can probiotics and prebiotics impact mucosal immunity?Gut Microbes, 1293300

54.   T. Ogita, M. Nakashima, H. Morita, Y. Saito, T. Suzuki, S. Tanabe, 2011Probiotics in Dairy Fermented ProductsJ. Biomed. Biotechnol., 2011:378417.

55.   L. Ong, A. Henriksson, N. P. Shah, 2006Probiotics in Dairy Fermented ProductsInternational Dairy Journal16446456

56.   L. Ong, A. Henrikssonb, N. P. Shaha, Probiotics in Dairy Fermented ProductsInternational Dairy Journal1720072007937945

57.   A. C. Ouwehand, S. Salminen, E. Isolauri, 2002Probiotics in Dairy Fermented ProductsAnton. Leeuw. Int. J. G., 82279289

58.   G. Perdigon, D. L. De Moreno, J. Valdez, M. Rachid, 2002Probiotics in Dairy Fermented ProductsEur. J. Clin. Nutr., 56 Suppl 3:S65S68.

59.   F. C. Prado, J. L. Parada, A. Pandey, C. R. Soccol, 2008Probiotics in Dairy Fermented ProductsFood Research International41111123

60.   R. A. Rastall, R. Fuller, H. R. Gaskins, G. R. Gibson, 2000Colonic functional foods. In Functional Foods, 7189GR Gibson and CM

Williams, editors]. Cambridge: Woodhead Publishing Limited.

61. B. S. Reddy, A. Rivenson, 1983Probiotics in Dairy Fermented ProductsCancer Research

62. G. Reid, A. Jessica, J. A. Younes, H. C. Van Der Mei, G. B. Gloor, R. Knight, H. J. Busscher, 2011Gut flora restoration: natural and supplemented recovery of human microbial communities. Nature Reviews. 92738

63. S. Rodgers, 2008Probiotics in Dairy Fermented ProductsTrends Food Sci. Tech. 19188197

64. R. P. Ross, G. Fitzgerald, K. Collins, C. Stanton, 2002Cheese delivering biocultures: probiotic cheese. Australian Journal of Dairy Technology71 EOF78 EOF

65. M. Saarela, G. Mogensen, R. Fonden, J. Matto, T. Mattila-Sandholm, 2000Probiotics in Dairy Fermented ProductsJournal of Biotechnology84197215

66. A. Samona, R. K. Robinson, 1994Probiotics in Dairy Fermented ProductsJournal of the Society of Dairy Technology475860

67. C. Santos-Rocha, O. Lakhdari, H. M. Blottiere, S. Blugeon, H. Sokol, L. G. Bermu'dez-Humara'n, V. Azevedo, A. Miyoshi, J. Dore, P. Langella, E. Maguin, G. M. Van De , 2012Probiotics in Dairy Fermented ProductsInflamm. Bowel. Dis., 18657666

68. E. Sendra, P. Fayos, Y. Lario, J. Fernandez-Lopez, E. Sayas-Barbera, J. Perez-Alvarez, 2008Probiotics in Dairy Fermented ProductsFood Microbiology251321

69. N. P. Shah, W. E. V. Lankaputhra, M. L. Britz, W. S. A. Kyle, 1995Probiotics in Dairy Fermented ProductsInternational Dairy Journal, 5515521

70. H. Sokol, B. Pigneur, L. Watterlot, O. Lakhdari, L. G. Bermudez-Humaran, J. J. Gratadoux, S. Blugeon, C. Bridonneau, J. P. Furet, G. Corthier, C. Grangette, N. Vasquez, P. Pochart, G. Trugnan, G. Thomas, H. M. Blottiere, J. Dore, P. Marteau, P. Seksik, P. Langella, 2008Faecalibacterium prausnitzii is an anti-inflammatory commensal bacterium identified by gut microbiota analysis of Crohn disease patients.Proc. Natl. Acad. Sci. U. S. A, 1051673116736

71. C. H. B. Souza, S. M. I. Saad, 2009Probiotics in Dairy Fermented ProductsLWT e Food Science and Technology, 42(2), 633 EOF640 EOF

72. C. H. B. Souza, F. C. A. Buriti, J. H. Behrens, S. M. I. Saad, 2008Probiotics in Dairy Fermented ProductsInternational Journal of Food Science and

Technology, 43(5), 871 EOF877 EOF

73. A. E. Stapleton, M. Au-Yeung, T. M. Hooton, D. N. Fredricks, P. L. Roberts, C. A. Czaja, Y. Yarova-Yarovaya, T. Fiedler, M. Cox, W. E. Stamm, 2011Probiotics in Dairy Fermented ProductsClinical Infectious Diseases1212 EOF1217 EOF

74. A. Talwalkar, K. Kailasapathy, 2004Probiotics in Dairy Fermented ProductsInternational Dairy Journal, 14142149

75. A. Talwalkar, C. W. Miller, K. Kailasapathy, M. H. Nguyen, (2004, 2004Effect of packaging materials and dissolved oxygen on the survivalof probiotic bacteria in yoghurt. International Journal of Food Science and Technology, 39(6), 605-611.

76. K. G. Thamer, A. L. B. Penna, 2005Probiotics in Dairy Fermented ProductsRevista Brasileira de Ciências Farmacêuticas

77. D. L. Topping, P. M. Clifton, 2001SHorty-chain fatty acids and human colonic function: roles of resistant starch and nonstarch polysaccharides. Physiological Reviews, 81(3), 1031-1064.

78. C. G. Vinderola, W. Prosello, D. Ghiberto, J. Reinheimer, 2000Viability of probiotic (Bifidobacterium, Lactobacillus acidophilus and Lactobacillus casei) and nonprobiotic microflora in argentinianFresh cheese. Journal of Dairy Science, 8319051911

79. M. Waser, K. B. Michels, C. Bieli, H. Floistrup, G. Pershagen, M. E. Von, M. Ege, J. Riedler, D. Schram-Bijkerk, B. Brunekreef, H. M. Van , R. Lauener, C. Braun-Fahrlander, 2007Probiotics in Dairy Fermented ProductsClin. Exp. Allergy, 37661670

80. S. Wohlgemuth, Loh. G. Gunnar, M. Blaut, 2010Probiotics in Dairy Fermented ProductsInternational Journal of Medical Microbiology300310

81. G. D. Wu, J. Chen, C. Hoffmann, K. Bittinger, Y. Y. Chen, S. A. Keilbaugh, M. Bewtra, D. Knights, W. A. Walters, R. Knight, R. Sinha, E. Gilroy, K. Gupta, R. Baldassano, L. Nessel, H. Li, F. D. Bushman, J. D. Lewis, 2011Linking long-term dietary patterns with gut microbial enterotypes.Science, 334105108

82. F. Yan, H. Cao, T. L. Cover, M. K. Washington, Y. Shi, L. Liu, R. Chaturvedi, R. M. Peek Jr, K. T. Wilson, D. B. Polk, 2011Probiotics in Dairy Fermented ProductsJournal of Clinical Investigation2242 EOF2253 EOF

83. H. Yang, M. C. Adams, 2004In vitro assessment of the upper gastrointestinal tolerance of potential probiotic dairy propionibacteria.

International Journal of Food Microbiology, 91253260

84.   T. Yatsunenko, F. E. Rey, M. J. Manary, I. Trehan, M. G. Dominguez-Bello, M. Contreras, M. Magris, G. Hidalgo, R. N. Baldassano, A. P. Anokhin, A. C. Heath, B. Warner, J. Reeder, J. Kuczynski, J. G. Caporaso, C. A. Lozupone, C. Lauber, J. C. Clemente, D. Knights, R. Knight, J. I. Gordon, 2012Probiotics in Dairy Fermented ProductsNature486222227

Chapter 3

ASSESSMENT OF THE CONCENTRATIONS OF VARIOUS ADVANCED GLYCATION END-PRODUCTS IN BEVERAGES AND FOODS THAT ARE COMMONLY CONSUMED IN JAPAN

Masayoshi Takeuchi[1], Jun-ichi Takino[2] , Satomi Furuno[3] , Hikari Shirai[3], Mihoko Kawakami[3] , Michiru Muramatsu[3] , Yuka Kobayashi[3] , Sho-ichi Yamagishi[4]

[1] Department of Advanced Medicine, Medical Research Institute, Kanazawa Medical University, Uchinadamachi, Ishikawa, Japan,

[2]Laboratory of Biochemistry, Faculty of Pharmaceutical Sciences, Hiroshima International University, Kure, Hiroshima, Japan,

[3]Department of Pathophysiological Science, Faculty of Pharmaceutical Science, Hokuriku University, Kanazawa, Ishikawa, Japan,

[4]Department of Pathophysiology and Therapeutics of Diabetic Vascular Complications, Kurume University School of Medicine, Kurume, Fukuoka, Japan

ABSTRACT

Dietary consumption has recently been identified as a major environmental source of pro-inflammatory advanced glycation end-products (AGEs) in humans. It is disputed whether dietary AGEs represent a risk to human health. N^{ϵ}-(carboxymethyl)lysine (CML), a representative AGE compound found in food, has been suggested to make a significant contribution to circulating CML levels. However, recent studies have found that the dietary intake of AGEs is not associated with plasma CML concentrations. We have shown that the serum levels of glyceraldehyde-derived AGEs (Glycer-AGEs), but not hemoglobin A1c, glucose-derived AGEs (Glu-AGEs), or CML, could be used as biomarkers for predicting the progression of atherosclerosis and future cardiovascular events. We also detected the production/accumulation of Glycer-AGEs in normal rats administered Glu-AGE-rich beverages. Therefore, we assessed the concentrations of various AGEs in a total of 1,650 beverages and foods that are commonly consumed in Japan. The concentrations of four kinds of AGEs (Glu-AGEs, fructose-derived AGEs (Fru-AGEs), CML, and Glycer-AGEs) were measured with competitive enzyme-linked immunosorbent assays involving immunoaffinity-purified specific antibodies. The results of the latter assays indicated that Glu-AGEs and Fru-AGEs (especially Glu-AGEs), but not CML or Glycer-AGEs, are present at appreciable levels in beverages and foods that are commonly consumed by Japanese. Glu-AGEs, Fru-AGEs, CML, and Glycer-AGEs exhibited concentrations of $\geq 85\%$, $2-12\%$, $<3\%$, and trace amounts in the examined beverages and $\geq 82\%$, $5-15\%$, $<3\%$, and trace amounts in the tested foods, respectively. The results of the present study indicate that some lactic acid bacteria beverages, carbonated drinks, sugar-sweetened fruit drinks, sports drinks, mixed fruit juices, confectionery (snacks), dried fruits, cakes, cereals, and prepared foods contain markedly higher Glu-AGE levels than other classes of beverages and foods. We provide useful data on the concentrations of various AGEs, especially Glu-AGEs, in commonly consumed beverages and foods.

INTRODUCTION

In humans, two major sources of advanced glycation end-products (AGEs) have been identified, exogenous and endogenous AGEs [1–5]. AGEs are formed by the Maillard reaction, a non-enzymatic reaction between the aldehyde or ketone groups of reducing sugars, such as glucose, fructose, and glyceraldehyde, and the terminal α-amino groups or ϵ-amino groups of protein lysine residues [2–5]. AGEs were originally characterized by their yellow-brown fluorescent color and their ability to form cross-links with and between amino groups; however, the term is now used for a broad range of advanced products of the glycation process, including N-(carboxymethyl)lysine (CML), N-(carboxyethyl) lysine (CEL), and pyrraline, which are colorless, do not fluoresce, and do not form cross-links with proteins [1–5]. The use of CML as a marker of AGE formation in food has recently led to the development of a database containing the CML concentrations of 549 foodstuffs [6,7]. However, the inconsistencies between the information in this database and data obtained with other methods highlight the considerable challenges associated with analyzing AGEs [8]. Moreover, dietary CML might pose a risk to human health, as it enhances oxidative stress and initiates inflammatory responses, which ultimately lead to atherosclerosis [9,10]. While previous studies have suggested that dietary CML makes a significant contribution to *in vivo* CML concentrations [10], two recent studies have reported that this is not true for humans [11] or rats [12]. Semba*et al.* suggested that the excessive consumption of foods considered to be high in AGEs might not have a major effect on serum CML concentrations [11]. We previously demonstrated that glucose, fructose, α-hydroxyaldehydes (glyceraldehyde and glycolaldehyde), and dicarbonyl compounds (glyoxal (GO), methylglyoxal (MGO), and 3-deoxyglucosone (3-DG)) were actively involved in protein glycation [13–16]. Seven immunochemically distinct classes of AGEs (glucose-derived AGEs, Glu-AGEs; fructose-derived AGEs, Fru-AGEs; glyceraldehyde-derived AGEs, Glycer-AGEs; glycolaldehyde-derived AGEs, Glycol-AGEs; GO-derived AGEs, GO-AGEs; MGO-derived AGEs, MGO-AGEs; and 3-DG-derived AGEs, 3-DG-AGEs) have been detected in the sera of type 2 diabetic patients on hemodialysis [13–16]. Based on these findings, we proposed an *in vivo* pathway for the formation of distinct AGEs involving the Maillard reaction, sugar autoxidation, and sugar metabolic pathways, as shown in Fig. 1.

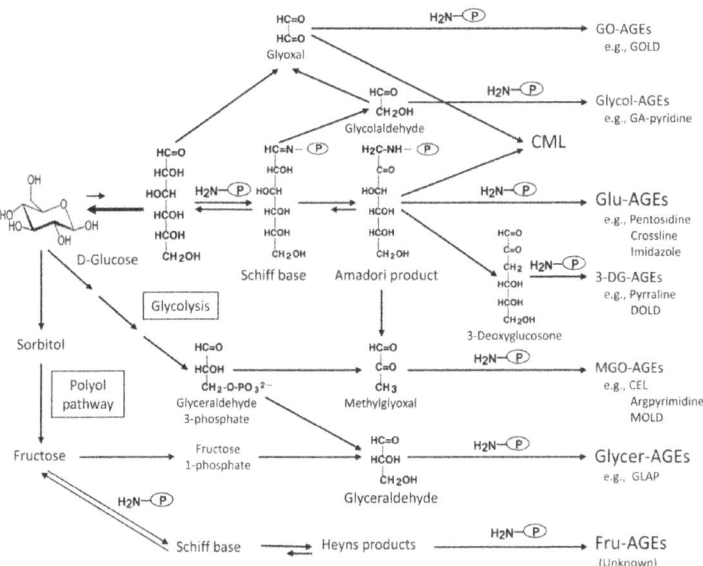

Figure 1: Alternative *in vivo* **AGE synthesis routes.** Reducing sugars, such as glucose, fructose, and glyceraldehyde, are known to react non-enzymatically with the amino groups of proteins to form reversible Schiff bases and Amadori/Heyns products. These early glycation products undergo further complex reactions such as rearrangement, dehydration, and condensation to become irreversibly cross-linked heterogeneous fluorescent derivatives, termed AGEs. Glu-AGEs, glucose-derived AGEs; Fru-AGEs, fructose-derived AGEs; Glycer-AGEs, glyceraldehyde-derived AGEs; Glycol-AGEs, glycolaldehyde-derived AGEs; MGO-AGEs, methylglyoxal-derived AGEs; GO-AGEs, glyoxal-derived AGEs; 3-DG-AGEs, 3-deoxyglucosone-derived AGEs; CML, N-(carboxymethyl)lysine; P-NH$_2$, free amino residue of protein; GOLD, GO-lysine dimer; GA-pyridine, 3-hydroxy-4-hydroxymethyl-1-(5-carboxy-

pentyl) pyridinium cation; DOLD, 3-DG-lysine dimer; CEL, N-(carboxyethyl)lysine; MOLD, MGO-lysine dimer; GLAP, glyceraldehyde-derived pyridinium compound. doi:10.1371/journal.pone.0118652.g001

We recently demonstrated that interactions between Glycer-AGEs and the receptor for AGEs (RAGE) affect intracellular signaling, gene expression, and the release of pro-inflammatory molecules and also induce oxidative stress in numerous types of cells, all of which can contribute to the pathological changes observed in various chronic diseases [17–19]. Furthermore, we detected increased hepatic RAGE expression and the enhanced production/ accumulation of Glycer-AGEs in normal rats administered Glu-AGE-rich beverages [20]. These findings indicate that Glu-AGEs, which are frequently found in beverages and foods, and hence, are taken into the body orally, enhance the production/accumulation of Glycer-AGEs, leading to Glycer-AGE-RAGE interactions. Our recent studies indicated that the serum levels of Glycer-AGEs, but not hemoglobin A1c (HbA1c), Glu-AGEs, or CML, could be used as biomarkers for predicting the progression of lifestyle-related diseases [21–25]. We previously demonstrated that the enhanced production/ accumulation of Glycer-AGEs after the oral consumption of Glu-AGEs plays an important role in the pathogenesis of vascular damage [26].

Therefore, the objective of the present study was to assess the concentrations of four kinds of AGEs; i.e., Glu-AGEs, Fru-AGEs, CML, and Glycer-AGEs, which have been detected in the sera of both non-diabetic and diabetic subjects [13–16,21–26], in beverages and foods that are commonly consumed in Japan.

MATERIALS AND METHODS

Preparation of various advanced glycation end-products (AGEs)

Glu-AGEs, Fru-AGEs, Glycer-AGEs, and CML proteins were prepared as described in previous studies [13–16]. Briefly, 25 mg/ml rabbit serum albumin (RSA, A0764, Sigma-Aldrich, St Louis, MO, USA) or bovine serum albumin (BSA, A0281, Sigma-Aldrich) were incubated under sterile conditions with 0.5 M D-glucose, D-fructose, or 0.1 M DL-glyceraldehyde and 5 mM diethylenetriaminepentaacetic acid (Dojindo Laboratories, Kumamoto, Japan) in 0.2 M phosphate buffer (pH 7.4) at 37°C for 8 weeks (or 7 days for the samples incubated with glyceraldehyde). Low molecular weight (LMW) reactants and the glucose, fructose, or glyceraldehyde were removed using a PD-10 chromatography column (GE Healthcare UK, Buckinghamshire, UK) and dialysis against phosphate buffered saline (PBS). CML-RSA, CML-BSA, CML-human serum albumin (HSA), CML-hemoglobin (Hb), and CML-ribonuclease A (RNase A) were prepared as described previously [13]. Briefly, 50 mg/ml each of RSA, BSA, HSA, Hb, and RNase A were incubated at 37°C for 24 h with 50 mM glyoxylic acid and 150 mM sodium cyanoborohydride in 0.2 M phosphate buffer (pH 7.4), before being subjected to PD-10 column chromatography and dialysis against PBS. Protein concentrations were determined with the DC protein assay reagent (Bio-Rad Laboratories, Richmond, CA, USA) using BSA as a standard.

Preparation of AGE-specific antibodies

Four kinds of immunoaffinity-purified antibodies were prepared as described in previous studies [13–16]. Briefly, 4 mg of three types of AGE-RSA (incubated with glucose, fructose, or glyceraldehyde) were emulsified in 50% Freund's complete adjuvant (Wako Pure Chemical Industries, Osaka, Japan) and then injected intradermally into Japanese white rabbits (Sankyo Labo Service Corporation, Tokyo, Japan). This procedure was repeated at weekly intervals for 6 weeks. After a 2-week break, the rabbits were given a booster injection of 4 mg of the appropriate antigen. The animals were bled on the 10th day after the last injection, and their sera were obtained for further affinity purification. CNBr-activated Sepharose 4B gels (GE Healthcare UK) were coupled to Glu-AGE-BSA, Fru-AGE-BSA, Glycer-AGE-BSA, or CML-BSA, as described previously [13–16]. The anti-Glu-AGE antiserum, which contained anti-Glu-AGE and anti-CML antibodies, was applied to a column (2.5 x 5.5 cm) containing Sepharose 4B coupled to Glu-AGE-BSA. After extensive washing with PBS, the adsorbed fractions were eluted with 20 mM sodium phosphate buffer containing 1 M potassium thiocyanate (pH 7.4) (elution buffer). The eluted fractions were pooled, concentrated using Centriprep-10 (Millipore Corporation, Billerica, MA, USA), and passed through a PD-10 column equilibrated with PBS. The eluted fraction was then loaded onto a column (1.5 x 5.5 cm) containing Sepharose 4B coupled with CML-BSA, which was washed with PBS to obtain the unadsorbed fraction (anti-Glu-AGE antibody). The adsorbed fraction (anti-CML antibody) was then eluted with elution buffer. The anti-Glu-AGE and anti-CML antibodies were pooled (separately), concentrated with Centriprep-10, and passed through a PD-10 column equilibrated with PBS, before being used in this study [13,15]. The anti-Fru-AGE or anti-Glycer-AGE antisera were applied to columns containing Sepharose 4B coupled to Fru-AGE- or Glycer-AGE-BSA. After extensive washing with PBS, the adsorbed fractions were eluted with elution buffer. Each of the eluted fractions was pooled, concentrated using Centriprep-10, and passed through a PD-10 column equilibrated with PBS. Then, they were loaded onto a column containing Sepharose 4B coupled to CML-BSA, which was washed with PBS to obtain the unadsorbed fraction (anti-Fru-AGE or anti-Glycer-AGE antibodies). The anti-Fru-AGE and anti-Glycer-AGE antibodies were pooled (separately), concentrated with Centriprep-10, and passed through a PD-10 column equilibrated with PBS and then used in this study [14,16].

The immunoaffinity-purified anti-Glu-AGE antibody did not recognize well-characterized AGE structures, such as CML, CEL, pyrraline, pentosidine, argpyrimidine, imidazolone, GO-lysine dimers (GOLD), and MGO-lysine dimers (MOLD). In addition, it did not recognize AGEs whose structures remain unknown, such as Fru-AGEs, Glycer-AGEs, Glycol-AGEs, GO-AGEs, MGO-AGEs, and 3-DG-AGEs [13,15,27]. Instead, the anti-Glu-AGE antibody specifically recognized unique unknown Glu-AGE structures. The immunoaffinity-purified anti-Fru-AGE antibody did not recognize well-characterized AGE structures, such as CML, CEL, pyrraline, pentosidine, and argpyrimidine. In addition, it did not recognize AGE whose structures remain unknown, such as Glu-AGEs, Glycer-AGEs, Glycol-AGEs, GO-AGEs, MGO-AGEs, and 3-DG-AGEs [16]. Instead, the anti-Fru-AGE antibody specifically recognized unique unknown Fru-AGE structures. The immunoaffinity-purified anti-Glycer-AGE antibody did not recognize well-characterized AGE structures, such as CML, CEL, pyrraline, pentosidine, argpyrimidine, imidazolone, GOLD, MOLD, and glyceraldehyde-derived pyridinium (GLAP). Furthermore, it did not recognize AGE whose structures are unknown, such as Glu-AGEs, Fru-AGEs, Glycol-AGEs, GO-AGEs, MGO-AGEs, and 3-DG-AGEs [14,23]. Instead, the anti-Glycer-AGE antibody specifically recognized unique unknown Glycer-AGE structures. The three types of AGE antibodies were able to detect both high-molecular weight (HMW) and LMW Glu-AGEs, Fru-AGEs, or Glycer-AGEs with unique unknown structures in serum [13–16,21–26]. On the other hand, the immunoaffinity-purified anti-CML antibody recognized a common epitope that is shared by protein-bound CML (such as CML-RSA, CML-BSA, CML-HSA, CML-Hb, and CML-RNase A) and free CML molecules, whereas it did not react with lysine, Amadori products (glycated HSA), unmodified proteins, pentosidine, pyrraline, argpyrimidine, imidazolone, or CEL [13]. The anti-CML antibody was able to detect both HMW and LMW CML structures in serum [13].

Competitive enzyme-linked immunosorbent assay (ELISA)

The concentrations of four kinds of AGEs (Glu-AGEs, Fru-AGEs, CML, and Glycer-AGEs) were measured with competitive ELISA using the immunoaffinity-purified antibodies described above. Briefly, a 96-well (flat bottomed without a lid, high binding) enzyme immunoassay / radioimmunoassay plate (Corning Incorporated, Corning, NY, USA) was coated with 1 µg/ml of

Glu-AGE-BSA, Fru-AGE-BSA, CML-BSA, or Glycer-AGE-BSA standard solution and incubated overnight at 4°C. The wells were washed three times with 0.3 ml of the washing solution (PBS containing 0.05% Tween-20), before being blocked *via* incubation for 1 h with 0.2 ml of a PBS solution containing 1% BSA. After the wells had been washed with the washing solution, test samples (50 µl) were added to each well as competitors for 50 µl of immunoaffinity-purified anti-Glu-AGE, anti-Fru-AGE, anti-CML, or anti-Glycer-AGE antibodies (1:1000~1:2500 dilution), and then the plates were incubated for 2 h at room temperature under gentle shaking in a horizontal rotary shaker (EYELA, MMS-1, Tokyo, Japan). The wells were then washed with washing solution and developed with an alkaline phosphatase-conjugated sheep anti-rabbit IgG (Millipore Corporation, Billerica, MA, USA) using p-nitrophenyl phosphate as the colorimetric substrate (Pierce, Rockford, IL, USA). The AGE concentrations of each sample were read from the calibration curves for Glu-AGE-BSA, Fru-AGE-BSA, CML-BSA, or Glycer-AGE-BSA standards and were expressed as Glu-AGE, Fru-AGE, CML, or Glycer-AGE units (U) per ml, where 1U corresponded to 1 µg of the Glu-AGE-BSA, Fru-AGE-BSA, CML-BSA, or Glycer-AGE-BSA standard [13–16].

Assessment of the concentrations of various AGEs

Commonly consumed beverages and foods were obtained from vending machines, convenience stores, supermarkets, fast food stores (including doughnut or hamburger stores), bento-ya (shops that sell lunch sets), or family restaurants. Samples of beverages and liquefied foods (mainly seasonings) were analyzed using competitive ELISA after they had been diluted. To prepare food samples for the AGE measurements, solid food was first crushed uniformly with a food processor (Cuisinart, Mini-Prep Processor/Little Pro Plus, Tokyo, Japan). In the case of the bento-ya/convenience store lunch boxes, any fish, meat, vegetables (including nimono/aemono), seasonings and spices, and rice within them were examined separately, whereas for hamburgers, any meat/fish, vegetables, sauce, and bread/rice were examined separately. We then weighed out 5 g of the uniformly crushed food, added 45 ml of the sample dilution buffer (Tris/HCl buffer containing 0.05% Tween-20, pH 7.4), and homogenized it for 1 min at 15,000 rpm using an Ace Homogenizer (Nippon Seiki Co., Ltd., Nagaoka, Japan). The homogenate was rotated with a tube rotator (AS ONE, AM-9, Osaka, Japan) for 3 h at room temperature. After being centrifuged for 20 min at 3,500 rpm at 4°C, the supernatant was used for the assessments of AGE levels. We purchased at least 2 of each beverage and food, and prepared at least 2 samples of each product for AGE measurement. The concentrations of the four abovementioned types of AGEs in each beverage or food extract were measured using competitive ELISA after the extract had been diluted 10- to 100,000-fold (after controlling for dilution) with sample dilution buffer. The AGE concentrations of each beverage and food item are shown as mean values of at least three measurements per sample and are expressed as AGE units (U). The tests for the four kinds of AGEs displayed sensitivity values of 0.1 U/ml. In the case of the bento-ya/convenience store lunch boxes, the AGE concentrations of any fish, meat, vegetables (including nimono/aemono), seasonings and spices, and rice were combined and expressed as mean units per meal/lunch box. For hamburgers, the AGE concentrations of any meat/ fish, vegetables, sauce, and bread/rice were combined and expressed as mean units per hamburger. The concentrations of the four kinds of AGEs were calculated based on standard serving sizes (65~500 ml/bottle) for beverages or on the standard serving sizes consumed by average Japanese people for foods (e.g., one bag/box of confectionery, one bag/box of prepared foods, 15 (~150 for noodle tsuyu) g/ml seasonings and spices, one hamburger, and one convenience store lunch box).

RESULTS

Assessment of the AGE concentrations of common beverages as determined by their AGE levels

We classified beverages according to the Japanese Agricultural Standard (JAS). The concentrations of the four kinds of AGEs in each type of beverage are shown in Fig. 2. Glu-AGEs, Fru-AGEs, CML, and Glycer-AGEs exhibited concentrations of $\geq 85\%$, 2–12%, <3%, and trace amounts, respectively. Table 1 shows a list of commonly consumed beverages that exhibited Glu-AGE concentrations of $\geq 100,000$ U/bottle. The highest Glu-AGE concentrations were detected in the lactic acid bacteria beverages Pil·Cres (two types; 264,090 and 240,870 U/65 ml bottle), Yakult (eight types; max: 243,890—min: 60,900 U/65 or 80 ml bottle), and lactic acid bacteria Power Peach (171,320 U/250 ml bottle). The lactic acid bacteria beverages containing large amounts of Glu-AGEs were all produced *via* processes in which high-fructose corn syrup (HFCS) and skimmed milk were mixed and reacted at high temperature, which would have altered the mixture's Glu-AGE levels, before the seeding and culturing of lactic acid bacteria. On the other hand, the lactic acid bacteria beverages that had been colored using caramel had low levels of Glu-AGEs. Moreover, beverages that contained artificial sweeteners and carbonated drinks displayed low levels of Glu-AGEs. Tea, black coffee, and oolong tea did not contain many Glu-AGEs. The numbers of beverages in each category that contained $\geq 100,000$; 50,000–99,999; 20,000–49,999; and <20,000 U/bottle of

Glu-AGEs are shown in Table 2. The concentration of Glu-AGEs was ≥20,000 U/bottle in *ca*. 45% of the beverages examined.

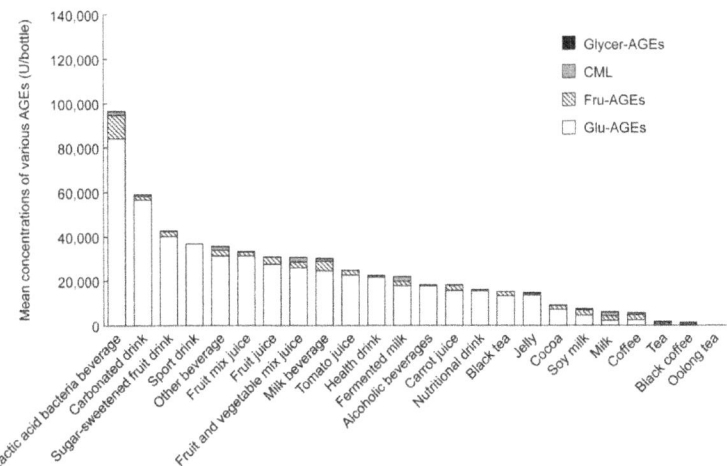

Figure 2: Mean concentrations of various AGEs in commonly consumed beverages. Beverages were classified according to the Japanese Agricultural Standard (JAS). The mean concentrations of four kinds of AGEs (Glu-AGEs, Fru-AGEs, CML, and Glycer-AGEs) in each beverage are expressed as AGE units (U) per bottle and are based on the standard serving size (65~500 ml/bottle) of each product in Japan. doi:10.1371/journal.pone.0118652.g002

Table 1: List of common beverages containing Glu-AGE levels of ≥100,000 U/bottle. doi:10.1371/journal.pone.0118652.t001

Name	AGE concentrations (U/bottle)*				Serving size(ml)
	Glu-AGEs	Fru-AGEs	CML	Total AGEs	
Pil Cre Probiotics	264,090	32,750	140	296,980	65
Pil Cre Sawayakajitate	240,870	24,450	130	265,450	65
Yakult	243,890	18,570	140	262,600	65
Fanta Fruit Punch	232,540	1,290	330	234,160	500
Natchan! Soda (Junnsuijitate Fruity Grape)	201,810	1,350	0	203,160	500
Fanta Grape	200,250	1,990	0	202,240	500
Yakult 300V LT	164,150	17,410	180	181,740	80
Prune 100	149,870	29,140	1,190	180,200	125
Lactic Acid Bacteria Power Peach	171,320	4,600	320	176,240	250
Canada Dry Ginger Ale Extra	173,130	1,930	0	175,060	500
Yakult 300V	145,940	17,250	170	163,360	80
Blue Ginger	151,350	2,120	480	153,950	500
Fanta Orange	146,830	850	130	147,810	500
Yakult SHEs	123,270	19,970	190	143,430	80
Yakult 400	124,990	14,050	140	139,180	80
Mountain Dew	130,310	2,400	90	132,800	500
Real KIAIDA Kiaida-!!	126,880	1,330	130	128,340	280
Lemons Lemon	121,050	5,670	850	127,570	140
Yakult 80 Ace	110,380	15,730	210	126,320	80
Fanta Green Apple	119,490	1,290	120	120,900	500
Lipton Sparkling Tea Soda (Cassis & White grape)	119,300	410	70	119,780	500
Mitsuya Cider (Sukattoshiroi)	100,470	12,920	3,420	116,810	500
Orange (TOPVALU)	108,320	5,210	1,120	114,650	500
Kokoichibann! Macano-gennki	104,460	3,150	30	107,640	100
Tottemo Orange (Qoo)	103,930	2,590	240	106,760	500
ORANGE ADE (Kobe-kyoryuuchi)	105,040	1,680	40	106,760	350
Junnsuimikann (Koiwai)	100,800	3,190	190	104,180	500

*the Glycer-AGE concentrations of these products are not shown because they were too low to assess accurately.

Table 2: The number of beverages that had their Glu-AGE concentrations tested (n = 885). doi:10.1371/journal.pone.0118652.t002

	Mean Glu-AGE concentrations	(U/bottle) ≥100,000	50,000–99,999	20,000–49,999	<20,000
Beverages (660):					
Carbonated Drinks (70)	56,670	12	19	21	18
Oolong Tea (7)	0				7
Black Tea (24)	13,240		1	7	16
Fruit Juice (184):					
Fruit Juice (64)	27,530	1	7	26	30
Mixed Fruit Juice (27)	31,350		4	12	11
Mixed Fruit and Vegetable Juice (28)	25,900		1	16	11
Sugar-sweetened Fruit Drinks (65)	40,200	4	16	21	24
Tomato Juice (16)	22,660			11	5
Carrot Juice (5)	15,620			2	3
Coffee (65):					
Black Coffee (23)	470				23
Coffee (42)	2,500			2	40
Soy Milk (22)	4,520			2	20
Other Beverages (267):					
Sports Drinks (14)	36,670		4	7	3
Health Drinks (13)	21,640		2	3	8
Nutritional Drinks (35)	15,540		1	10	24
Cocoa (10)	7,190		1		9
Jelly (53)	13,630			19	34
Tea (36)	570				36
Other Beverages (106)	31,260	1	20	45	40
Milk and Dairy Products (90):					
Milk (11)	2,270				11
Dairy Products (79):					
Fermented Milk (18)	17,880			6	12
Milk-based Beverages (35)	24,570		6	12	17
Lactic Acid Bacteria-containing Beverages (26)	84,100	9	13		4
Alcoholic Beverages (135)	17,750		6	50	79
(Number of beverages)		(27)	(101)	(272)	(485)

Assessment of the AGE concentrations of common foods as determined by their AGE levels

We classified the foods according to the Standard Tables of Food Composition in Japan (5th revised and enlarged edition, 2009). The concentrations of the four kinds of AGEs were measured in over 750 commercially available foods based on the standard servings consumed by average Japanese people. The concentrations of the four kinds of AGEs in each type of food are shown in Fig. 3. Glu-AGEs, Fru-AGEs, CML, and Glycer-AGEs exhibited concentrations of ≥82%, 5–15%, <3%, and trace amounts, respectively. Commonly consumed foods that demonstrated Glu-AGE levels of ≥100,000 U/meal are shown in Table 3. The highest Glu-AGE levels in the small food group were detected in snacks that had been prepared *via* dry-heat processing. These products contained high levels of Glu-AGEs because the processes used to produce them involved the heating of raw materials containing large amounts of reducing sugars (such as HFCS and dried fruit) and soybean flour or flour (which contain large quantities of lysine) at a high temperature for a long period of time. On the other hand, prepared foods such as bentos containing many vegetables and broiled fish contained low levels of Glu-AGEs. Moreover, the algae, vegetables, and pulses did not contain high levels of Glu-AGEs. The numbers of foods in each category that contained ≥100,000; 50,000–99,999; 20,000–49,999; and <20,000 U/meal of Glu-AGEs are shown in Table 4. Approximately 23% of the foods examined contained ≥20,000 U of Glu-AGEs per meal.

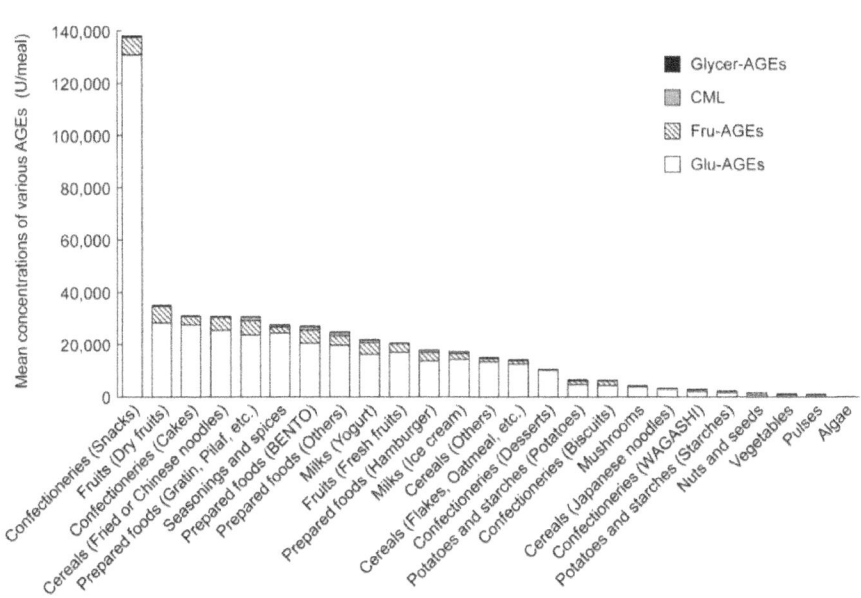

Figure 3: Mean concentrations of various AGEs in commonly consumed foods. Food items were classified according to the Standard Tables of Food Composition (5th revised and enlarged edition, 2009). The mean concentrations of four kinds of AGEs (Glu-AGEs, Fru-AGEs, CML, and Glycer-AGEs) in each food item are expressed as AGE units (U) per meal and are based on the standard servings consumed by average individuals in Japan. doi:10.1371/journal.pone.0118652.g003

Table 3: List of common foods with Glu-AGE levels of $\geq 100,000$ U/meal. doi:10.1371/journal.pone.0118652.t003

Name	AGE concentrations (U/meal)*				Serving size (g)
	Glu-AGEs	Fru-AGEs	CML	Total AGEs	
SOYJOY Apricot	893,780	25,340	500	919,620	30
SOYJOY Mango + Coconut	834,100	22,100	500	856,700	30
SOYJOY Raisin + Almond	828,010	21,450	530	849,990	30
SOYJOY Hawthorn	788,720	20,200	450	809,370	30
SOYJOY Prune	540,530	14,110	520	555,160	30
SOYJOY Strawberry	506,520	25,090	520	532,130	30
SOYJOY Cocoa +Orange	456,650	23,050	520	480,220	30
SOYJOY Apple	453,590	21,280	520	475,390	30
Light Meal Soy Bar Raisin + Almond Flavor	433,320	19,190	480	452,990	27
Rice au Gratin (Joutou-youshoku)	335,710	15,360	2,340	353,410	250
Light Meal Block Fruit Flavor	282,780	37,060	600	320,440	41
Light Meal Soy Bar Pineapple Flavor	277,760	12,210	470	290,440	27
SOYJOY Blueberry	270,210	18,500	460	289,170	30
Old Fashion	250,770	3,900	760	255,430	62
Light Meal Soy Bar Cocoa + Orange Flavor	206,450	11,570	420	218,440	27
Eel Bento	203,340	7,540	4,430	215,310	(a meal)
Chocolate Fashion	206,990	3,180	460	210,630	65
Light Meal Block Blueberry Yogurt Flavor	192,430	12,940	610	205,980	41
Calorie Mate Block Fruit Flavor	178,720	10,310	270	189,300	40
Old Fashion Maccha Green Tea	179,710	2,890	430	183,030	62
SOYJOY Banana	174,920	4,990	120	180,030	30
Old Fashion Maccha Green Tea Chocolate	161,010	4,410	400	165,820	64
SOYJOY Orange	157,750	5,770	550	164,070	30
Hamburger Steak with Sauce (Ogawaken)	145,610	10,430	5,040	161,080	210
Natsukashino Sauce Yakisoba	144,150	8,660	440	153,250	184
Tomato Ketchup (Kagome)	150,040	3,040	130	153,210	15
Pudding Super BIG (Meiji)	137,340	12,120	250	149,710	200
Hayashi Rice (Ogawaken)	135,400	7,500	1,150	144,050	350
Organic Tomato Ketchup (Del Monte)	140,710	2,220	80	143,010	15
Haagen-Dazs Bitter Caramel	126,210	3,990	90	130,290	120
Honnkaku Okonomi Sauce (Otafuku)	125,290	1,580	150	127,020	15
Ippei-chan Yomiseno Yakisoba	115,920	4,920	1,210	122,050	265
Hiyashichuuka Soup Soy Sauce	101,590	7,670	2,580	111,840	150
GYU-DON (Kinno-donburi)	104,790	2,660	1,500	108,950	180
HAYASHI (Curry-ya)	103,260	4,540	400	108,200	210

*the Glycer-AGE concentrations of these products are not shown because they were too low to assess accurately.

Table 4: The number of food items that had their Glu-AGE concentrations tested (n = 767). doi:10.1371/journal.pone.0118652.t004

	Mean Glu-AGE concentrations	(U/meal)			
		≥100,000	50,000–99,999	20,000–49,999	<20,000
Cereals (87):					
Fried or Chinese Noodles (22)	25,510	2	3	3	14
Flakes, Oatmeal, and Rice (10)	12,520		1	2	7
Japanese Noodles (14)	2,970			1	13
Others (41)	13,300		3	6	32
Potatoes and Starches (27):					
Potatoes (8)	4,550				8
Starches (19)	1,680				19
Pulses (15)	770				15
Nuts and Seeds (8)	900				8
Vegetables (39)	740				39
Fruits (29):					
Fresh Fruit (21)	16,910			7	14
Dried Fruit (8)	28,400		1	2	5
Mushrooms (12)	3,800				12
Algae (22)	160				22
Milks (44):					
Yogurt (16)	16,310		3	1	12
Ice Cream (28)	14,300	1		5	22
Confectionery (197):					
Snacks (60)	130,840	17	1	4	38
Cakes (49)	28,060	4	1	6	38
Desserts (45)	10,110	1		1	43
Biscuits (27)	4,420			1	26
Wagashi (16)	2,080				16
Seasonings and Spices (78)	24,540	4	4	24	46
Prepared Foods (209):					
Gratin, Pilaf, or Fried Rice (27)	23,650	1	1	3	22
Bento (48)	20,430	1	1	16	30
Hamburger (41)	13,670			11	30
Others (93)	19,630	4	5	21	63
(Number of food items)		(35)	(24)	(114)	(594)

DISCUSSION

A previous study indicated that a significant proportion of pro-inflammatory AGEs are derived from dietary components [28]; however, it is disputed whether such AGEs are a health risk [8,28–30]. ELISA or liquid chromatography-mass spectrometry (LC-MS) are usually used to assess the levels of AGEs in bodily secretions, foods, and beverages. However, it is not possible to assess the levels of both HMW- and LMW-AGEs using any of the current approaches. Subjecting foods to heating results in the formation of various AGE molecules, with the types of AGEs produced depending on the heat treatment method employed and the food involved. Unfortunately, most previous studies only examined the levels of a small number of AGE molecules because of methodological issues [31,32]. The majority of studies of AGEs have focused on representative molecules, particularly CML. Based on assessments of CML levels obtained with ELISA, a database of food products and their AGE concentrations has been produced [6,7]. Foods containing elevated levels of protein and fat, meat substitutes, and processed meats were found to display markedly increased AGE concentrations. However, discrepancies have been detected between the information in the abovementioned database and the data obtained using other approaches, which indicates that more accurate methods for analyzing AGE levels are required [33]. As AGEs are produced when food is heated, eating processed foods, which are subjected to high temperatures during their production, results in greater exposure to AGEs. The Maillard reaction, in which sugar groups react with proteins (which causes foods to turn brown) and cross-links form between proteins, is responsible for AGE synthesis during the heating of foodstuffs [31]. Conversely, CML is colorless and does not promote protein cross-linking or fluoresce [2–5].

There are no established methods for evaluating the serum concentrations of AGEs. Whilst most studies of AGEs have examined CML levels, the role played by CML in pathological conditions is poorly understood. Although numerous other AGE structures are known to exist, little is known about which AGEs make significant contributions to disease. In previous studies, we found that the serum concentrations of Glycer-AGEs, but not those of Glu-AGEs, CML, or HbA1c, could be useful biomarkers for predicting the development of cardiovascular events and atherosclerosis [21–25]. Hence, in the present study we subjected a range of foods and beverages that are commonly consumed in Japan to testing in order to determine their AGE concentrations (their CML, Glu-AGE, Glycer-AGE, and Fru-AGE concentrations). The present study indicates that CML is not a suitable biomarker of dietary AGE consumption (Figs. 2 and 3 and Tables 1 and 3). The findings of the present study indicate that people should not consume excessive amounts of particular types of food such as cereals, dried fruits, seasonings, prepared foods, and confectionery (snacks) or certain beverages, particularly sugar-sweetened fruit drinks, lactic acid bacteria beverages, mixed fruit juices, sports drinks, and carbonated drinks.

Interestingly, the present study demonstrated that a markedly greater number of foods and beverages contain Glu-AGEs than Fru-AGEs, even among dried fruit products and fruit juices, which contain large amounts of fructose. During glycation, the initial kinetics of the reaction are affected by the protein involved, the temperature at which the reaction occurs, the concentration of the reducing sugar, and the percentage of the reducing sugar that possesses an open-chain structure. Compared with glucose, a greater proportion of fructose exhibits an open-chain structure. Studies of glycation have found that the initial rates of fructose-adduct formation are increased in Hb; however, glucose and fructose display similar levels of reactivity with RNase A, and glucose reacts with albumin 8 times more readily than fructose. The discrepancies between these findings might be due to differences in the local conditions at the reaction sites [34]. A Japanese study examined the amounts of glucose and fructose in various fruits. As a result, it was found that the levels of glucose and fructose in 100 g of fruit were as follows: 1.5 g and 1.5 g, respectively, in oranges; 3.0 g and 5.5 g, respectively, in apples; 7.5 g and 8.0 g, respectively, in grapes; and 4.0 g and 2.5 g, respectively, in bananas. Therefore, the detection of Glu-AGEs in fruit seems reasonable, although they contain even more Fru-AGEs. Irrespective of the proteins involved, AGEs are defined as molecules that contain AGE structures. It has been demonstrated that Glu-AGE structures often develop in casein that has reacted with glucose as well as in RNase A, Hb, and albumin. In the present study, the IC_{50} values of three types of AGE-BSA or CML-BSA standards were found to range from 1.0–1.5 U/mL in competition experiments involving an anti-CML antibody and three anti-AGE antibodies (data not shown); therefore, it is considered that all three anti-AGE antibodies and the anti-CML antibody had similar ability to recognize AGEs in beverages and foods.

When assessing the levels of AGEs using ELISA, it is essential to be aware of the cross-reactivity of the antibodies employed. Detailed assessments of the relevant epitopes are necessary to ensure this, together with precise knowledge of the target structure. In addition, it is necessary to validate the ELISA in each matrix, e.g., using spiking, as the local chemical milieu can affect antigen-antibody binding [8]. At present, our knowledge of the various complex pathways involved in AGE synthesis is insufficient. As AGE synthesis pathways are complex and exhibit a great degree of diversity, a large number of molecules are defined as AGEs. A significant number of dietary AGEs, including argpyrimidine, CEL, pentosidine, imidazolones, CML, pyrraline, GOLD, DOLD, and MOLD, *etc.*, have been detected *in vivo*[35,36]. In previous studies, we have found that anti-Fru-AGE, anti-Glycer-AGE, and anti-Glu-AGE antibodies recognize epitopes that differ from previously described AGE structures [13–16,23,27]; i.e., it is suggested that these AGE antibodies recognize novel AGE structures.

In humans, it has been demonstrated that roughly 10% of the AGEs in foods and beverages are taken up into the body. Of these, ~33% are excreted in urine within 48 h of their consumption, while ~67% accumulate within the body [37]. In a study conducted in the USA, Semba *et al.* stated that adults exhibit median (25th/75th percentile) serum CML concentrations of 0.69 (0.60/0.80) μg/ml [11]. In addition, non-diabetic individuals with normal renal function were found to have a mean serum Glu-AGE level of 10.5±1.3 U (one unit equals 1 μg of Glu-AGE-BSA)/ml, whereas diabetic patients and diabetic patients on hemodialysis exhibited serum Glu-AGE concentrations that were >2 times greater (24.7±2.4 U/ml) and ~8 times greater (79.4±9.9

U/ml), respectively, than those seen in the non-diabetic subjects [38]. We examined the serum Glu-AGE levels of healthy subjects and Japanese diabetic nephropathy patients and found that they were 10–20 U/ml and 30–50 U/ml, respectively. The consumption of 100,000 U of dietary Glu-AGEs results in a blood Glu-AGE level of ~2.0 U/ml [100,000 (U) × 0.1 (the proportion that is absorbed) x 1/5,000 (ml of blood)]. The dietary intake of food products containing <20,000 U Glu-AGEs has little effect on the body. On the other hand, care should be taken when mixing Glu-AGE-containing products or consuming large amounts of Glu-AGE-containing beverages or foods as this can result in elevated concentrations of Glu-AGEs and sugar in the blood and promote the hepatic build-up of Glu-AGEs [20,26].

In a previous study, we detected elevated hepatic RAGE expression in normal rats that had been given a Glu-AGE-rich beverage (Yakult) [20]. In addition, the rats' hepatic cells were found to contain Glycer-AGEs and Glu-AGEs, despite the fact that the beverage administered to the rats did not contain Glycer-AGEs. The above findings suggest that the synthesis and hepatic accumulation of Glycer-AGEs are promoted by Glu-AGEs, which are often found in foods and beverages, resulting in increased Glycer-AGE-RAGE binding [20,26]. In another study, AST-120 (Kremezin, Kureha-Chemical Co., Tokyo, Japan), an oral adsorbent that slows the development of chronic renal failure (CRF) by promoting the removal of uremic toxins, reduced the serum Glu-AGE and Glycer-AGE concentrations of non-diabetic CRF patients [26]. In addition, in an examination of the expression profiles of endothelial cells extracted from the latter patients' serum samples the mRNA expression levels of monocyte chemoattractant protein-1, vascular cell adhesion molecule-1, and RAGE were found to be significantly downregulated in the cells acquired after AST-120 (Kremezin) treatment compared with those seen in the endothelial cells obtained prior to treatment [26]. The latter results indicated that the consumption of Glu-AGEs might contribute to the development of vascular damage in pathological conditions linked to Glycer-AGE-RAGE interactions. Furthermore, they suggest that reducing the absorption of dietary Glu-AGEs might be a useful strategy for treating lifestyle-related conditions. Further clinical studies might help to elucidate whether lifestyle-related conditions can be prevented by encouraging people to reduce their consumption of Glu-AGEs.

In conclusion, we have presented useful information regarding the AGE concentrations of numerous beverages and foods that are commonly consumed in Japan. As dietary AGEs are derived from numerous precursors, they include a broad range of compounds with different structures and molecular weights. There is insufficient detailed data about the functions and molecular structures of AGEs. Furthermore, little is known about the *in vivo* activity of AGEs or about their bioavailability and absorption. This is partly because of a dearth of accurate analytical techniques for assessing the concentrations of AGEs in food and tissues. The structures of the epitopes recognized by anti-Fru-AGE, anti-Glycer-AGE, and anti-Glu-AGE antibodies were not examined in the present study; however, we were able to determine that they differ from those of well-defined AGEs as well as those of AGEs derived from carbonyl or sugar molecules with unknown structures. Accordingly, it is possible that Glycer-AGEs, Fru-AGEs, and Glu-AGEs have unique structures, but studies involving spectroscopic and biochemical analyses are required to confirm this.

AUTHOR CONTRIBUTIONS

Conceived and designed the experiments: MT SY. Performed the experiments: MT JT SF HS MK MM YK. Analyzed the data: MT JT. Contributed reagents/materials/analysis tools: MT JT. Wrote the paper: MT.

REFERENCES

1. Monnier VM, Cerami A. Non-enzymatic browning in vivo: possible process for aging of long-lived proteins. Science. 1981;211: 491–493. pmid:6779377 doi: 10.1126/science.6779377

2. Bucala R, Cerami A. Advanced glycosylation: chemistry, biology, and implications for diabetes and aging. Adv Pharmacol. 1992;23: 1–34. pmid:1540533 doi: 10.1016/s1054-3589(08)60961-8

3. Vlassara H, Bucala R, Striker L. Pathogenic effects of advanced glycosylation: biochemical, biologic, and clinical implications for diabetes and aging. Lab Invest. 1994;70: 138–151. pmid:8139257

4. Brownlee M. Advanced protein glycosylation in diabetes and aging. Ann Rev Med. 1995;46: 223–234. pmid:7598459 doi: 10.1146/annurev.med.46.1.223

5. Takeuchi M, Makita Z. Alternative routes for the formation of immunochemically distinct advanced glycation end-products in vivo. Curr Mol Med. 2001;1: 305–315. pmid:11899079 doi: 10.2174/1566524013363735

6. Goldberg T, Cai W, Peppa M, Dardaine V, Baliga BS, Uribarri J, et al. Advanced glycoxidation end products in commonly consumed foods. J Am Diet Assoc. 2004;104: 1287–1291. pmid:15281050 doi: 10.1016/j.jada.2004.05.214

7. Uribarri J, Woodruff S, Goodman S, Cai W, Chen X, Pyzik R, et al.

Advanced glycation end products in foods and a practical guide to their reduction in the diet. J Am Diet Assoc. 2010 110: 911–916. doi: 10.1016/j. jada.2010.03.018. pmid:20497781

8. Poulsen MW, Hedegaard RV, Andersen JM, de Courten B, Bugel S, Nielsen J, et al. Advanced glycation end products in food and their effects on health. Food Chem Toxicol. 2013;60: 10–37. doi: 10.1016/j. fct.2013.06.052. pmid:23867544

9. Sebekova K, Somoza V. Dietary advanced glycation endproducts (AGEs) and their health effects—PRO. Mol Nutr Food Res. 2007;51: 1079–1084. pmid:17854003 doi: 10.1002/mnfr.200700035

10. Uribarri J, Cai W, Peppa M, Goodman S, Ferrucci L, Striker G, et al. Circulating glycotoxins and dietary advanced glycation endproducts: two links to inflammatory response oxidative stress, and aging. J Gerontol A Biol Sci Med Sci. 2007;62: 427–433. pmid:17452738 doi: 10.1093/gerona/62.4.427

11. Semba RD, Ang A, Talegawkar S, Crasto C, Dalal M, Jardack P, et al. Dietary intake associated with serum versus urinary carboxymethyllysine, a major advanced glycation end product, in adults: the Energetics Study. Eur J Clin Nutr. 2012;66: 3–9. doi: 10.1038/ejcn.2011.139. pmid:21792213

12. Alamir I, Niquet-Leridon C, Jacolot P, Rodriguez C, Orosco M, Anton PM, et al. Digestibility of extruded proteins and metabolic transit of N-ε-carboxymethyllysine in rats. Amino Acids. 2013;44: 1441–1449. doi: 10.1007/s00726-012-1427-3. pmid:23160731

13. Takeuchi M, Makita Z, Yanagisawa K, Kameda Y, Koike T. Detection of noncarboxymethyllysine and carboxymethyllysine advanced glycation end products (AGE) in serum of diabetic patients. Mol Med. 1999;5: 393–405. pmid:10415164

14. Takeuchi M, Makita Z, Bucala R, Suzuki T, Koike T, Kameda Y. Immunological evidence that non-carboxymethyllysine advanced glycation end-products are produced from short chain sugars and dicarbonyl compounds in vivo. Mol Med. 2000;6: 114–125. pmid:10859028

15. Takeuchi M, Yanase Y, Matsuura N, Yamagishi S, Kameda Y, Bucala R, et al. Immunological detection of a novel advanced glycation end-product. Mol Med. 2001;7: 783–791. pmid:11788793

16. Takeuchi M, Iwaki M, Takino J, Shirai H, Kawakami M, Bucala R, et al. Immunological detection of fructose-derived advanced glycation end-products. Lab Invest. 2010;90: 1117–1127. doi: 10.1038/

labinvest.2010.62. pmid:20212455

17. Takeuchi M, Yamagishi S. TAGE (toxic AGEs) hypothesis in various chronic diseases. Med Hypotheses. 2004;63: 449–452. pmid:15288366 doi: 10.1016/j.mehy.2004.02.042

18. Takeuchi M, Yamagishi S. Involvement of toxic AGEs (TAGE) in the pathogenesis of diabetic vascular complications and Alzheimer's disease. J Alzheimers Dis. 2009;16: 845–858. doi: 10.3233/JAD-2009-0974. pmid:19387117

19. Takeuchi M, Takino J, Yamagishi S. Involvement of the toxic AGEs (TAGE)-RAGE system in the pathogenesis of diabetic vascular complications: A novel therapeutic strategy. Curr Drug Targets. 2010;11: 1468–1482. pmid:20583971 doi: 10.2174/1389450111009011468

20. Sato T, Wu X, Shimogaito N, Takino J, Yamagishi S, Takeuchi M. Effects of high-AGE beverage on RAGE and VEGF expressions in the liver and kidneys. Eur J Nutr. 2009;48: 6–11. doi: 10.1007/s00394-008-0753-4. pmid:19083041

21. Tsunosue M, Makiko N, Ohta Y, Matsuo Y, Ueda K, Ninomiya M, et al. An α-glucosidase inhibitor, acarbose treatment decreases serum levels of glyceraldehyde-derived advanced glycation end products (AGEs) in patients with type 2 diabetes. Clin Exp Med. 2010;10: 139–141. doi: 10.1007/s10238-009-0074-9. pmid:19834782

22. Tahara N, Yamagishi S, Matsui T, Takeuchi M, Nitta Y, Kodama N, et al. Serum levels of advanced glycation end products (AGEs) are independent correlates of insulin resistance in non-diabetic subjects. Cardiovasc Ther. 2012;30: 42–48. doi: 10.1111/j.1755-5922.2010.00177.x. pmid:20626403

23. Tahara N, Yamagishi S, Takeuchi M, Honda A, Tahara A, Nitta Y, et al. Positive association between serum level of glyceraldehyde-derived advanced glycation end products (AGEs) and vascular inflammation evaluated by ^{18}F-fluorodeoxyglucose positron emission tomography (FDG-PET). Diabetes Care. 2012;35: 2618–2625. doi: 10.2337/dc12-0087. pmid:22912424

24. Ueda S, Yamagishi S, Matsui T, Noda Y, Ueda SI, Jinnouchi Y, et al. Serum levels of advanced glycation end products (AGEs) are inversely associated with the number and migratory activity of circulating endothelial progenitor cells in apparently healthy subjects. Cardiovasc Ther. 2012;30: 249–254. doi: 10.1111/j.1755-5922.2011.00264.x. pmid:21884000

25. Fukushima Y, Daida H, Morimoto T, Kasai T, Miyauchi K, Yamagishi SI, et al. Relationship between advanced glycation end products and plaque

progression in patients with acute coronary syndrome: The JAPAN-ACS Sub-study. Cardiovasc Diabetol. 2013;12: 5. Available: http://www.cardiab.com/content/12/1/5.   doi: 10.1186/1475-2840-12-5. pmid:23289728

26.   Ueda S, Yamagishi S, Takeuchi M, Kohno K, Shibata R, Matsumoto Y, et al. Oral adsorbent AST-120 decreases serum levels of AGEs in patients with chronic renal failure. Mol Med. 2006;12: 180–184. pmid:17088950

27.   Yamagishi S, Inagaki Y, Okamoto T, Amano S, Koga K, Takeuchi M, et al. Advanced glycation end product-induced apoptosis and overexpression of vascular endothelial growth factor and monocyte chemoattractant protein-1 in human-cultured mesangial cells. J Biol Chem. 2002;277: 20309–20315. pmid:11912219 doi: 10.1074/jbc.m202634200

28.   Vlassara H, Striker GE. AGE restriction in diabetes mellitus: a paradigm shift. Nat Rev Endocrinol. 2011;7: 526–539. doi: 10.1038/nrendo.2011.74. pmid:21610689

29.   Semba RD, Gebauer SK, Baer DJ, Sun K, Turner R, Silber HA, et al. Dietary intake of advanced glycation end products did not affect endothelial function and inflammation in healthy adults in a randomized controlled trial. J Nutr. 2014;44: 1037–1042. doi: 10.3945/jn.113.189480

30.   Kellow NJ, Savige GS. Dietary advanced glycation end-product restriction for the attenuation of insulin resistance, oxidative stress and endothelial dysfunction: a systematic review. Eur J Clin Nutr. 2013;67: 239–248. doi: 10.1038/ejcn.2012.220. pmid:23361161

31.   Henle T. Protein-bound advanced glycation endproducts (AGEs) as bioactive amino acid derivatives in foods. Amino Acids. 2005;29: 313–322. pmid:15997413 doi: 10.1007/s00726-005-0200-2

32.   Zhang Q, Ames JM, Smith RD, Baynes JW, Metz TO. A perspective on the Maillard reaction and the analysis of protein glycation by mass spectrometry: probing the pathogenesis of chronic disease. J Proteome Res. 2009;8: 754–769. doi: 10.1021/pr800858h. pmid:19093874

33.   Assar S, Moloney C, Lima M, Magee R, Ames J. Determination of N$^\varepsilon$-(carboxymethyl)lysine in food systems by ultra performance liquid chromatography-mass spectrometry. Amino Acids. 2009;36: 317–326. doi: 10.1007/s00726-008-0071-4. pmid:18389168

34.   Schalkwijk CG, Stehouwer CD, van Hinsbergh VW. Fructose-mediated non-enzymatic glycation: sweet coupling or bad modification. Diabetes Metab Res Rev. 2004;20: 369–382. pmid:15343583 doi: 10.1002/dmrr.488

35.   Ahmed N, Argirov OK, Minhas HS, Cordeiro CA, Thornalley PJ.

Assay of advanced glycation endproducts (AGEs): surveying AGEs by chromatographic assay with derivatization by 6-aminoquinolyl-N-hydroxysuccinimidylcarbamate and application to N-epsilon-carboxymethyl-lysine- and Nepsilon-(1-carboxyethyl)lysine-modified albumin. Biochem J. 2002;364: 1–14. pmid:11988070

36.   Ahmed N, Mirshekar-Syahkal B, Kennish L, Karachalias N, Babaei-Jadidi R, Thornally PJ. Assay of advanced glycation endproducts in selected beverages and food by liquid chromatography with tandem mass spectrometric detection. Mol Nutr Food Res. 2005;49: 691–699. pmid:15945118 doi: 10.1002/mnfr.200500008

37.   Koschinsky T, He CL, Mitsuhashi T, Bucala R, Liu C, Buenting C, et al. Orally absorbed reactive glycation products (glycotoxins): an environmental risk factor in diabetic nephropathy. Proc Natl Acad Sci USA. 1997;94: 6474–6479. pmid:9177242 doi: 10.1073/pnas.94.12.6474

38.   Makita Z, Vlassara H, Cerami A, Bucala R. Immunochemical detection of advanced glycosylation end products in vivo. J Biol Chem. 1992;267: 5133–5138. pmid:1371995

Chapter 4

TRADITIONAL FERMENTED FOODS OF LESOTHO

Tendekayi H. Gadaga[1], Molupe Lehohla[2], VictorNtuli[3]

[1]Department of Environmental Health, University of Swaziland,
[2]Department of Pharmacy, National University of Lesotho
[3]Department of Biology, National University of Lesotho

ABSTRACT

This paper describes the traditional methods of preparing fermented foods and beverages of Lesotho. Information on the preparation methods was obtained through a combination of literature review and face to face interviews with respondents from Roma in Lesotho. An unstructured questionnaire was used to capture information on the processes, raw materials and utensils used. Four products; motoho (a fermented porridge), Sesotho (a sorghum based alcoholic beverage), hopose (sorghum fermented beer with added hops) and mafi (spontaneously fermented milk), were found to be the main fermented foods prepared and consumed at household level in Lesotho. Motoho is a thin gruel, popular as refreshing beverage as well as a weaning food. Sesotho is sorghum based alcoholic beverage prepared for household consumption as well as for sale. It is consumed in the actively fermenting state. Mafi is the name given to spontaneously fermented milk with a thick consistency. Little research has been done on the technological aspects, including the microbiological and biochemical characteristics of fermented foods in Lesotho. Some of the traditional aspects of the preparation methods, such as use of earthenware pots, are being replaced, and modern equipment including plastic utensils are being used. There is need for further systematic studies on the microbiological and biochemical characteristics of these these products.

INTRODUCTION

Lesotho is a landlocked country covering 30 350sq km with a population of about 1.8m. The terrain is characterized by mountains and valleys with the altitude ranging from 1388m (lowlands) to 3482m above sea level (Majara, 2005). The livelihoods of the communities differ depending on whether they live in the mountains, foothills, valleys or urban areas. However, many people live in the rural areas, especially in the mountains and maintain a traditional lifestyle. The people consume diverse traditional foods including wild vegetables and traditional fermented foods. Fermentation is one of the oldest methods of preserving foods. It has benefits in that it is inexpensive, provides much needed nutrients and destroys undesirable components including natural toxins and pathogenic microorganisms, and enhances the flavour (Steinkraus, 2002; Blandino et al., 2003).

Traditional fermentation is usually done spontaneously without the addition of a commercial starter culture. The fermenting microorganisms come from the raw materials, utensils or from a previous batch of the fermented product (back slopping) (Steinkraus, 2002). Several studies have shown the important role of fermentation in preventing food borne illnesses, especially childhood diarrhoea (Motarjemi, 2002) and enhancing dietary diversity. The predominant microorganisms in most of these foods are lactic acid bacteria, especially Lactococcus spp. (Blandino et al., 2003). Many African cereal fermented foods are characterised by the presence of lactic acid, the main end product of fermentation by Lactoccocus spp. Examples of African fermented foods include mahleu/mahewu (sour sorghum or maize meal non-alcoholic beverage, consumed in South Africa and Zimbabwe) (Gadaga et al., 1999), togwa (thin sour maize meal porridge, consumed in Tanzania) (Mugula et al., 2003), and kenkey (thick sour maize meal porridge, consumed in Ghana) (Halm et al., 1993).

Spontaneously fermented milk is consumed in many Southern African countries and is known by various names including amasi (South Africa), emasi (Swaziland) and madila (Botswana). Sorghum is a staple crop in Lesotho. It is drought tolerant and has short season varieties. It is therefore used to make many traditional foods including fermented alcoholic beverages and unfermented thin and thick porridges. According to available information, very little research has been done on traditional fermented foods of Lesotho, yet some of the traditional foods are widely consumed by the population. The objective of this article, therefore, was to document the preparation methods of traditional fermented foods of Lesotho. The paper will help stimulate new research on the technological characteristics (including the microbiology and biochemistry) of the fermented foods and encourage modernization of the products

BACKGROUND ON FOOD FERMENTATION IN LESOTHO

Information about the traditional processing methods was obtained through a combination of literature review and interviewing women who prepared the foods in the Roma area of Lesotho. A semi-structured questionnaire was used for the interviews. Sorghum and maize are staple cereal crops in Lesotho. However sorghum is preferably used to prepare traditional fermented foods. Interviews with women in the Roma area of Lesotho revealed that four products; motoho (a fermented porridge), sesotho (a sorghum based alcoholic beverage), hopose (sorghum fermented beer with added hops) and mafi (spontaneously fermented milk), were the main fermented foods prepared and consumed at household level. Leting, ting, seqhaqhabola (sekhakabolo) and sekumukumu are other traditional fermented foods consumed in Lesotho as documented in literature. The preparation methods are outlined below and comparisons with descriptions of other similar products consumed in other countries are made.

Non-alcoholic fermented foods

Motoho

Motoho is a non-alcoholic sour porridge. Some Basotho (people of Lesotho) like to think of it as a beverage because of its slurry-like consistency. The red type sorghum is usually used for making motoho, hence the product has a brownish colour. Figure 1 is a schematic diagram of the production process for motoho as described by women preparing the product in Lesotho. The sorghum meal is mixed with warm water (1 part sorghum meal to 3 parts water) to make a thin slurry. A traditional starter culture called tomoso (described below) is then added (1 part tomoso to 20 parts of the slurry) and the mixure is allowed to ferment. In summer, the ambient temperature varies between 25-30oC and fermentation takes 24 h. Normal monthly winter minimum temperatures range from – 6.3°C in the lowlands to 5.1°C in the highlands. It is therefore common in winter, that the fermentation vessel is covered with a blanket to retain warmth and speed up the fermentation. Fermentation takes about 48-72 h in winter. The fermented product is then boiled for 20-30 minutes and allowed to cool to about 25-30oC, and served. Grinding of the coarse grains before boiling is optional, and this is often omitted (Figure 1). Households usually prepare motoho on a weekly basis. However, some prepare it daily for sale. Motoho can sometimes be prepared for occasions like funerals or other family feasts and is also served to visitors as a beverage. The product is consumed by the whole family, in the same way as non-fermented porridge called lesheleshele.

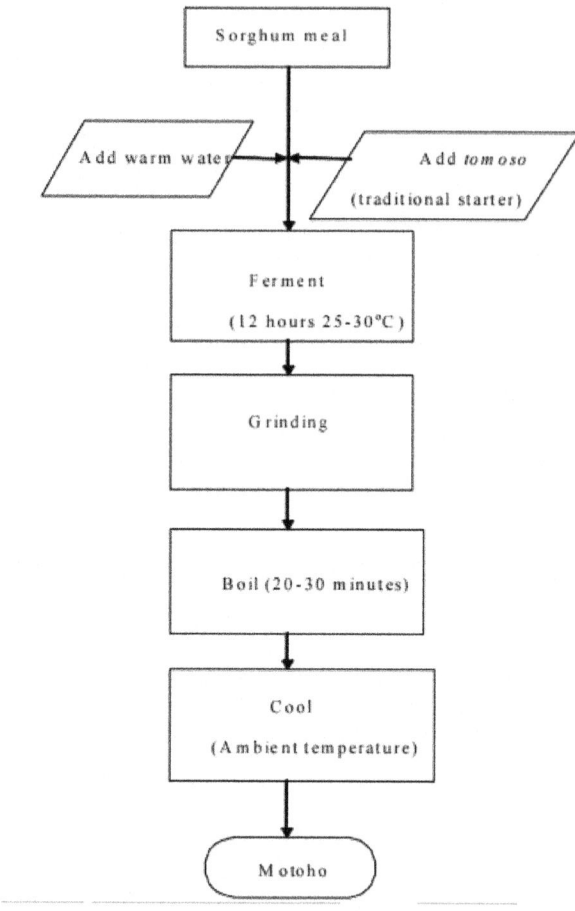

Figure 1: Traditional preparation of motoho in Lesotho

It has a shelf-life of about 5 days at 25oC and therefore can be prepared in large quantities for future use. This is probably the reason why motoho has now been successfully produced commercially. It can be bought from street vendors or the supermarket. The boiling of motoho after fermentation is also similar to the process described in the commercial production of commercialised instant mahewu before drum drying (Heseltine and Wang, 1980). However, scientific literature on the microbiology of motoho and the standardization of its preparation is scanty. Sakoane and Walsh (1987) carried out a study to assess the safety of motoho as a weaning food. It was observed that motoho effectively inhibited strains of enteropathogenic Escherichia coli, Salmonella Typhii and Shigella boydii 3 hours after fermentation. The pH decreased from

7 to 6.4 during the 12 hours of fermentation. Although the product was boiled after fermentation, it was still found to effectively inhibit these pathogens, suggesting that either the acid or a heat stable inhibitory substance prevented the growth of the microorganisms.

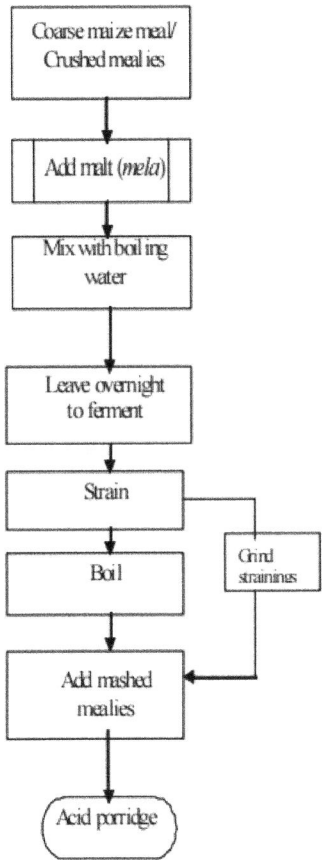

Figure 2: Preparation of sekhakabolo (Source: From description of Orpen (1902))

Sakoane and Wash (1987) also observed that some colonies of Salmonella Typhii inoculated at low dilution were able to grow in autoclaved motoho, raising concerns about the safety of the product. Motoho needs to be handled with very high standards of hygiene to prevent post fermentation contamination. Lesheleshele, a non-fermented sorghum porridge consumed in Lesotho had no inhibitory effects (Sakoane and Walsh, 1987). In a similar study on the safety of mahewu (a maize based fermented gruel), Simango and Rukure

(1991) reported that some strains of Escherichia coli survived for 24 h in the fermented product, but they did not increase in number. Such microorganisms were thought to have found their way into the mahewu from the leftover maize porridge used in the preparation of the product, and from contaminated hands and utensils. The pathogens can proliferate in the product before acidity has sufficiently developed.

Tomoso used in initiating fermentation of motoho is a liquid starter culture made by mixing a small amount of sorghum flour with a minimal amount of warm water such that the water just covers the flour. It is then left to ferment spontaneously for a day. Alternatively, it is obtained from a previous successful fermentation (back slopping option). Although the types of microorganisms in tomoso have not been systematically studied, it is expected to contain high numbers of lactic acid bacteria that then ferment the slurry into a sour product.

Seqhaqhabola (Sekhakabolo)

An early paper by Orpen (1902) described another sour porridge called seqhaqhabola (sekhakabolo) that was prepared by the Basotho. According to the report, coarse maize meal or crushed mealies were mixed with sorghum malt (Orpen, 1902). The sorghum malt acted both as an inoculum and a source of enzymes. Boiling water was then added and mixed well. The mixture was then cooled and left to ferment for about 24 h. It was then strained using baskets and boiled. The strainings were not discarded, but were ground to finer particle size and mixed with the boiled mixture to give a sour porridge as shown in Figure 2. Seqhaqhabola is similar to ilambazi lokubilisa, a sour porridge prepared by the Ndebele speaking people in Zimbabwe (Gadaga et al., 1999). In making ilambazi lokubilisa, maize meal is thoroughly mixed with a little amount of water and allowed to ferment in a closed vessel for 2-4 days. The fermented meal is then used to make sour porridge (Simango, 1997). Ilambazi lokubilisa has been shown to be batericidal to strains of enteric pathogens belonging to the genera Aeromonas, Campylobacter and Salmonella, and bacteriostatic to strains of Shigella and Escherichia coli (Simango and Rukure, 1992). This makes the product relatively safe and suitable for use as a weaning food. The process for traditional malting of sorghum used in the preparation of seqhaqhabola and other fermented products is shown in Figure 3.

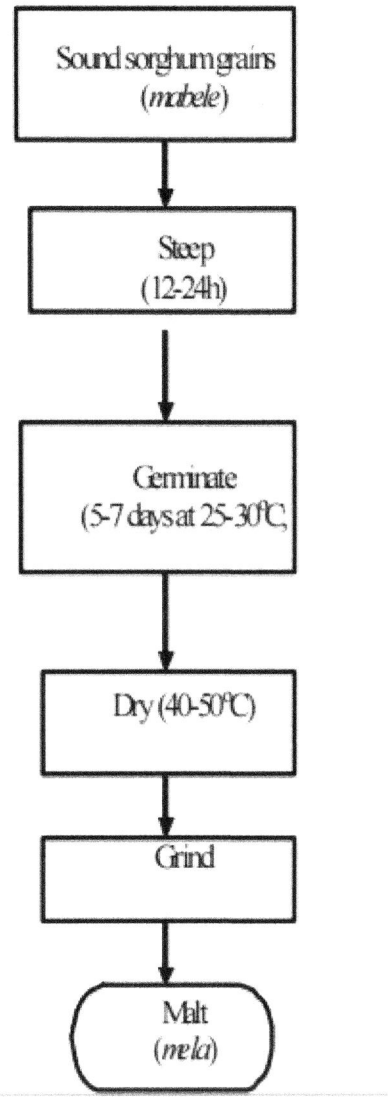

Figure 3: Traditional preparation of sorghum malt (mela) in Lesotho

Ting

Franz (1971) reported that ting was a sour porridge consumed by the TswanaSotho community in the present day Limpopo province of South Africa and possibly the Sotho in Lesotho. Ting could be prepared in either of two ways: (a) sorghum (mabele) meal was mixed with milk whey and stirred until a thick consistency was obtained and served; OR (b), sorghum meal (40-45%) was mixed with warm water in an earthenware pot and the slurry was allowed to stand in a warm place (30-37oC) for 2 days until it had 'risen' and turned sour (fermented) (Franz, 1971; Sekwati-Monang and Gänzle, 2011). The fermented mixture (pH 3.5-4.0) was then gradually added to boiling water while heating sufficiently to make two types of porridge, bogobe and motogo (stiff and soft porridge respectively) (Sekwati-Monang and Gänzle, 2011) (Figure 4).

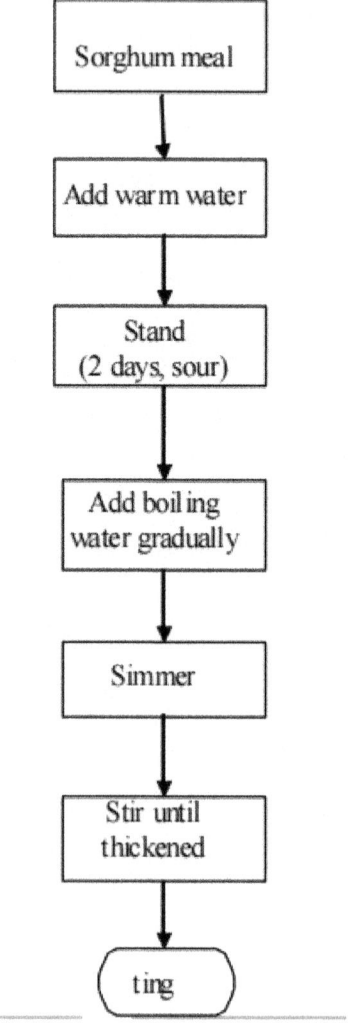

Figure 4: Preparation of ting (From description of Franz (1971)

The second method described for preparation of ting is similar to that of mutwiwa, a sorghum based sour porridge in Zimbabwe (Gadaga et al., 1999). In the preparation of mutwiwa, dried maize grains are first dehulled using pestle and mortar. The dehulled grains are washed and steeped in clean water and left to ferment until gas production ceases (Gadaga et al., 1999). The fermented maize is then dried and pounded into a meal which is then used to make a thin or thick porridge. A study on ting by Madoroba et al. (2009; 2011) found that Lactobacillus fermentum, Lb. plantarum and Lb. Rhamnosus, Lb. parabuchneri, Lb. casei, Lb. coryniformis, Lb. curvatus and Weissella cibaria were the predominant lactic acid bacteria in the product. These species are also commonly isolated from a wide range of African traditional fermented foods and beverages including fufu, kenkey and ogi (Steinkraus, 1996; Mugula et al., 2003). Lb. plantarum in particular is known to be acid tolerant and can break down fermentable substrates in plant based materials (Steinkraus, 2002). The presence of these microorganisms and the low pH can inhibit the proliferation of pathogens and enhance the safety of the food. In another study, Kunene et al. (1999) in South Africa reported that adults consumed the thick form of the porridge from ting at ceremonies such as weddings and funerals, while a diluted form/thin porridge was used as weaning food.

Leting

Leting is described in literature as another fermented sour mash consumed in Lesotho. According to the description by Orpen (1902), malted Sorghum bicolor was dried and ground into a fine meal. A portion of this meal was then mixed thoroughly with hot water until no particles of the malt remained at the bottom. It was then ready for serving. However, when left at room temperature (c.a. 25oC), it fermented and became sour. Orpen (1902) noted that leting was similar to sweet wort, but was acidic, refreshing and nourishing with up to 10.5% alcohol. Despite the alcohol content, it was consumed by all age groups. The reported alcohol content suggests that the product was left to ferment for a considerable time. Novellie (1981) also reported that the alcohol content of leting increased by volume from 0.2 to 3.1% from day 1 to day 6, indicating that as it was left to ferment it became more alcoholic. Goldberg and Thorp (1946) observed that the souring of the sorghum meal and malt during the making of leting resembled the first stage of sorghum beer brewing, and hence if left to ferment for a long time it would became alcoholic on storage. Hesseltine and Wang (1980) put leting in the same category as sorghum beer and described it as a thick, acidic, weakly alcoholic drink. However, Orpen (1902) mentioned that leting had some health and/or nutritional benefits. For example, during the early part of the 20th century, leting was successfully used to prevent scurvy among soldiers stationed at Bloemfontein, South Africa (Orpen 1902). It was argued that leting should be given to workers on a regular basis so as to add to their health and vitality (Orpen 1902; Harford, 1905; Goldberg and Thorp, 1946). Goldberg (1946) and Rooney (1985) also compared leting to mahewu. Goldberg (1946) reported that it contained 9.5% solids, while Rooney (1985) described a souring process that was similar to that described by Orpen (1902). According to Rooney (1985), ground sorghum, sometimes mixed with sprouted sorghum, was exposed to boiling water and held at elevated temperatures for up to 20 min and then allowed to cool. The product was left to sour with very little alcohol being produced (Rooney 1985). The product was strained before serving. The strainings were then used to make a light fermented drink by adding water, or the dregs were mixed with more sorghum meal to make a type of bread. This description of leting suggests that it may be similar to sorghum or millet based Zimbabwean product called masvusvu (Zvauya et al., 1997) and not mahewu, which is made from maize (Gadaga et al., 1999). Briefly, mahewu is prepared in the following way: Left-over thick maize meal porridge is cleaned, broken into small pieces, and mixed with water. A little sorghum or millet malt is then added and mixed thoroughly and left to ferment for 24 to 48 h. The fermentation is due to the natural flora of the maize meal, malt and the utensils (Gadaga et al., 1999). Backslopping is rarely practiced but the same pot is usually used every time a fresh product is made. Mahewu is now produced commercially in several countries in southern Africa. On the other hand, in preparing masvusvu, finger millet is malted then milled and the flour is mixed with water. The mixture is slowly heated to almost boiling for 80 minutes. The resulting mash is called masvusvu, which is cooled, strained and served or allowed to fermented to make other products like alcoholic beverages.

Alcoholic beverages

Traditional alcoholic beverages in Lesotho are generally referred to as joala. There are a number of such beverages, which are all intoxicating and are meant for consumption by adults during feasts or just as refreshing drinks. These beverages include hopose, sekumukumu and sesotho beer.

Hopose

This is a home-made alcoholic beverage which derives its name from the use of hops in its preparation. This makes it unique among African traditional alcoholic beverages. The hops are mixed with warm water, to which brown wheat meal/flour is added to make a thin gruel. Brown sugar (about 1kg for 20 liter preparation) is then added and the mixture is left for about 20 to 30 min.

A traditional starter culture (or commercial yeast, Saccharomyces cerevisiae) and malt are then added to the lukewarm mixture. The mixture is left overnight after which it is sieved to remove large pieces, and another portion of brown sugar is then added. The mixture is then ready for consumption.

Sekumukumu

This alcoholic beverage is made from bread, brown sugar and tomoso. Traditionally, this is prepared from sorghum meal bread made from the spent grains from the processing of leting. Water is boiled then cooled to ambient temperature (25-30oC). The sorghum meal bread is broken into small pieces and added into the water, together with tomoso and/or yeast. The mixture is then left to ferment for 24-48 h. It is then sieved to remove the dregs before consumption.

Sesotho (sorghum beer)

Traditional sorghum beer in Lesotho is called sesotho (Asita et al., 2011). According to information obtained from women who prepare Sesotho, sorghum (or occasionally maize meal) and wheat meal are mixed together in equal amounts and cold water is added to make a stiff consistency. Boiling water is then added to make a thin gruel, which is then cooled to about 30-35oC. A traditional liquid starter, tomoso, is added and the container is covered with a blanket to retain warmth. The mixture is left to ferment overnight or for 24 h to 48 h, depending on the ambient temperature. It is then boiled for 2-3 h and then cooled to 30-35oC. A solid starter culture called moroko (spent dregs from a previous fermentation) is added and the mixture is fermented for a further 24-48 h. It is then filtered to remove coarse particles to give a refreshing alcoholic drink (Figure 5).

Orpen (1902) reported that the sorghum beer in Lesotho was prepared by first boiling sweet wort, similar to the process in making leting. Sorghum meal was then added and the mixture was allowed to ferment for 24-48 h. The product was alcoholic with a raw taste and gritty texture. While the method used by the traditional brewers in Lesotho today can take up to 5 days, the method described by Orpen (1902) only took about 2 days. The characteristics of these two products are bound to be different. The Northern Sotho in South Africa also used to make a sorghum beer called bjalwa (Franz, 1971), which is probably the same name as the Sesotho joala, only spelt differently.

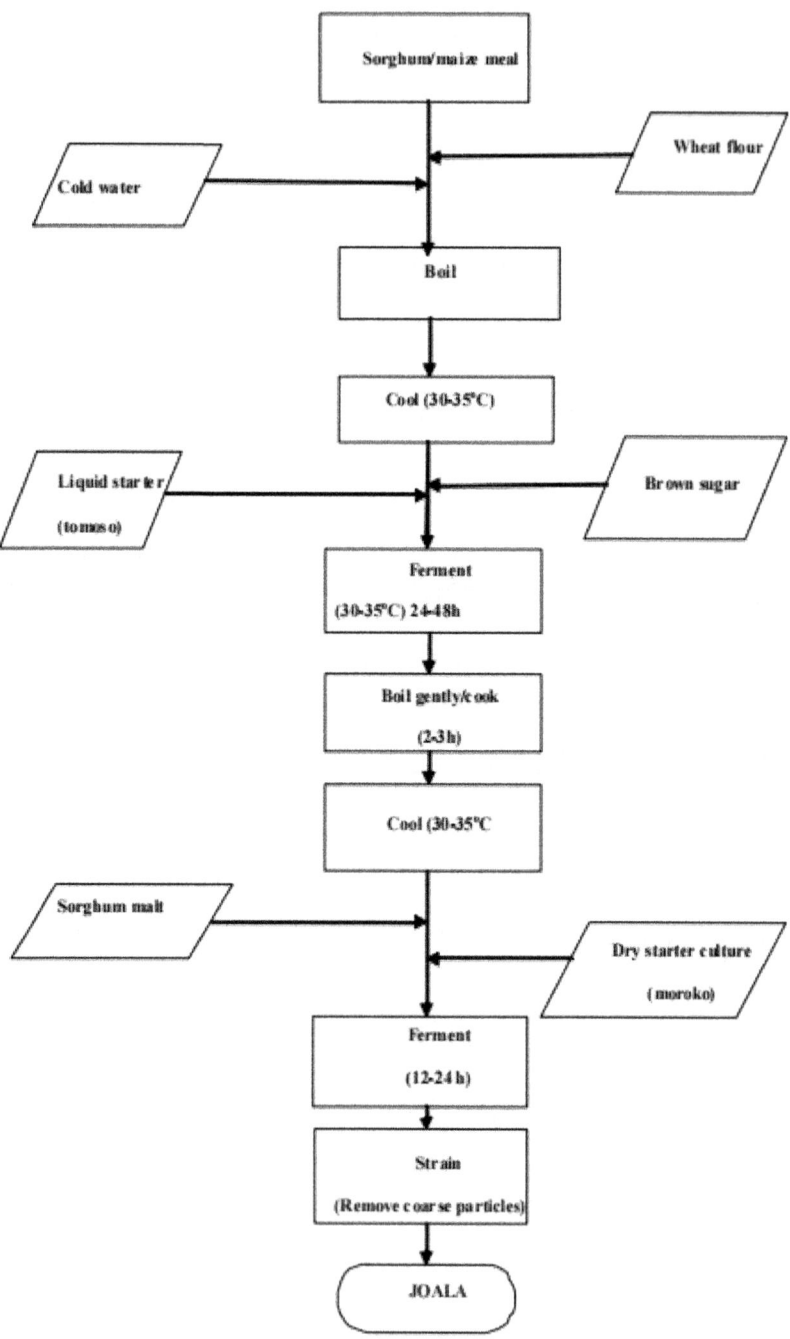

Figure 5: Traditional preparation of sesotho (sorghum beer) in Lesotho

Bjalwa was made by mixing sorghum malt with clean water in earthenware pots and allowed to ferment for a day or more. The mixture was then boiled and cooled to ambient temperature. More malt was then added and again allowed to ferment for a day or more (24h- 48 h). This fermented product was then strained and cooled to give bjalwa. The process is very similar to that described by women making sorghum beer in Lesotho today, except that no tomoso was added (Figure 5). Apart from the description of the preparation process, there is no additional information on the microbiological, biochemical or nutritional information of joala in literature.

The processing of sesotho is similar to other sorghum beers of the Bantu people in southern Africa. For example, it is similar to doro/uthwala, a traditional beer in Zimbabwe, which is also produced over 5 to 7 days (Madovi, 1981; Gadaga et al., 1999). Umqombothi is the Shangaan name given to the South African sorghum beer while the Sotho speaking people call it joala (Katongole, 2008). It has a pinkish colour and a yoghurty flavour and is consumed in the active state of fermentation with a shelf life of 2-3 days (Novellie, 1966). Other studies on similar products in southern Africa have shown that the alcohol content in fermented sorghum beer can range from 3-5% (Hesseltine, 1979; Parawira and Muchuweti, 2008), and is rich in B-vitamins. Umqombothi was found to contain up to 2.9% alcohol (Katongole, 2008). Yeasts and lactic acid bacteria are thought to be the predominant microorganisms during the fermentation, with Saccharomyces cerevisiae and Candida krusei being some of the most common yeasts isolated. Lactic acid bacteria isolated from other sorghum beers include Lactobacillus helveticus, Lb. salivarius, Pediococcus damnosus, and P. partulus (Katongole, 2008).

Other wild yeasts and bacteria from the ingredients and the beer pots should be present and contribute to the flavour characteristics of the beer as well as inhibit pathogenic microorganisms. There is therefore need to isolate and identify the diversity of microorganisms in sesotho beer. Sorghum beer prepared in metal containers is also high in iron (Mandishona et al., 1999; Choma and Alberts, 2007). In South Africa, traditional beer fermented in either iron pots or plastic containers was found to have iron levels ranging from 15 mg/l to 67.8 mg/l and 6 mg/l to 17 mg/l, respectively. Consumption of the traditional beer seemed to prevent iron deficiency in those at risk of developing such deficiency, but appeared to precipitate iron overloading those at risk of developing iron overload. Moyo et al. (1997) also observed that the mean concentration of iron in the supernatants of nine samples of traditional beer was 46 mg/L. The men (14.3%) in the study showed a combination of an elevated serum ferritin and a transferrin saturation of more than 70%, suggestive of substantial iron overload. Mandishona et al. (1999) also concluded that the consumption of traditional beer, rich in iron, protected women against iron deficiency as none of the women drinkers in the study had iron deficiency anaemia.

Fermented milk

The traditional fermented milk in Lesotho is called mafi. It is traditionally prepared by allowing raw milk to ferment spontaneously in clay pots until thick curds form. This may take 2 to 3 days at 25-30oC. The thick curds are then consumed with thick porridge called papa or on its own as a refreshing drink. This is similar to the other differently named fermented milks in Southern Africa, such as amasi/emasi (South Africa, Zimbabwe/Swaziland). The South Sotho in South Africa also called spontaneously fermented milk mafi (Beukes et al., 2000). In this case the whey is usually drained from the milk curds, but it is also consumed as a side relish with thick sorghum or maize meal porridge (papa or bohobe). Franz (1971) described a product similar to mafi that was prepared by the Northern Sotho or Sotho-Tswana community in the Petersburg region in South Africa. The product was called mange. Milk was curdled in large earthenware pots. The butterfat was then skimmed off with the aid of a calabash vessel. The remaining milk was then sieved through sisal mats to obtain a thick product. Earthenware pots are no longer popular as fermentation vessels, even in Lesotho. The women interviewed in the Roma area of Lesotho reported that they now use plastic or metal containers. The weather in Lesotho can be very cold in winter and it is common to find that the fermenting vessel is covered with a blanket to maintain a warm temperature. Mafi is also comparable to madila, the traditional fermented milk produced in Botswana, (Ohiokpehai and Jagow, 1998). To prepare madila fresh milk is filtered through a strainer and placed in an enamel/metal bucket. This is then kept in a warm place (c.a. 30oC) for 24 h to initiate fermentation. The soured milk is then poured into a woven polypropylene sack and a further bucket of one day old soured milk is added each day over a seven or eight day period. During this period the madila continues to ferment. The bag is then hung from a beam for three or four days during which time the whey drains away through the woven bag. Finally the madila is removed from the bag and mixed with fresh milk in a ratio of 4:1 before consumption or sale (Ohoikpehai and Jagow, 1998). Although the microbial and biochemical properties of mafi are yet to be studied in detail, previous work with amasi in Zimbabwe showed that several species of lactic acid bacteria, yeasts and even some coliforms were present during fermentation (Gadaga et al., 2000; Mutukumira, 1995). The pH of the product is usually below 4.5, sufficient to inhibit many pathogenic microorganisms such as Salmonella spp. and Escherichia coli strains. However, Gran et al. (2003)

found that pathogenic strains of E. coli inoculated into spontaneously fermented milk or that produced by back-slopping had unacceptably high numbers of the E. coli surviving, even 48 h after fermentetion. In a separate study, Nyatoti et al., (1997) reported that out of 12 samples of naturally soured milk used as weaning foods, two contained enteropathogenic E. coli. Strains of E. coli O157:H7 are of particular concern and they have been isolated from raw milk. They can also contaminate milk post-pasteurization if there is poor handling of the milk. However, in a study with madila (spontaneously fermented milk in Botswana), Patty-Hanson et al. (2009) found that E. coli O157:H7 could be inhibited when the lactoperoxidase (LP) system in the raw milk was activated. The counts were reduced by more than 5.0 log cfu mL-1 . These observations, however, suggest that if not hygienically handled, fermented milk products such as mafi, madila and amasi can pose a food safety risk.

CONCLUSION

The traditional fermented foods, motoho, sesotho beer, hopose and mafi are commonly prepared and consumed at household level in Lesotho. Motoho is now produced commercially and is a popular weaning food. However, some pathogenic microorganisms were able to grow in motoho, and fermented milk products similar to mafi, demonstrating the need to pay attention to hygiene and food safety issues for these fermented products. Sesotho beer (joala) and hopose bring much needed income for poor families. They are principally made from sorghum meal and wheat meal, respectively. Despite the fact that traditionally brewed sorghum beer is regarded as safe due to its inhospitable environment (pH 3-4), saftey may be compromised by microbial contaminants from raw materials, ingredients and the food handler, thereby jeopardising the health of the consumer. Some traditional fermented foods that were consumed in the past by the Basotho may no longer be available as the population adapts to new food habits. There is need for further research on the technical, microbial and biochemical characteristics of these fermented products.

ACKNOWLEDGMENTS

The authors wish to thank the women who were interviewd in Roma, Lesotho for the time they took to explain the different preparation methods for the fermented foods.

REFERENCES

1.   ASITA, A.O., TANOR, E.B., MAGAMA, S., KHOABANE, N.M. 2011. Assessment of sorghum beer for alcohol and metal ions content

and genotoxicity in mice bone marrow. Journal of Toxicology and Environmental Health Sciences, 3(11), 317-327.

2.   BEUKES, E.M., BESTER, B.H., MOSTERT, J.F. 2000. The microbiology of South African traditional fermented milks. International Journal of Food Microbiology, 63(3), 189–197.

3.   BLANDINO, A., AL-ASEERI, M.E., PANDIELLA, S.S., CANTERO, D., WEBB, C. 2003. Cereal based fermented foods and beverages. Food Research International, 36(6), 527-543.

4.   FRANZ, H.C. 1971. The traditional diet of the Bantu in the Pietersburg district. South African Medical Journal, 45, 1232-1235.

5.   GADAGA, T.H., MUTUKUMIRA, A.N., NARVHUS, J.A., FERESU, S.B. 1999. A review of traditional fermented foods and beverages of Zimbabwe. International Journal of Food Microbiology, 53(1), 1-11

6.   GADAGA, T.H., MUTUKUMIRA, A.N., NARVHUS, J.A. 2000. Enumeration and identification of yeasts isolated from Zimbabwean traditional fermented milk. International Dairy Journal, 10(7), 459-466

7.   GOLDBERG, L., THORP, J.M. 1946. A survey of vitamins in African foodstuffs. VI: Thiamin, riboflavin and nicotinic acid in sprouted and fermented cereals and foods. South African Journal of Medical Science, 11(4), 177-185

8.   Gran, H.M., Gadaga, H.T., Narvhus J.A. 2003. Utilization of various starter cultures in the production of amasi, a Zimbabwean naturally fermented raw milk product. International Journal of Food Microbiology, 88(1), 19-28

9.   HALM, M., LILLIE, A., SØRENSEN, A.K., JAKOBSEN, M. 1993. Microbiological and aromatic characteristics of fermented maize doughs for kenkey production in Ghana. International Journal of Food Microbiology, 19(2), 135-143

10.   HARFORD, C.F. 1905. The drinking habits of uncivilized and semi-civilized races. British Journal of Inebriety, 2(3), 92-103.

11.   HESSELTINE, C.W., WANG, H.L. 1980. The Importance of Traditional Fermented Foods. BioScience, 30(6), 402-404.

12.   HESSELTINE, C.W. 1979. Some important fermented foods of middle Asia, the Middle East and Africa. Journal of the American Oil Chemistry Society, 56(3), 367-374.

13.   CHOMA, S.S.R., ALBERTS, M. 2007. Effect of traditional beer consumption on the iron status of a rural South African population. South African Journal of Clinical Nutrition, 20(2), 62-68

14. KATONGOLE, J.N. 2008. The Microbial Succession in indigenous Fermented Maize products. MSc Dissertation, University of the Free State, South Africa.

15. KUNENE, N.F., HASTINGS, J.E., von HOLY, A. 1999. Bacterial populations associated with a sorghum-based fermented weaning cereal. International Journal of Food Microbiology, 49, 75-83

16. MADOROBA, E., STEENKAMP, E.T., THERON, J., HUYS, G., SCHEIRLINCK, I., CLOETE, T.E. 2009. Polyphasic taxonomic characterization of lactic acid bacteria isolated from spontaneous sorghum fermentations used to produce ting, a traditional South African food. African Journal of Biotechnology, 8(3), 458-463.

17. MADOROBA, E., STEENKAMP, E.T., THERON, J., SCHEIRLINCK, I., CLOETE T.E., HUYS, G. 2011. Diversity and dynamics of bacterial populations during spontaneous sorghum fermentations used to produce ting, a South African food. Systematic and Applied Microbiology, 34(3), 227-234.

18. MADOVI, P.B. 1981. Food handling in Shona villages of Zimbabwe. Ecology of Food and Nutrition, 11(3), 133-144.

19. MAJARA, N. 2005. Land degradation in Lesotho: A synoptic perspective. Dissertation, Master of Natural Science, University of Stellenbosch, South Africa.

20. MANDISHONA, E.M., MOYO, V.M., GORDEUK, V.R., KHUMALO, H., SAUNGWEME, T., GANGAIDZO, I.T., GOMO, Z.A.R., ROUAULT, T., MACPHAIL, A.P. 1999. A traditional beverage prevents iron deficiency in African women of child bearing age. In European Journal of Clinical Nutrition, 53(9), 722-725

21. MOTARJEMI, Y. 2002. Impact of small-scale fermentation technology on food safety in developing countries. International Journal of Food Microbiology, 75, 213-229

22. MOYO, V.M., GANGAIDZO, I.T., GOMO, Z.A.R., KHUMALO, H., SAUNGWEME, T., KIIRE, C.F., ROUAULT, T., GORDEUK, V.R. 1997. Traditional beer consumption and the iron status of spouse pairs from a rural community in Zimbabwe. Blood, 89(6), 2159-2166.

23. MUGULA, J.K., NNKO, S.A., NARVHUS, J.A., SØRHAUG, T. 2003. Microbiological and fermentation characteristics of togwa, a Tanzanian fermented food. International Journal of Food Microbiology, 80(3), 187–199.

24. MUTUKUMIRA, A.N. 1995. Properties of amasi, a natural fermented milk produced by smallholder milk producers in Zimbabwe.

Milchwissenschaft, 50, 201-205.

25. NOVELLIE, L. 1966. Bantu beer: Popular drink in South Africa. International Journal of Brewing and Distillation, 1, 27-31.

26. NOVELLIE, L. 1981. Fermented porridges. Proceedings of the International Symposium on Sorghum Grain Quality (An International Symposium), Patancheru, India, 113-120.

27. NYATOTI, V.N., MTERO, S., Rukure, G. 1997. Pathogenic Escherichia coli in traditional African weaning foods. Food Control, 8, 51-54.

28. OHIOKPEHAI, O., JAGOW, J. 1998. Improving madila - a traditional fermented milk from Botswana. Intermediate Technology Food Chain, 23, 6

29. ORPEN, J.M. 1902. Diet of native labourers: A lecture read before the Rhodesia Scientific Association, Argus Printing and Publishing Company, Salisbury, Rhodesia, 1-12p

30. PARAWIRA, W., MUCHUWETI, M. 2008. An overview of the trend and status of food science and technology research in Zimbabwe over a period of 30years. Scientific Research Essays, 3(12), 599-612.

31. PARRY-HANSON, A., JOOSTE, P., BUYS, E. 2009. Inhibition of Escherichia coli O157:H7 in commercial and traditional fermented goat milk by activated lactoperoxidase. Dairy Science and Technology, 89(6), 613-625

32. ROONEY, L.W. 1985. Food and nutritional quality of sorghum. Fighting hunger with research (A Five-Year Technical Research Report of the Grain Sorghum/Pearl Millet Collaborative Research Support Program), INTSORMIL, 131-139

33. SAKOANE, A.L., WALSH, A. 1987. Bacteriological properties of traditional sour porridges in Lesotho. Improving young child feeding in Eastern and Southern Africa- Household Level Food Technology (Proceeding of a workshop) Nairobi, Kenya, 261-269.

34. SEKWATI-MONANG, B., GÄNZLE, M.G. 2011. Microbiological and chemical characterisation of ting, a sorghum-based sourdough product from Botswana. International Journal of Food Microbiology, 150(2/3), 115-121.

35. SIMANGO, C., RUKURE, G. 1991. Survival of Campylobacter jejuni and pathogenic Escherichia coli in mahewu, a fermented cereal gruel. In Transactions of the Royal Society of Tropical Medicine and Hygiene, 85(3), 399-400

36. SIMANGO, C., RUKURE, G. 1992. Survival of bacterial enteric

pathogens in traditional fermented foods. Journal of Applied Bacteriology, 73(1), 37-40.

37. SIMANGO, C. 1997. Potential use of traditional fermented foods for weaning in Zimbabwe. Journal of Social Science and Medicine, 44(7), 1065-1068.

38. STEINKRAUS, K. H. 2002. Fermentations in world food processing. Comprehensive Reviews in Food Science and Food Safety, 1(1), 23-32.

39. STEINKRAUS, K.H. 1996. Handbook of indigenous fermented foods. Marcel Dekker, New York, 213-23. ISBN 0-8247-9352-8

40. ZVAUYA, R., MUGOCHI, T., PARAWIRA, A. 1997. Microbial and biochemical changes occurring during production of masvusvu and mangisi, traditional Zimbabwean beverages. Plant Foods for Human Nutrition, 51(1), 43- 51.

Chapter 5

THE ROLE OF YEAST AND LACTIC ACID BACTERIA IN THE PRODUCTION OF FERMENTED BEVERAGES IN SOUTH AMERICA

Fábio Faria-Oliveira[1], Raphael H.S. Diniz[1], Fernanda Godoy-Santos[1], Fernanda B. Piló[1], Hygor Mezadri[1], Ieso M. Castro[1] and Rogelio L. Brandão[1]

[1]Laboratório de Biologia Celular e Molecular, Núcleo de Pesquisas em Ciências Biológicas, Universidade Federal de Ouro Preto, Brazil

ABSTRACT

Fermentation is one of the oldest forms of food preservation in the world. In South America, most fermented beverages are nondairy products featuring several other food raw materials such as cereals, fruits, and vegetables. Generally, natural fermentations are carried out by yeast and lactic acid bacteria forming a complex microbiota that acts in cooperation. Yeast have a prominent role in the production of beverages, due to the ability to accumulate high levels of ethanol and to produce highly desirable aroma compounds, but lactic acid bacteria are particularly important in fermentation because they produce desirable acids, flavor compounds, and peptides that inhibit the growth of undesirable organisms. Among the South America beverages based on cereals and vegetables, the fermented beverages chicha, caxiri, cauim and champús, and cachaça, a fermented and distilled beverage, could be cited. Genetic and physiological analyses of Saccharomyces cerevisiae strains isolated from cachaça have been shown to present interesting traits for beer production, such as flocculation and production of aroma compounds, fundamental to high-quality beer. The study of these traditional beverages allows the identification of new microorganism strains displaying enhanced resistance or new flavor and aroma profiles that could lead to applications in several industries and ultimately new products.

INTRODUCTION

Alcoholic beverages have been consumed by mankind since ancient times. These products of fermented sugar-rich goods, namely, cereals, roots, and fruits, are present worldwide since the oldest records [1,2]. In fact, several of mankind's milestones, such as the dawn of agriculture, are closely linked with the production of some type of alcoholic beverages. Similar processes of fermentation emerged independently in many civilizations across the globe. Interestingly, the main players of the whole process are relatively few, mostly yeast from the *Saccharomyces* genus and lactic acid bacteria (LAB) [3, 4]. Nowadays, such microorganisms have a significant role in several industrial relevant processes, including the production of beer, wine, cheese, and bread. Importantly, the popularity of fermented beverages, namely, beer and wine, is such that their worldwide consumption is second only to nonalcoholic drinks as water, tea, and coffee [5].

This chapter aims to contribute to a comprehensible analysis of the role of yeast and LAB on the production of fermented beverages from South America. The microbiological diversity associated with the fermentation of a wide diversity of raw materials, from sugarcane to cassava, as well as new potential biotechnological applications will be addressed.

Ethanol and Lactic Acid Fermentation

Yeast Diversity and Metabolism

Yeast are unicellular fungi, being the simplest eukaryotes. Present in a great number of environments, yeast can be found not only in decomposing fruit, trees, and soils but also in commensal relationships with higher eukaryotes, humans included, and even saltwater. The high diversity of species, almost 1500 species have been described [6], is closely related to this wide distribution. Some of these yeast are adapted to extreme environments, such as high salt concentrations [7], low pH [8], or extremely cold temperatures [9, 10]. The genus *Saccharomyces*, particularly *Saccharomyces cerevisiae*, is strongly associated with the production of fermented products for human consumption, namely, bread, wine, and beer [2]. After several millennia of close coexistence, through phenotypic selection, these species evolved to produce goods with organoleptic properties pleasant to humans. However, given the high degree of diversity found in nature, it is expected to find yeast with new and more interesting characteristics for the industry in new and unexplored niches [11, 12]. Yeast, as other heterotrophic organisms, have the anabolism coupled with catabolism. In one hand, the oxidation of organic molecules, as sugars, yields adenosine 5-triphosphate (ATP) that, in turn, is used as an energy resource for the cell. On the other hand, such organic molecules can also be used as building blocks or to generate intermediary compounds for the synthesis of other compounds, some of which with high commercial value.

The high diversity of environments where yeast can be found is closely related to the variety of carbon sources that can be used. Hexoses such as glucose, fructose, galactose, or mannose are the most common substrates, but some species can use pentoses like xylose or arabinose. Several industrial relevant species can metabolize disaccharides as maltose, lactose, or sucrose, and some, as *Saccharomyces diastaticus*, can even metabolize dextrins (glucose polymers) [13, 14]. Nevertheless, glucose and fructose, to a lesser extent, are the preferred substrates.

In order to use glucose as carbon source, first and foremost, yeast have to sense the presence of this sugar in the surrounding environment and then express the adequate proteins to transport it across the plasma membrane [15, 16]. Whenever glucose is sensed in the medium, changes in the cell proteome will occur. Several processes contribute to the overall change in enzymes levels, including alteration of mRNA translation rates, mRNA stability, or protein synthesis and/or degradation. However, the major response is the extensive upregulation of a large number of genes required for the metabolism of glucose, such as genes encoding glycolytic pathway enzymes, leading to the adaptation to the fermentative metabolism. Moreover, in genes encoding for proteins involved in the metabolism of alternative substrates, gluconeogenic and respiratory pathways are repressed strongly by glucose (for reviews, see [17, 18]). In *S. cerevisiae*, a glucose concentration as little as 15 mM is enough to induce such changes [19].

S. cerevisiae presents an extensive family of hexose transporters, including more than 20 members: (i) 18 genes encoding transporters (*HXT1-HXT17*, *GAL2*) and (ii) at least two genes encoding sensors (*SNF3*, *RGT2*). Some studies suggest that Gpr1p and Hxk2p may sense glucose levels [17, 20]. The transporters can be divided in two classes regarding glucose affinity: (1) low affinity for glucose and high transport capacity, the most important proteins are Hxt1p and Hxt3p, and (2) high affinity and low transport capacity, the key proteins being Hxt2p, Hxt4p, and Hxt7p.

Following uptake by the hexose transporters, glucose enters the glycolytic pathway in order to be metabolized to pyruvate (Figure 1, steps from glucose to pyruvate), whereby the production of energy in the form of ATP is coupled to the generation of intermediates and reducing power in the form of NADH for biosynthetic pathways [21, 22]. The phosphorylation of glucose to glucose-6-phosphate, requiring ATP, is the initial step of glycolysis, by the action of the hexokinases (Hxk1/2p) and the glucokinase (Glk1p), which are linked to high-affinity glucose uptake. The glucose-6-phosphate is then isomerized to fructose-6-phosphate by the phosphoglucose isomerase, encoded by *PGI1* gene. The next step, done by the phosphofructokinase (Pfk1/2p), also requires energy. The fructose-6-phosphate molecule is converted into fructose 1,6-biphosphate through the transfer of inorganic phosphate from ATP. In turn, yeast aldolase (fructose 1,6-bisphosphate aldolase—Fba1p) is responsible for the reversible cleavage of fructose 1,6-bisphosphate to glyceraldehyde 3-phosphate and dihydroxyacetone phosphate.

These two resulting compounds can be interconverted, in a reversible way, by the action of the triosephosphate isomerase (Tpi1p). Glyceraldehyde 3-phosphate is further metabolized to ultimately yield pyruvate, while some of the dihydroxyacetone phosphate follows gluconeogenesis. This step is fundamental for the osmotic and redox homoeostasis, as the dihydroxyacetone can be converted to glycerol yielding NAD^+. Glyceraldehyde 3-phosphate is first oxidized by NAD^+ and then phosphorylated under the catalysis of the 3-phosphate dehydrogenase (Tdh1/2/3p). The resulting 1,3-diphosphoglycerate is, in turn, converted to 3-phosphoglycerate by the action of phosphoglycerate kinase (Pgk1p), yielding 1 molecule of ATP. The enzyme phosphoglycerate mutase (Pgm1p) promotes the relocation of the phosphate group from C3 to C2, allowing the dehydration by the enolase (Eno1/2 p), resulting in the phosphoenolpyruvate. Then the pyruvate kinase (Pyk1p) converts this highly energetic molecule to pyruvate, yielding a second molecule of ATP.

The pyruvate molecule can be further processed through different metabolic alternatives, the respiratory or the fermentative pathways (Figure 2). The selection of one of the route depends greatly on the expression or repression of some genes, which in turn are tightly regulated on the environmental conditions [24]. The genus to which the yeast belongs also plays a role in the prevalence of one route over the other.

The fermentative pathway is particularly relevant to industry, as several important commodities are produced through this process (characteristic of particular organisms). In *S. cerevisiae*, the first step is the decarboxylation of pyruvate to yield acetaldehyde and carbon dioxide (CO_2), through the action of the pyruvate decarboxylase (Pdc1/5/6p). The acetaldehyde can be further reduced to form ethanol by the enzyme alcohol dehydrogenase (Adh1p), allowing the reoxidation of NADH to NAD^+. Besides the direct products of fermentation, ethanol and CO_2, several other by-products are generated during the process, including cell biomass, glycerol, and some organic acids. Overall, the ethanol fermentation is a redox-neutral process since the reduced coenzyme NADH produced during glycolysis, in the oxidation of glyceraldehyde

3-phosphate, is latter reoxidized in the reduction of acetaldehyde to ethanol [25]. Nevertheless, given that biomass is a product of fermentation, and it is in a more oxidized state than glucose, an excess of reducing equivalents may be generated. As mentioned above, glycerol production plays an important role in the redox balance restoration. The glycolytic intermediate dihydroxyacetone is reduced to glycerol 3-phosphate, oxidizing NADH to NAD^+, in a reaction catalyzed by the NAD^+-dependent glycerol 3-phosphate dehydrogenase (Gpd1/2p). Glycerol 3-phosphate is then dephosphorylated to glycerol due to the action of glycerol 3-phosphatase (Gpp1/2p) [5, 26, 27]. The presence of glycerol may contribute to the organoleptic properties in the final product of fermentation, such as wine.

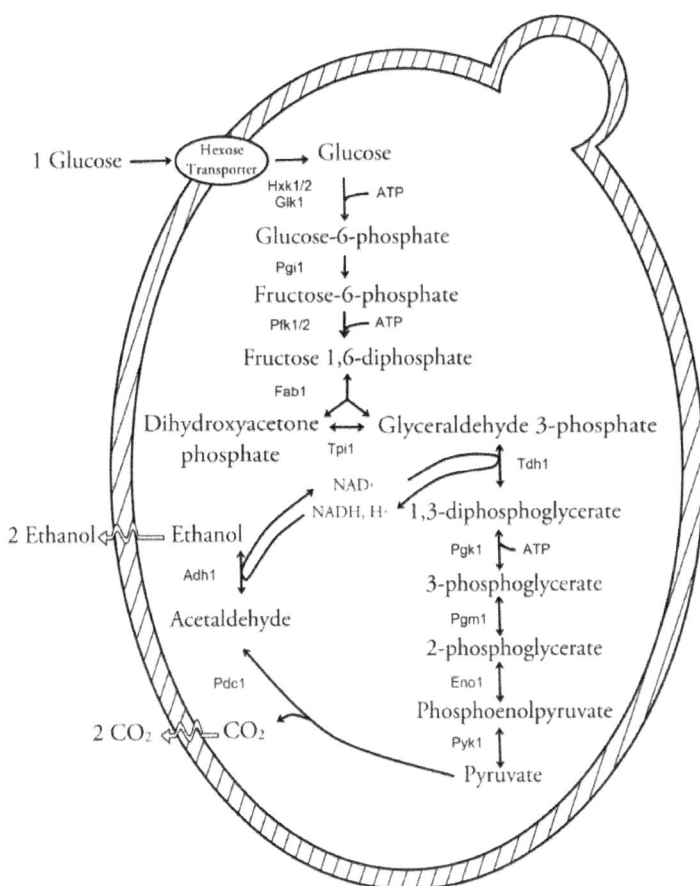

Figure 1: Glycolysis and alcoholic fermentation steps on *S. cerevisiae* (adapted from [23]).

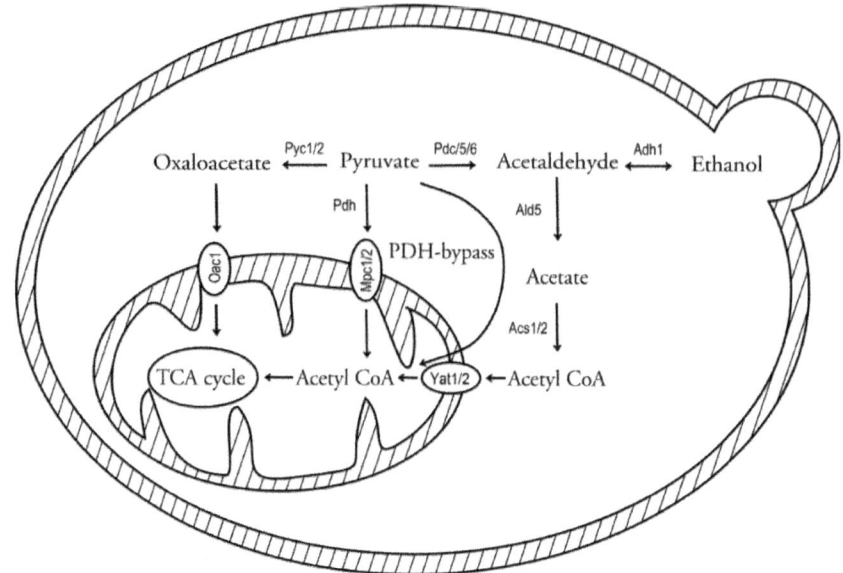

Figure 2: Pyruvate metabolic fates. The pyruvate yielded during glycolysis can be converted into two intermediates of TCA cycle: acetyl-CoA, by the pyruvate dehydrogenase complex (PDH), and/or oxaloacetate, by pyruvate carboxylases (Pyc1/2p). These molecules can be transported into the mitochondria by the pyruvate carriers (Mpc1p or Mpc2p) and the oxaloacetate carrier (Oac1p), respectively. Another alternative is the decarboxylation to acetaldehyde, by the pyruvate decarboxylase (Pdc1/5/6p), which ultimately can yield ethanol. Adh1p—alcohol dehydrogenase; Ald5p—acetaldehyde dehydrogenase; Acs1/2p—acetyl-CoA synthases; Yat1/2p—carnitine acetyltransferases (adapted from [22]).

Although most microorganisms ferment in the absence of oxygen, this is not always the case. Even if oxygen is available, high concentrations of sugars present in the environment will lead yeast to choose fermentation over respiration. This inhibition of aerobic metabolism if glucose is available, both in the presence or absence of oxygen, is denominated the Crabtree effect [28]. *S. cerevisiae* is known as Crabtree positive since it will produce ethanol aerobically if the glucose available is higher than 15 mM [19]. The availability of high sugar concentrations in the surrounding environment stimulates glycolysis, which in turn leads to the production of increasing amounts of ATP, through substrate-level phosphorylation. At the same time, the availability of additional ATP will reduce the respiration and ATP synthesis, through oxidative phosphorylation, leading to a decrease in oxygen consumption. On the other hand, Crabtree-negative yeast do not present a glucose inhibition of aerobic respiration, so these microorganisms resort to this more efficient form of energy metabolism, producing biomass via tricarboxylic acid (TCA) cycle. Nevertheless, these species are able to ferment, but mainly in anaerobic conditions. Importantly, Crabtree is not exclusive to yeast, as it has been detected in many mammalian tumor cells [29–31].

During aerobic respiration (Figure 3), acetyl-CoA is produced by the decarboxylation of the glycolytic pyruvate, by the action of the pyruvate dehydrogenase complex. Then acetyl-CoA will enter the tricarboxylic acid (TCA) cycle, where it will be used to generate reducing equivalents, NADH and $FADH_2$. These molecules will fuel the oxidative phosphorylation, through the highly conserved electron transport chain. Besides the production of reducing coenzymes, the TCA cycle provides intermediates to several other biochemical pathways, including the synthesis of amino acids and nucleotides (for reviews, see [22, 32]).

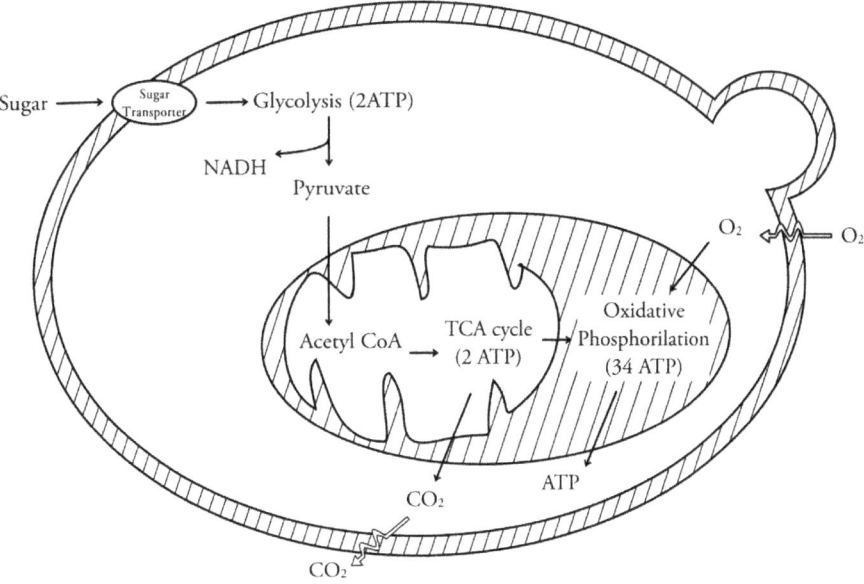

Figure 3: Aerobic respiration in *S. cerevisiae* (adapted from [33]).

Lactic Acid Bacteria

Lactic acid bacteria (LAB) constitute an ubiquitous and heterogeneous group capable of fermenting carbohydrate with the production of lactic acid as a major end product [34]. LAB are found in diverse nutrient-rich habitats associated with plant and animal's matter, as well as in respiratory, gastrointestinal, and genital tracts of humans [35, 36]. A typical LAB is Gram positive, present a GC content below 55%, generally nonsporulating, usually nonmotile, fastidious, catalase negative (pseudocatalase may occur in some LAB), aerotolerant, and acid tolerant [34]. Taxonomic parameters have distributed LAB members into two phyla, *Firmicutes* and *Actinobacteria*. Within the *Firmicutes*phylum, LAB members belong to the order *Lactobacillales* and comprise the following genera:*Aerococcus, Alloiococcus, Carnobacterium, Enterococcus, Lactobacillus, Lactococcus, Leuconostoc,Oenococcus, Pediococcus, Streptococcus, Symbiobacterium, Tetragenococcus, Vagococcus,* and*Weissella*. Regarding LAB members belonging to the *Actinobacteria* phylum, the only species belongs to the *Bifidobacterium* genus [35, 37, 38]. Nevertheless, it is worth mentioning that *Bifidobacterium* is poorly phylogenetically related to typical LAB. These bacteria have been considered as LAB given its physiological similarity and the shared biochemical properties [39, 40].

Usually, LAB members are nonpathogenic organisms with a reputed generally recognized as safe (GRAS) status. The *Lactobacillus* genus includes some of the most important GRAS species involved in food microbiology and human nutrition [41, 42]. The remarkable ability of these bacteria to adapt to different environments resulted in a large number of industrially relevant strains. Among these are *Aerococcus, Carnobacterium, Enterococcus, Lactobacillus, Lactococcus, Leuconostoc, Pediococcus,Streptococcus,* and *Bifidobacterium* [35, 43, 44]. Furthermore, given that LAB greatly contribute to the effective acidification of the matrix and consume rapidly fermentable sugars, these bacteria are frequently predominant in the natural fermentation microbiota [44].

Pathway of Homolactic and Heterolactic Acid Fermentation in Lab

LAB are able to live in the presence of oxygen; however, they obtain their energy by substrate-level phosphorylation. These bacteria do not present a functional respiratory system, as they lack the ability to synthesize cytochromes and porphyrins, key components of respiratory chains [45, 46]. Therefore, an important parameter used in the differentiation of the LAB species is the type of lactate fermentations: homofermentative and heterofermentative [35]. As a general rule, homofermentative lactic acid bacteria use the Embden–Meyerhof–Parnas pathway (EMP pathway or glycolysis) to produce pyruvate, while heterofermentative lactic acid bacteria use the pentose phosphate pathway (PPP). However, a third pathway, the Bifidum pathway, presents distinct reactions (Figure 4) [45, 46].

In the homofermentative lactate fermentation, as the name implies, the major end product generated is lactate. Initially, two ATP molecules are produced per mole of glucose via the oxidation of phosphoglyceraldehyde. In a second stage, NADH molecules resulting from the previous oxidative stage are used to reduce the pyruvate, forming lactate [45, 46]. The overall reaction is as follows:

glucose + 2 ADP + 2 Pi → 2 lactate + 2 ATPglucose + 2 ADP + 2 Pi → 2 lactate + 2 ATP

Some representative homolactic LAB genera include *Lactobacillus*, *Lactococcus, Enterococcus,Streptococcus*, and *Pediococcus* species [38].

Conversely, in the heterofermentative lactate fermentation pathway, lactate is not the only end product; significant amounts of CO_2 and ethanol, or acetate, are also produced. In this pathway, lactate is produced by the decarboxylation and isomerization reactions of the PPP. Glucose is oxidized to ribulose-5-phosphate that is isomerized to xylulose-5-phosphate, which in turn is cleaved to form phosphoglyceraldehyde and acetyl phosphate. The phosphoglyceraldehyde molecule is oxidized to pyruvate by reactions of glycolytic pathway, whereas the acetyl phosphate is reduced to ethanol [45,46]. The overall reaction is as follows:

glucose+ ADP + Pi→ ethanol + lactate + CO2 + ATPglucose+ ADP + Pi→ ethanol + lactate + CO2 + ATP

Some representative heterolactic LAB genera include *Leuconostoc*, *Oenococcus*, and *Weissella* [38]. It is worth mentioning that heterofermentative lactate fermentation produces only one ATP molecule per glucose, while the homofermentative lactate fermentation produces two ATP molecules per glucose.

Bifidum Pathway

The Bifidum pathway is a particular metabolic route found in *Bifidobacterium bifidum*, which uses reactions of the PPP and homofermentative pathway, producing primarily acetate and lactate [38, 45]. In this pathway, 2.5 ATP molecules are produced per glucose. As such, ATP yields are greater than for the homofermentative or heterofermentative pathways, due to the presence of key enzymes, fructose-6-phosphate phosphoketolase and xylulose-5-phosphate phosphoketolase. These proteins catalyze two important steps: the cleavage of one molecule of fructose-6-phosphate, yielding one molecule of erythrose-4-phosphate and one of molecule acetyl phosphate, and the cleavage of two xylulose-5-phosphate into two glyceraldehyde 3-phosphate and two acetyl phosphate, respectively [45, 46]. The overall reaction is as follows:

2 glucose +5 ADP + 5 Pi → 3 acetate + 2 lactate+5 ATP2 glucose +5 ADP + 5 Pi → 3 acetate + 2 lactate+5 ATP

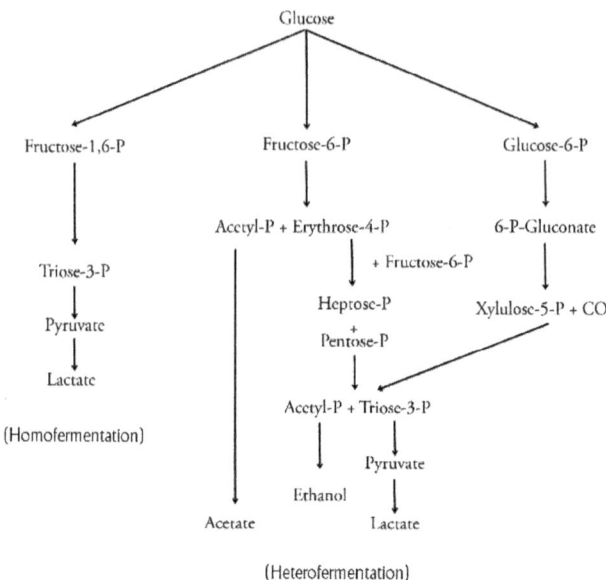

Figure 4: Schematic representation of the metabolism of hexoses by lactic acid bacteria (adapted from [47]).

Lab—Beverage Industry Applications

Over the years, LAB has been explored on a large scale in several food industry segments (processing of meats, vegetables, and beverages) occupying a central role in these niches [43, 48–50]. Withal, there are some reasons that explain their use in the food production industry. Among these are the following: the production of antimicrobial substances, which restricts the growth of harmful microorganisms, and the production of metabolites, which influences the nutritional, texture, and organoleptic qualities of the end products [36, 51]. Moreover, LAB have also been used as probiotics, which shows several potential health benefit [52]. Thus, in general, LAB enhances the shelf life and microbial safety of end products [43]. However, based on the microorganisms profile present in the raw material, their effects may be either beneficial or disadvantageous to the food processing. For instances, malolactic fermentation (MLF) is a secondary fermentation in wine normally carried out by LAB, especially by *Leuconostoc oenos,* which usually occurs at the end of alcoholic fermentation by yeast [53]. In this metabolism, L-malic acid is decarboxylated to L-lactic acid and CO_2, a reaction catalyzed by the malolactic enzyme without the release of intermediates. As a consequence of this pathway, the acidity is reduced which turn it a crucial process in wine and cider production [53].

However, it is noteworthy that MLF is not only important as a deacidification process in wine, but for the aroma and microbial stability of wine [53–55]. Additionally, this fermentation prevents the malic acid utilization by other undesirable microorganisms. Another industrial application of LAB is the use of starter cultures as inoculants during the malting process, a complex biological process essential to the production of fermented beverages, in order to improve the malt quality and safety. In these conditions, LAB can improve the extraction, fermentability, and nitrogen yield of wort and the foam stability, color, and flavor of beer [50]. Moreover, another important effect of the LAB in the malting process is their ability of antimicrobial substances production (e.g., bacteriocins) that restricts the growth of harmful bacteria to malting [50, 56–58].

FERMENTED BEVERAGES OF SOUTH AMERICA

The traditional foods, mainly those produced by spontaneous fermentation, are present in the daily life of the population and play an important role in the cultural identity of different communities [59]. Indigenous or traditional fermented foods refer to the products that, since the beginning of history, are an integral part of the diet and can be prepared in household or cottage industry, using simple techniques and equipments [60].

In South America, there are various traditional fermented beverages, mainly produced by fermentation of cereals, vegetables, and root tubers. Among these beverages could be cited the traditional beverages *cauim* and *caxiri*, made by Brazilian natives and the traditional beers *chicha* and *champús*, typical of the Andes [59, 61–64]. *Cachaça* is also a typical beverage, from Brazil, where the fermented sugarcane juice is then distilled to produce a spirit. Table 1 shows some characteristics of these traditional beverages.

Table 1: Catalogue of traditional fermented beverages of South America.

Beverage	Raw material	Microorganisms	Country
cauim	cassava, rice, peanuts	Lactic acid bacteria (LAB),*Saccharomyces cerevisiae*, other yeast	Brazil
caxiri	cassava	LAB, *Bacillus*spp., *S. cerevisiae*(predominant yeast), other yeast	Brazil
champús	maize	*S.cerevisiae*, other yeast	Colombia
chicha	maize	LAB, *S.cerevisiae*(predominant yeast),*Aspergillus* spp.	Peru

cachaça	sugarcane	*S. cerevisiae*; LAB, other yeast	Brazil

[i] - Source: [59, 61-64]

Caxiri

Caxiri is an alcoholic fermented beverage made from cassava (*Manihot esculenta* Crantz) and produced by Brazilian natives. The *Yudjá-Pakaya* native tribe is located in the Xingu Park in the Mato Grosso state, Brazil [64]. Initially, cassava roots are incubated for 2 days in running water to soft the skin. After this stage, the cassava is peeled, cut into small pieces, and placed in handmade straw press for removing the excess of water. The resulting paste from pressing is grated into flour. Then the flour is roasted for about 2 hours. Subsequently, the flour is mixed with water and grated sweet potato. The mixture is placed in special open vessels for fermentation to occur. After 24–48 hours, the beverage is ready for consumption [64]. In the course of the fermentation process, a change of pH occurs, decreasing from pH ~5.8 to ~3.1–3.2. The final beverage has an ethanol concentration of approximately 80 g/L, and 28 g/L of lactic acid.

Ramos et al. (2010) investigated the microbiota involved in *caxiri* by culture-dependent and culture-independent methods of cultivation. These authors found that the bacterial populations varied from 3.05 to 5.33 log CFU/mL, and the populations of yeast from 3.27 to 7.34 log CFU/mL. Yeast dominate fermentation after 48 hours, being *S. cerevisiae* the dominant yeast species. Other yeast species, such as *Rhodotorula mucilaginosa, Pichia membranifaciens, Pichia guilliermondii* and *Cryptococcus luteolus* were also found. The bacteria belonging to the genus *Bacillus* were the most prevalent in *caxiri*. LAB species were found at the beginning of the fermentation and after 24 hours of fermentation. The microbiota found in *caxiri* may be from different sources, such as raw materials, environmental, or utensils used in the preparation of the beverage [64].

Cauim

Cauim (Kawi) is a nonalcoholic fermented beverage produced from various substrates, such as banana, cassava, cotton seed, maize, pumpkin, peanuts, and rice. This drink is produced in Brazil, in the Mato Grosso state, by *Tapirapé-Tapi'itãwa* native tribe [59, 63]. *Cauim* represents great importance to the tribe since it is consumed by adults and children as part of their daily meal [59, 63]. For the production of *cauim* from cassava, initially this substrate is fermented for 3 to 5 days in running water. Afterward, the cassava is peeled, cut into pieces and placed on sun to dry. The dried pieces are crushed, and the obtained flour is added to water. The mixture is boiled for 2 hours and then cooled at room temperature. An inoculum obtained from the chewed sweet potato (*Ipomoea batatas*) prepared by indigenous women is then added to the porridge to start the fermentation. After 24–48 hours, the beverage is ready for consumption [59].

For the preparation of beverages with other substrates, the procedure is similar to *cauim* made from cassava. The main substrate is cooked for approximately 2 hours and cooled at room temperature. After cooled, the inoculum is added to start the fermentation [63]. During fermentation, a progressive acidification occurs, and the pH of beverage decrease from ~5.5 to ~3.4 [59].

Almeida et al. (2006) found LAB as the dominant microorganisms during *cauim* fermentation. Among the LAB obtained in that work, *Lactobacillus pentosus* and *Lactobacillus plantarum* were the most prevalent species. Gram-positive sporulating bacteria, as the genus *Bacillus* spp., were found in small amounts. In a recent work, Ramos et al. (2010) monitored the community of yeast and LAB in *cauim*prepared from rice and peanuts, by culture-dependent and culture-independent methods. LAB were found in higher counts, ranging from 7.4 to 8.4 log CFU/mL, while yeast were found at 4.0 to 6.6 log CFU/mL. The most prevalent species of yeast found were *Pichia guilliermondii*, *Kluyveromyces lactis*,*Candida* sp., *Rhodosporidium toruloides*, and *S. cerevisiae*. The most prevalent species of LAB belonged to genus *Lactobacillus*: as *L. plantarum, L. fermentum, L. paracasei*, and *L. brevis* [63].

Champús

Champús is a cereal-based fermented beverage, sweet and sour, and with low alcohol content. This drink, typically found in Colombia, can also be found in other countries in South America, such as Ecuador and Peru [62]. *Champús* can be prepared from different cereals (maize, rye, and wheat) alone or in a mixture, with other raw materials, such as pineapple, *lulo* (*Solanum quitoense* Lam.), *panela*(brown sugar paste), herbs (orange leaves), and spices (cloves and cinnamon) [62].

In Colombia, the beverage is produced by boiling the kernels of maize, for about 2 hours. Thereafter, the beans are cooled to room temperature, and then fruits, *panela*, and other ingredients are added. The beverage is cooled to 12°C–15°C, and after 24–48 hours of a spontaneous fermentation process, it is ready for consumption. The final beverage has a low alcohol content (2.5%–4.2%), and the pH is between 3.5 and 4.0 but can vary according to the ingredients [62]. The microorganisms responsible for*champús* fermentation (during storage at low temperatures), such as yeast and LAB, come from the fruit since the microbial derived from corn grains are eliminated during the period of boiling [62]. Osorio-Cadavid et al. (2008) found seven genera of yeast when twenty samples of *champús* from Colombia were analyzed. The most prevalent species of yeast founded in this study were *Pichia fermentans*, *S. cerevisiae*, and *Issatchenkia orientalis* (*Candida krusei*).

Chicha

Chicha is an alcoholic beverage, clear, yellowish, and sparkling, which resembles the taste of cider and that has been consumed by Andean indigenous population for hundreds of years. This beverage is produced in regions of the Andes and sometimes in low-lying regions of Ecuador, Brazil, Peru, Bolivia, Colombia, and Argentina [65]. *Chicha* is a generic name that comprises a series of fermented or nonfermented beverages that can be prepared from various raw materials such as cereals and fruits [66].

Chicha can be produced in different ways. Although the recipes pass from generation to generation, all of them use the conversion of starch into sugar, followed by fermentation of sweet wort. As the production process resembles the brewing process, the traditional *chicha*, made from maize, can be named as the Andean indigenous beer [67].

In the Andean region, the most common maize *chicha* is the *chicha de jora*. This *chicha* is prepared from yellow corn grain (*maíz amarillo*) malted (germinated and dried) or chewed. In *chicha de jora* production process, hydrolysis of starch is obtained by the malting process of the kernels of maize or the action of salivary amylase in the case of chewing [65].

The production process of *chicha de jora* is laborious. After malting, it proceeds to the wort boiling (consisting of maize flour plus water), a process that takes many hours. During the wort boiling, sugar and *panela* as well as herbs and spices may be added. Subsequently, the wort is cooled and filtrated. Then the beverage is placed in special vessels for the occurrence of fermentation. The *chicha* is ready for consumption when its sweet taste disappears and the flavor becomes a little stronger. However, if not consumed immediately, the beverage becomes bitter, and after 7 days, it usually is converted into vinegar [65, 68]. During the *chicha de jora* production, a wort acidification occurs, with a decreasing pH values of ~5.7–5.3 to ~3.7–3.5. The final beverage presents an ethanol concentration between 9 and 10 g/L [69].

In some Andean countries is produced the *chicha morada*, a beverage prepared with purple maize. Purple corn is a pigmented variety of *Zea mays* L., originating mainly from Peru and Bolivia. This drink is prepared by boiling purple maize with pineapple, quince peel, cinnamon, and cloves [70, 71]. In Ecuador, in addition to *chicha de jora*, other kind of *chichas* like *chicha de morocho*, prepared with white maize, *chicha de Yamor* or seven-grain *chicha*, and *chicha de yuca* (cassava *chicha*) are also produced. The seven-grain *chicha* is produced from seven varieties of maize as *jora*, yellow corn, white corn, black corn, *chulpi* corn, *morocho* corn, and popcorn [68]. The *chicha de yuca*, produced by the indigenous and mestizo population in the Amazon region of Ecuador, is produced in a peculiar way since chewing is used. After the chewing process and fermentation of cassava, a mixture of Ungurahua palm (*Oenocarpus bataua* subsp. *bataua*, Arecaceae) fruit juice with the fermented mass is made, and thus the beverage is ready for consumption. For the preparation of the juice, the fruits are first harvested and then they are soaked in hot water for the removal of mesocarp. The seeds are dropped and then pieces fermented cassava are added to the mixture of mesocarp and water [68, 72].

In *chichas* from Ecuador and Brazil, LAB, yeast, and *Bacillus* species were found as the microorganisms associated with this beverages [73, 74]. Blandino et al. (2003) found yeast, bacteria, and filamentous fungi in *chichas* of Peru. Elizaquível et al. (2015) found the *Lactobacillus* genus as the most prevalent and the one with the highest diversity of species in *chichas* of Argentina, by culture-dependent and culture-independent methods. In another work, the yeast obtained from *chichas de jora* collected in 10 *chicherías* (*chicha* producers) in Peru were identified as belonging to the species *S. cerevisiae* [66]. Vallejo et al. (2013) considered *S. cerevisiae* as the responsible yeast for the fermentation of Peruvian *chichas de jora*.

The *chicha* microbiota may come from different environments. The LAB found in *chichas* may have been introduced from the raw materials, as many species are commonly found in vegetables and plants, and also transferred from humans and animals, natural hosts of these bacteria [73]. The yeast involved in *chichas* production process come from different sources, such as handlers, raw materials, utensils, and equipment used in the preparation of these beverages or can be carried by insects. The clay vessels and wooden spoons, used in the preparation of *chichas*, provide an ideal microhabitat for yeast, infiltrating into tiny cavities of such utensils [75]. In Quito, Ecuador, two isolates of yeast from old vessels obtained from deep tombs of La Florida archaeological site were recovered, which were identified as *Candida theae*, a new species belonging to the clade *Lodderomyces* [76].

Cachaça

In Brazil, *cachaça* (ka.ʃa.s) was the name given to (i) waste of sugar production (beginning of XVI century), (ii) waste of sugar production when fermented (around XVI–XVII centuries), and finally (iii) product of the distillation of the fermented sugarcane (XVII century to nowadays). The first mention of *cachaça* occurs in 1622 with the name of "augoa ardente" [aqua vitae] or spirit, in Bahia State (Brazil), and the first use of the name *cachaça* instead of aqua vitae occurs in 1660. Considering the three ethnic groups that formed the Brazilian nation (Native Brazilians, Africans, and Europeans), scarce information is available about the real contribution of each group to the initial production of fermented sugarcane. However, *cachaça*, which is the result of the distillation of fermented sugarcane juice, was certainly "discovered" by Europeans, the most technologically advanced group, who had knowledge and equipment to do so [77]. Nowadays, *cachaça* is the typical and exclusive denomination to Brazilian spirit produced from sugarcane juice with alcohol content ranging 38%–48% (v/v) at 20°C (68°F), which present unique characteristics (Table 2) [78]. Brazil has an estimated installed capacity of *cachaça* production ranging from 1.2 to 1.5 billion litters/year; however, the production is less than 800 million litters/year. According to the Brazilian Institute of Geography and Statistics (IBGE), almost 15,000 establishments are currently producing *cachaça*. In 2014, 10.2 million liters of *cachaça* were exported to 66 countries, generating US$ 18.33 million in revenue [79].

Table 2: Components present in *cachaça* and its limits in accordance with the Brazilian law.

Compounds	Units	Limits
Copper	mg/L	5.0
Ethyl carbamate	μg/L	210.0
Volatile acidity	mg/100mL anhydrous ethanol	150.0
Total esters	mg/100mL anhydrous ethanol	200.0
Aldehydes	mg/100mL anhydrous ethanol	30.0
Total higher alcohols*	mg/100mL anhydrous ethanol	360.0
Furfural+HM[F]+	mg/100mL anhydrous ethanol	5.0
Methanol	mg/100mL anhydrous ethanol	20.0
Acrolein	mg/100mL anhydrous ethanol	5.0
Particles in suspension	-	Absent
Dry extract	g/L	6.0
Total sugars	g/L	6.0
		38-48

[i] - Source: Ministry of Agriculture, Livestock and Supply - Brazil

[ii] - *Sum of isobutyl (2-methyl-1-propanol), isoamyl (2-methyl-1-butanol and 3-methyl-1-butanol), and n-propyl (1-propanol) alcohols.

[iii] - + 5-(Hydroxymethyl)furfural

Production

The main raw material for the production of *cachaça* is the juice of sugarcane (*Saccharum* spp.). The first step to prepare of the fermentation medium is the extraction of sugarcane juice. Small producers extract the juice by crushing the sugarcane using a mill. Yet, the large producers use a more complex system of extraction: (i) crushing system, (ii) cutting machines, and (iii) shredders [80]. The resulting sugarcane juice is an opaque (color ranging from brown to dark green), viscous, and sweet liquid. The color is due to different pigments such as chlorophyll and polyphenols, while the viscosity is due mainly to the presence of colloidal proteins. The sugarcane content in fermentable carbohydrates is sucrose (11%–18%), glucose (0.2%–1%), and fructose (0%–0.6%) [81]. A great number of microorganisms is associated with the sugarcane plant, and during the extraction of juice, these microorganisms can be transferred to the fermentation medium. The yeast present in the juice belong to the genus: *Candida*, *Cryptococcus*, *Kluyveromyces*, *Hansenula*, *Rhodotorula*, *Saccharomyces*, and *Torulopsis*. The main bacteria genus are *Leuconostoc*, *Streptotococcus*, *Lactobacillus*, and *Bacillus*[82]. In order to reduce the number of microorganisms and/or optimize the fermentation, several processes to improve the quality of the fermentation medium can be implemented. The most common procedure is the decantation of the sugarcane juice, thus eliminating coarse particles as soil/sand that can damage the *cachaça* manufacturing equipment. A dilution or concentration step (most unusual process) may be performed, particularly if the crop was harvested at a nonoptimal time. The juice supplementation with nutrients can provide a more robust fermentation; however, such practice is not common among the majority of *cachaça* producers. Studies have shown that addition of ammonium sulfate, magnesium sulfate, cobalt sulfate, and vitamins (especially B complex) provide huge productivity gains. Other actions such as filtration, addition of antibiotics, use of disinfectants, temperature control, and must acidity correction can also be used to improve the quality of fermentations [83].

Microorganisms

As important as the preparation of the medium is the preparation of microorganisms that will ferment the sugarcane juice, the so-called foot-of-vat.

Traditionally, *S. cerevisiae* is the most used yeast, but some other species are also utilized, namely, *Pichia* sp. [11, 84]. The cell concentration usually found at the beginning is about 10^6–10^7 cells/mL and about 10^8 cells/mL at the end of the fermentation. Traditionally, the majority of the *cachaça* is produced through self-inoculation, using different protocols. Usually, a homemade mixture is prepared: (i) diluted sugarcane juice (carbon source), (ii) rice bran and/or corn bran (nutrient sources), and (iii) lemon/orange juice (to reduce initial pH) [85]. All these ingredients are mixed and kept at rest for 24 hours, during which wild yeast present in sugarcane juice will multiply—microorganism multiplication is verified through foaming. At the end of the first 24 hours, a new volume of diluted sugarcane juice is added, and the process is restarted [85]. The process is repeated until it reaches 20% of the volume of the fermentation vat. At the end of the process, a wild yeast community is obtained, which is adapted to the physical and chemical conditions of the fermentation. Producers, which aim to improved organoleptic characteristics, tend to use selected yeast in fermentative process. There are two basic types of selected yeast: (i) yeast that are used to produce ethanol or other beverages without being specific for *cachaça* production and (ii) yeast selected from their own vats of *cachaça* fermentation. Yeast selected in their own fermentation tanks have advantages over the first because they are more adapted to the *cachaça* fermentation specific conditions and are able to provide unique features to the product [86, 87]. There are numerous advantages to using selected yeast, including (i) rapid and (ii) homogeneity of the fermentation, (iii) higher fermentation yields, (iv) higher quality of the final product, (v) lowest risk of contaminations, and (vi) highest resistance to stress [88]. The improved aroma obtained in the final product is another advantage of using the selected yeast [86]. Given that *cachaça* fermentation develops in an environment with few controls, these vats represent a unique ecological niche. Interesting phenotypes, such as the production of flavor and aroma compounds, can be assessed by the detailed characterization of isolated strains [89]. The use of yeast strains producing aroma compounds, as esters (ethyl acetate, isoamyl acetate, ethyl hexanoate, ethyl octanoate, and ethyl decanoate) and higher alcohols (propyl, isoamyl, and isobutyl alcohol), can significantly increase the final quality of *cachaça*. Thus, such profile can be a differential characteristic between several *cachaça* producers or even distinct producing regions—if all producers in a particular region start using a selected yeast strain.

Fermentation

The *cachaça* fermentation can occur in simple batch system (most common process) and fed batch or continuous culture (less common). The batch system with cell recycling at the end of each fermentation is the most widespread in the *cachaça* production. In general, the fermentation occurs in 24/36 hours depending on the system efficiency [90]. After fermentation, the product is generally removed by pump systems, or via a valve near the base of the fermenter, where about 80% of fermented must is removed. The remainder medium, approximately 20%, is composed primarily of sedimented yeast [90]. Some producers can perform specific treatments to decrease the contamination load of the inoculum: (i) acid treatment (Melle–Boinot method and its modifications); (ii) by stirring and spray air with diluted sugarcane juice, or (iii) only add new medium for fermentation and restart a new fermentation cycle, without any special treatment [91]. *Cachaça* production occurs mainly on small farms where financial resources are generally scarce. Control of the fermentation conditions is rarely held, including (i) maintaining the fermentation temperature, (ii) standardization of sugarcane juice, or (iii) fermentation in closed vat, in order to prevent contamination. Moreover, hygiene conditions are not always in conformity to national or international standards for the production of beverages [92]. Quite often, the fermentation vats are kept next to animal facilities or made from improvised materials (such as wood, rubber tires, etc.). Thus, the standardization of the final product is not achieved, and the sale of *cachaça*for more demanding/rigorous markets like Europe or the US is greatly impaired. Thereby, government agencies, research laboratories, and private companies together, and some individual initiatives are trying to change the *cachaça* production outlook in order to not only improve the quality of the final product but also regulate unlicensed producers.

Distillation

After fermentation of sugarcane juice, the medium is taken to steel distillation columns (industrial*cachaça*) or special copper-made distillers called "alembics" (majority of the *cachaça* producers). In this last process, the distillation product can be divided into three parts: (i) head or "strong water" (5%–10%), (ii) heart (approximately 80%), and (iii) tail or "weak water" (5%–20%) [19]. The head is the first distilled portion of a fermented must, and it is rich in substances such as aldehydes, methanol, and esters. The head portion has alcohol content of 65/70% v/v. The next distillate volume is called the heart and displays ethanol concentration ranging from 35/55% v/v. This fraction has lower content of chemical contaminants that affect negatively the sensory characteristics of the product. The last fraction is known as tail, where several acids and furfurals are found, presenting an alcohol content around 14% v/v. The best quality *cachaça* are produced only with the heart fraction of the distillate, and the remainder is discarded or used to produce other products such as liqueurs or bioethanol for fuel [93].

Aging

The *cachaça* aging process includes the storage of the distillation product in wood barrels under specific conditions (temperature, humidity, aeration, etc.) for a period not less than 1 year. Numerous biochemical reactions occur during the aging process, the main being oxidation and esterification reactions [94]:

R−COH (alcohol)+ O2 → R−HCO (aldehyde)(oxidation reaction)R−HCO (aldehyde)+ O2→ R−COOH (acid)(oxidation reaction)R−HCO (aldehyde) + R−COOH (acid)→ R−COO−R (ester) (esterification reaction)R−COH (alcohol)+ O2 → R−HCO (aldehyde)(oxidation reaction)R−HCO (aldehyde)+ O2→ R−COOH (acid)(oxidation reaction)R−HCO (aldehyde) + R−COOH (acid)→ R−COO−R (ester) (esterification reaction)

Alcohols are relatively stable to oxidation but can form significant amounts of aldehydes in the presence of phenol and water. Aldehydes are highly reactive and may oxidize to form the corresponding organic acid. Through esterification reactions, acids react with alcohols to form esters, which soften the odor of aldehydes, giving a pleasant odor to the *cachaça*. Beside the aldehydes, some sulfur compounds also decrease during maturation, such as sulfide and disulfide. In addition, alcohol and water, through capillary penetration and osmosis, pass through the interstices of the timber hydrolysing the hemicellulose and lignin [95]. The hydrolysis products are extracted, enriching the distillate and enhancing the quality of the drink [93]. Sensory gains of the beverage varies according to the chemical composition of wood, the aging time, the capacity of the barrel, the porosity, and the thickness of the timber [96]. Some studies have shown that blending aged with fresh *cachaça* is feasible and reasonable since the final product has better organoleptic and commercial characteristics. Many types of woods can be used for the manufacture of high-quality aging barrels, such as amburana (*Amburana cearencis*), jequitibá (*Cariniana estrellensis* and *Cariniana legalis*), ipê (*Tabebuia* genus), balsam wood (*Myrocarpus frondosus avium* and *Prunus cerasus*)—Brazilian trees, and oak (*Quercus* genus) [97].

YEAST AND LAB NEW POTENTIAL APPLICATIONS

South America presents a wide variety of fermented and distilled beverages, which have several unique characteristics, greatly influenced by the fermentative metabolism of microorganisms. Therefore, those microorganisms present a large potential for utilization in the development of new beverages, or even in new biotechnological applications. In this context, several scientific works have focused in the isolation and characterization of such microorganisms [11, 86, 87].

Wild Yeast

During fermentation, yeast and LAB cells are submitted to several stress factors, such as: high osmotic pressure and hydrostatic pressure, high concentrations of ethanol, anaerobic atmosphere, temperature, and nutrient limitation [98]. Such pressures promote the genetic adaptation of the individuals, leading to the survival of only the fittest cells. The increasing number of such alterations will lead to changes in the fermentation subproducts, some of which contribute to the organoleptic properties of the final products. Consequently, some of those subproducts may contribute to improve the beverages and, in this way, increasing the diversification of this industrial niche. Furthermore, the utilization of microorganisms isolated from traditional products, as *chicha* or *cachaça*, in the fermentation or maturation of new drinks production can lead to development of new promising products.

Recently, wild yeast isolates from *cachaça* fermentation vats in Brazil were innovatively evaluated in the production of beer [99]. First, in that study, 21 isolates belonging to the collection of the Laboratory of Cell and Molecular Biology /UFOP were surveyed for the production of aroma compounds in the beers. For that, compounds known for their influence in aroma and flavor, such as higher alcohols, esters, acetaldehyde, diacetyl, and ethanol, were analyzed by gas chromatography. After a careful analysis of each aromatic profile, two superior isolates were chosen (LBCM78 and LBCM45). In the same study, LBCM78 and LBCM45 were shown to be suitable to the production of ale and lager beers, respectively. The physicochemical composition of the produced beers were analyzed and compared to similar beers produced by commercial brewer yeast strains [99]. As a result, in the production of a wheat beer, the commercial strain WB-06 showed higher values of n-propanol than LBCM78. However, no significant differences were observed in the isobutanol and isoamyl alcohol levels. Similarly, the LBCM45 and the commercial strain W-34/70 were used to produce a lager beer, and the final products were analyzed as before. Between those two beers, the productions of isobutanol or isoamyl alcohol were similar. Moreover, no differences were observed in the ethyl acetate and diacetyl content of all the different beers produced by the four strains. As a final step of that study, beers were submitted to a sensorial evaluation by a group of trained tasters, from the Craft Brewers Association of Minas Gerais. The wild yeast showed a similar production of ethanol to the commercial strains (3.41–4.80% v/v), and the sensory analysis of the beers produced from LBCM45 and LBCM78 strains showed good acceptance in the evaluation panel [99]. These results suggest that yeast strains isolated from*cachaça* vats have a great potential for the production of new beers since a good production of volatile compounds and ethanol were observed.

Mixed Fermentation Yeast/Lab

Recently, our research group started a work to approach the utilization of both yeast and LAB in the fermentation of *cachaça* and beer [100]. In that study, it was verified that the presence of LAB in the fermentation of such beverages appears to contribute to the production of organic acids, which may contribute to the formation of aromatic compounds, and to the reduction of the pH, inhibiting the growth of spoiling microorganisms. In a recent collaboration between our group and a known Minas Gerais craft brewery, several LAB were isolated from beers with high acidity level. These isolates were characterized molecularly and identified as *Lactobacillus brevis*, which are mandatory hetero-fermentative microorganisms that produce lactate, ethanol, or acetic acid and CO_2 as final metabolites [40]. Moreover, in the next step of that study, those isolates were tested in fermentations at different temperatures in order to assess their applicability in the production of certain beer styles. The selected isolate LBCM718 showed a good growth in temperatures above 18°C, and then it was tested in beer mixed fermentations with two ale yeast strains, LBCM78 and WB-06. The viability of both yeast and LAB as well as the wort final pH were analyzed [100]. The ethanol production was not affected by the presence of *L. brevis*. Such fermentations are frequent in beers from the lambic or fruit beer style, where LAB contribute with acidity and lactate, yielding ethyl lactate— an important aromatic compound. Moreover, a test was conducted to test the LAB resistance to iso-α-acids, in the concentration range generally found in beer—17 to 55 ppm [101]. There was no formation of inhibition zones, which suggest that isolate can be used in the production of beers.

In a study from another research group, a mixed fermentation of *S. cerevisiae* and *Lactococcus lactis*led to the improvement of *cachaça* quality. When the concentrations of higher alcohols, such as propanol, isobutanol, and isoamyl alcohol from three fractions (head, tail, and heart), were compared, the researchers found higher ethanol levels in the *cachaça* from mixed fermentation than in the *cachaça*produced from a pure *S. cerevisiae* inoculum.

Finally, the evaluation comparing the two products was performed by 40 trained tasters, men and women aged between 22 and 50 years old. In that evaluation, both *cachaças* were evaluated for flavor, color, and overall acceptability. The *cachaça* produced by mixed culture obtained higher scores in the categories aroma and appearance, while the *cachaça*produced from pure culture yeast showed higher global acceptance. Both beverages showed similar flavor [102].

Both studies show that the use of bacteria and yeast simultaneously in fermentation apparently does affect the growth of both cultures. Similarly, the ethanol production in these mixed fermentations was the same. Furthermore, the use of mixed fermentations appears to improve the aroma of both beer and*cachaça*, a potential alternative to the development of new products.

New Spirits

Brazil is the country with the world's largest fruit production; however, there is a huge postharvest waste of raw material that generates losses to the farmer. Therefore, there is the necessity to develop new processes and products to reduce these losses. In this context, an alternative is the use of these fruits for the production of alcoholic beverages [103].

In a previous study, a research group developed a fermentation process from *cajá (Spondias mombin)* pulp for the production of a new beverage. In that study, *cajá* pulp was inoculated with *S. cerevisiae*and then fermented at 22°C during 10 days. Analyses of amount of alcohol and higher alcohols were carried out to determine the compounds present in the beverage. Simultaneously, the final product was subject to sensory analysis. The amount of alcohol found in the *cajá* beverage was in averaged 12°GL, comparable to those found in wines. The total amounts of higher alcohols found were about 0.7 g/L, while the values for these in wines ranging from 0.1 to 0.3 g/L [104]. Sensory analysis showed a good acceptance by the tasters [103].

In another study, it was evaluated the quality of fruit spirits produced through different treatments [105]. Mango, grape, and passion fruit were used as raw materials, and the fermentation was performed using *S. cerevisiae* cultures. Distillation was performed in copper still with controlled temperature between 85°C and 90°C, and the amount of alcohol in beverage was standardized at 40°GL. After that, oak chips and umburana chips were added to the spirits for 60, 90, and 120 days. At the end of this period, the samples were conducted to sensory analysis using 10 trained panelists, using the quantitative descriptive analysis method. All three products obtained were well accepted by the tasters, being the passion fruit distillate the best evaluated. Additionally, 90 days was the best period of aging for those particular spirits [105].

As noted in these studies, alcoholic beverages obtained from tropical fruits were well accepted in sensory tests, demonstrating the potential application of these substrates in the production of new beverages.

Another study had as objective to obtain and characterize a new spirit from the fermentation of cheese whey. The cheese whey is a by-product of the dairy industry that has a high impact in the environment. The researchers used the yeast *Kluyveromyces fragilis*, due to its ability to grow in medium containing lactose, in high yields, and without the production of toxins. In order to achieve that objective, the whey powder was acidified and deproteinized, and the resulting supernatant was used for fermentation [106]. The fermentation of the whey with a high concentration of lactose (200 g/L), after 92 hours, obtained a final product with an ethanol content of 9.6% (v/v). After distillation of the fermented beverage, the heart fraction was diluted to 40% (v/v) ethanol content. The chemical analysis revealed that the higher alcohols were the most abundant group of volatile compounds present in this fraction, containing isoamyl, isopentyl, isobutyl, and 1-propanol, all present in large quantities. Among the esters, the ethyl acetate was found the highest concentration. This compound has a significant effect on the organoleptic characteristics of wines and spirits. Furthermore, the authors concluded that it was possible to obtain a spirit with pleasant smell and taste from cheese whey, containing high concentration of lactose, this being an alternative to by-product of dairy industry [106].

From these studies, we can see distinct possibilities for the production of new beverages, by changing the yeast strain/species, or using blends of different microorganisms, such as yeast and LAB. Moreover, it is possible to use several different substrates for the production of these beverages, such as fruit and cheese whey.

CONCLUSION

Studies on South American beverages are scarce when compared to other beverages like wine, beer, or even sake. This is mainly due to years of neglect to research in these countries. Until recently, the economic difficulties of the South American countries prevented investments in scientific research. Nowadays, with the economic stability, these countries increased the scientific funding, and a new reality seems to arise. In this context, the understanding of the microorganisms present in typical South American beverages opens the door to the development of new technologies, contributing to the overall scientific and economic development of such countries. For example, the isolation of yeast in *cachaça*fermentation vats may lead to the discovery of new strains resistant to different stresses, which can be used not only to produce *cachaça,* but also to produce bioethanol. Moreover, lactic acid bacteria can promote the appearance of new products such as beers with unique or regional flavors. Thus, studies of these microorganisms diversity, present in such unique environments as the traditional beverages, help uncovering new potential applications. Furthermore, the knowledge of those microorganisms can promote the revival of traditional beverages, as *cauim* that was the most consumed beverage in South America in the centuries XVI–XVII. Nevertheless, despite the increased investment in research, few laboratories have the know-how necessary or the availability of resources to invest in the "screening" of yeast or lactic acid bacteria in traditional beverages. Research on South American traditional beverages is important to improve the quality of those beverages, but also to develop new products from these microorganisms.

ACKNOWLEDGEMENTS

The authors were supported by grants from the following Brazilian agencies: Conselho Nacional de Desenvolvimento Científico e Tecnológico (CNPq), Coordenação de Aperfeiçoamento de Pessoal de Nível Superior (CAPES), and Fundação de Amparo à Pesquisa do Estado de Minas Gerais

REFERENCES

1. McGovern PE, Zhang J, Tang J, Zhang Z, Hall GR, Moreau RA, Nuñez A, Butrym ED, Richards MP, Wang C. Fermented beverages of pre- and proto-historic China. Proc Natl Acad Sci USA. 2004;101:17593.

2. Faria-Oliveira F, Puga S, Ferreira C. Yeast: World's finest Chef. In: Muzzalupo I, editor. Food Industry. Rijeka: Intech; 2013. p. 519–547. DOI: 10.5772/53156

3. Erten H, Ağirman B, Gündüz C, Çarşanba E, Sert S, Bircan S, Tangüler H. Importance of yeast and lactic acid bacteria in food processing. In: Malik A, Erginkaya Z, Ahmad S, Erten H, editors. Food Processing: Strategies for Quality Assessment. New York: Springer; 2014. p. 351–378. DOI: 10.1007/978-1-4939-1378-7_14

4. Katina K, Poutanen K. Nutritional aspects of cereal fermentation with lactic acid bacteria and yeast. In: Gobbetti M, Gänzle M, editors. Handbook on Sourdough Biotechnology. Springer US; 2013. p. 229–244. DOI: 10.1007/978-1-4614-5425-0_9

5. Larsson K, Ansell R, Eriksson P, Adler L. A gene encoding a glycerol 3-phosphate dehydrogenase (NAD+) complements an osmosensitive mutant ofSaccharomyces cerevisiae. Mol Microbiol. 1993; 10:1101–1111.

6. Kurtzman CP, Fell JW, Boekhout T. The Yeast: A Taxonomic Study. Amsterdam: Elsevier; 2011.

7. Kejžar A, Gobec S, Plemenitaš A, Lenassi M. Melanin is crucial for growth of the black yeast Hortaea werneckii in its natural hypersaline environment. Fungal Biol. 2013; 117:368–379. DOI: 10.1016/j.funbio.2013.03.006

8. Gadanho M, Libkind D, Sampaio JP. Yeast diversity in the extreme acidic environments of the Iberian pyrite belt. Microbial Ecol. 2006; 52:552–563. DOI: 10.1007/s00248-006-9027-y

9.   Hashim N, Bharudin I, Nguong D, Higa S, Bakar F, Nathan S, Rabu A, Kawahara H, Illias R, Najimudin N, Mahadi N, Murad A. Characterization of Afp1, an antifreeze protein from the psychrophilic yeast Glaciozyma antarctica PI12. Extremophiles. 2013; 17:63–73. DOI: 10.1007/s00792-012-0494-4

10. Tsuji M, Yokota Y, Shimohara K, Kudoh S, Hoshino T. An application of wastewater treatment in a cold environment and stable lipase production of antarctic basidiomycetous yeast Mrakia blollopis. PLoS One. 2013; 8:e59376. DOI: 10.1371/journal.pone.0059376

11. Conceição LEFR, Saraiva MAF, Diniz RHS, Oliveira J, Barbosa GD, Alvarez F, da Mata Correa LF, Mezadri H, Coutrim MX, Afonso RJdCF, Lucas C, Castro IM, Brandão RL. Biotechnological potential of yeast isolates from cachaça: the Brazilian spirit. «»Journal of Industrial Microbiology and Biotechnology. 2015; 42:237–246. DOI: 10.1007/s10295-014-1528-y

12. Steensels J, Verstrepen KJ. Taming wild yeast: potential of conventional and nonconventional yeast in industrial fermentations. Ann Rev Microbiol. 2014; 68:61–80. DOI: 10.1146/annurev-micro-091213-113025

13. Kongkiattikajorn J. Production of amylase from Saccharomyces diastaticus sp. and hydrolysis of cassava pulps for alcohol production. J Agric Sci Technol B. 2012; 2:909–918.

14. Laluce C, Mattoon JR. Development of rapidly fermenting strains of Saccharomyces diastaticus for direct conversion of starch and dextrins to ethanol. App Environ Microbiol. 1984; 48:17–25.

15. Spencer-Martins I. Transport of sugars in yeast: implications in the fermentation of lignocellulosic materials. Biores Technol. 1994; 50:51–57.

16. Kruckeberg AL, Dickinson JR. Carbon metabolism. In: Dickinson JR, Schweizer M, editors. The metabolism and molecular physiology of Saccharomyces cerevisiae. London: CRC; 2004. p. 42–103.

17. Gancedo JM. The early steps of glucose signalling in yeast. FEMS Microbiol Rev. 2008; 32:673–704.

18. Galdieri L, Mehrotra S, Yu S, Vancura A. Transcriptional regulation in yeast during diauxic shift and stationary phase. Omics. 2010; 14:629–638.

19. Meijer MM, Boonstra J, Verkleij AJ, Verrips CT. Glucose repression in Saccharomyces cerevisiae is related to the glucose concentration rather than the glucose flux. J Biol Chem. 1998; 273:24102–24107.

20. Rintala E, Wiebe M, Tamminen A, Ruohonen L, Penttilä M. Transcription of hexose transporters of Saccharomyces cerevisiae is affected by change in oxygen provision. BMC Microbiol. 2008; 8:53.

21. Nelson DL, Cox MM. Lehninger Principles of Biochemistry. New York: W. H. Freeman; 2008.

22. Rodrigues F, Ludovico P, Leão C. Sugar metabolism in yeast: an overview of aerobic and anaerobic glucose catabolism. In: Rosa CA, Peter G, editors. Biodiversity and Ecophysiology of Yeast. 2006. p. 101–121.

23. Pretorius IS. Tailoring wine yeast for the new millennium: novel approaches to the ancient art of winemaking. Yeast. 2000; 16:675–729.

24. Pronk JT, Steensma HY, Van Dijken JP. Pyruvate metabolism in Saccharomyces cerevisiae. Yeast. 1996; 12:1607–1633.

25. van Dijken JP, Scheffers WA. Redox balances in the metabolism of sugars by yeast. FEMS Microbiol Lett. 1986; 32:199–224.

26. Eriksson P, André L, Ansell R, Blomberg A, Adler L. Cloning and characterization of GPD2, a second gene encoding a DL-glycerol 3-phosphate dehydrogenase (NAD+) in Saccharomyces cerevisiae, and its comparison with GPD1. Mol Microbiol. 1995; 17:95–107.

27. Norbeck J, Påhlman AK, Akhtar N, Blomberg A, Adler L. Purification and characterization of two isoenzymes of DL-glycerol-3-phosphatase from Saccharomyces cerevisiae. J Biol Chem. 1996; 271:13875–13881.

28. Crabtree HG. Observations on the carbohydrate metabolism of tumours. Biochem J. 1929; 23:536.

29. De Deken R. The Crabtree effect: a regulatory system in yeast. J Gen Microbiol. 1966; 44:149.

30. Golshani-Hebroni SG, Bessman SP. Hexokinase binding to mitochondria: a basis for proliferative energy metabolism. J Bioenerg Biomemb. 1997; 29:331–338.

31. Skinner C, Lin SJ. Effects of calorie restriction on life span of microorganisms. Appl Microbiol Biotechnol. 2010; 88:817–828.

32. Murray DB, Haynes K, Tomita M. Redox regulation in respiring Saccharomyces cerevisiae. Biochim Biophys Acta. 2011; 1810:945–958.

33. Feldmann H. Yeast metabolism. In: Feldmann H, editor. Yeast Molecular Biology—A Short Compendium on Basic Features and Novel Aspects. Munich: Adolf-Butenandt-Institute, University of Munich; 2005.

34. Boone DR, Castenholz RW, Garrity GM, Brenner DJ, Krieg NR, Staley

JT. Bergey›s Manual® of Systematic Bacteriology. Springer Science & Business Media; 2005.

35.   Liu W, Pang H, Zhang H, Cai Y. Biodiversity of Lactic Acid Bacteria. In: Zhang H, Cai Y, editors. Lactic Acid Bacteria. Springer Netherlands; 2014. p. 103–203. 10.1007/978-94-017-8841-0_2

36.   Hoover DG, Steenson LR. Bacteriocins of lactic acid bacteria. Academic Press; 2014.

37.   Horvath P, Coute-Monvoisin AC, Romero DA, Boyaval P, Fremaux C, Barrangou R. Comparative analysis of CRISPR loci in lactic acid bacteria genomes. Int J Food Microbiol. 2009; 131:62–70. DOI: 10.1016/j.ijfoodmicro.2008.05.030

38.   Stiles ME, Holzapfel WH. Lactic acid bacteria of foods and their current taxonomy. Int J Food Microbiol. 1997; 36:1–29.

39.   Klein G, Pack A, Bonaparte C, Reuter G. Taxonomy and physiology of probiotic lactic acid bacteria. Int J Food Microbiol. 1998;41:103–125.

40.   Felis GE, Dellaglio F. Taxonomy of lactobacilli and bifidobacteria. Curr Issues Intest Microbiol. 2007;8:44.

41.   Giraffa G, Chanishvili N, Widyastuti Y. Importance of lactobacilli in food and feed biotechnology. Res Microbiol. 2010;161:480–487. DOI: 10.1016/j.resmic.2010.03.001

42.   König H, Fröhlich J. Lactic acid bacteria. In: Biology of Microorganisms on Grapes, in Must and in Wine. Springer; 2009. p. 3–29.

43.   Ananou S, Maqueda M, Martínez-Bueno M, Valdivia E. Biopreservation, an ecological approach to improve the safety and shelf-life of foods. In: Communicating Current Research and Educational Topics and Trends in Applied Microbiology. Formatex; 2007. p. 475–486.

44.   Howarth GS, Wang H. Role of endogenous microbiota, probiotics and their biological products in human health. Nutrients. 2013;5:58–81.

45.   White D, Drummond JT, Fuqua C. The physiology and biochemistry of prokaryotes. New York: Oxford University Press; 2007.

46.   Romano AH, Eberhard SJ, Dingle SL, McDowell TD. Distribution of the phosphoenolpyruvate: glucose phosphotransferase system in bacteria. J Bacteriol. 1970;104:808–813.

47.   Kandler O. Carbohydrate metabolism in lactic acid bacteria. Antonie van Leeuwenhoek. 1983;49:209–224.

48.   Leroy F, De Vuyst L. Lactic acid bacteria as functional starter cultures for the food fermentation industry. Trends Food Sci Tech. 2004;15:67–78.

49.   Justé A, Malfliet S, Waud M, Crauwels S, De Cooman L, Aerts G, Marsh TL, Ruyters S, Willems K, Busschaert P. Bacterial community dynamics during industrial malting, with an emphasis on lactic acid bacteria. Food Microbiol. 2014;39:39–46.

50.   Bokulich NA, Bamforth CW. The microbiology of malting and brewing. Microbiol Mol Biol Rev. 2013;77:157–172. DOI: 10.1128/ MMBR.00060-12

51.   Smid EJ, Kleerebezem M. Production of aroma compounds in lactic fermentations. Annu Rev Food Sci Technol. 2014;5:313–326. DOI: 10.1146/annurev-food-030713-092339

52.   Soomro AH, Masud T, Kiran A. Role of lactic acid bacteria (LAB) in food preservation and human health–a review. Pak J Nutr. 2002;

53.   Liu SQ. A review: malolactic fermentation in wine—beyond deacidification. J Appl Microbiol. 2002;92:589–601.

54.   Swiegers JH, Bartowsky EJ, Henschke PA, Pretorius IS. Yeast and bacterial modulation of wine aroma and flavour. Aust J Grape Wine Res. 2005;11:139–173.

55.   Nedovic VA, Durieuxb A, Van Nedervelde L, Rosseels P, Vandegans J, Plaisant AM, Simon JP. Continuous cider fermentation with co-immobilized yeast and Leuconostoc oenos cells. Enzyme Microb Tech. 2000;26:834–839.

56.   Rouse S, Sun F, Vaughan A, Sinderen D. High-throughput isolation of bacteriocin-producing lactic acid bacteria, with potential application in the brewing industry. J Inst Brew. 2007;113:256–262.

57.   Vaughan A, Eijsink VGH, O›Sullivan TF, O›Hanlon K, Van Sinderen D. An analysis of bacteriocins produced by lactic acid bacteria isolated from malted barley. J Appl Microbiol. 2001;91:131–138.

58.   Lowe DP, Arendt EK. The use and effects of lactic acid bacteria in malting and brewing with their relationships to antifungal activity, mycotoxins and gushing: a review. J Inst Brew. 2004;110:163–180.

59.   Almeida EG, Rachid CC, Schwan RF. Microbial population present in fermented beverage ‹cauim› produced by Brazilian Amerindians. Int J Food Microbiol. 2007;120:146–151. DOI: 10.1016/j. ijfoodmicro.2007.06.020

60.   Aidoo KE, Nout MJ, Sarkar PK. Occurrence and function of yeast in Asian indigenous fermented foods. FEMS Yeast Research. 2006;6:30–39. DOI: 10.1111/j.1567-1364.2005.00015.x

61.   Blandino A, Al-Aseeri M, Pandiella S, Cantero D, Webb C. Cereal-

based fermented foods and beverages. Food Res Int. 2003;36:527–543.

62. Osorio-Cadavid E, Chaves-Lopez C, Tofalo R, Paparella A, Suzzi G. Detection and identification of wild yeast in Champús, a fermented Colombian maize beverage. Food Microbiol. 2008;25:771–777. DOI: 10.1016/j.fm.2008.04.014

63. Ramos CL, de Almeida EG, Pereira GVdM, Cardoso PG, Dias ES, Schwan RF. Determination of dynamic characteristics of microbiota in a fermented beverage produced by Brazilian Amerindians using culture-dependent and culture-independent methods. Int J Food Microbiol. 2010;140:225–231. DOI: 10.1016/j.ijfoodmicro.2010.03.029

64. Santos CC, Almeida EG, Melo GV, Schwan RF. Microbiological and physicochemical characterisation of caxiri, an alcoholic beverage produced by the indigenous Juruna people of Brazil. Int J Food Microbiol. 2012;156:112–121. DOI: 10.1016/j.ijfoodmicro.2012.03.010

65. Steinkraus K. Handbook of Indigenous Fermented Foods, revised and expanded. CRC Press; 1995.

66. Vallejo JA, Miranda P, Flores-Félix JD, Sánchez-Juanes F, Ageitos JM, González-Buitrago JM, Velázquez E, Villa TG. Atypical yeast identified asSaccharomyces cerevisiae by MALDI-TOF MS and gene sequencing are the main responsible of fermentation of chicha, a traditional beverage from Peru. Syst Appl Microbiol. 2013;36:560–564. DOI: 10.1016/j.syapm.2013.09.002

67. Gomes F, Lacerda I, Libkind D, Lopes C, Carvajal E, Rosa C. Traditional foods and beverages from South America: microbial communities and production strategies. In: Krause J, Fleischer O, editors. Industrial Fermentation: Food Processes, Nutrient Sources and Production Strategies. 2009. p. 79–114.

68. Piló FB. Leveduras e bactérias lácticas associadas à chicha, uma bebida tradicional produzida no Equador [thesis]. Belo Horizonte, MG: Universidade Federal de Minas Gerais; 2014.

69. Elizaquível P, Pérez-Cataluña A, Yépez A, Aristimuño C, Jiménez E, Cocconcelli PS, Vignolo G, Aznar R. Pyrosequencing vs. culture-dependent approaches to analyze lactic acid bacteria associated to chicha, a traditional maize-based fermented beverage from Northwestern Argentina. Int J Food Microbiol. 2015;198:9–18.

70. Lago C, Landoni M, Cassani E, Doria E, Nielsen E, Pilu R. Study and characterization of a novel functional food: purple popcorn. Mol Breed. 2013;31:575–585.

71. Schwarz M, Hillebrand S, Habben S, Degenhardt A, Winterhalter P.

Application of high-speed countercurrent chromatography to the large-scale isolation of anthocyanins. Biochem Eng J. 2003;14:179–189. DOI: 10.1016/S1369-703X(02)00219-X

72.   Miller C. Fruit production of the ungurahua palm (Oenocarpus bataua subsp. bataua, Arecaceae) in an indigenous managed reserve. Econ Bot. 2002;56:165–176.

73.   Cox L, Caicedo B, Vanos V, Heck E, Hofstaetter S, Cordier J. A catalogue of some ecuadorean fermented beverages, with notes on their microflora. World J Microbiol Biotechnol. 1987;3:143–153.

74.   Puerari C, Magalhães-Guedes KT, Schwan RF. Physicochemical and microbiological characterization of chicha, a rice-based fermented beverage produced by Umutina Brazilian Amerindians. Food Microbiol. 2015;46:210–217.

75.   Carvajal Barriga EJ. Arqueología Microbiana. Nuestra Ciencia. 2012;14:3–7.

76.   Chang CF, Lin YC, Chen SF, Carvajal Barriga EJ, Barahona PP, James SA, Bond CJ, Roberts IN, Lee CF. Candida theae sp. nov., a new anamorphic beverage-associated member of the Lodderomyces clade. Int J Food Microbiol. 2012;153:10–14. DOI: 10.1016/j.ijfoodmicro.2011.09.012

77.   Cavalcante MS. A verdadeira história da cachaça. São Paulo: Sá editora; 2011. 608 p.

78.   Lei 10958/04. Presidency of the Republic. Brasilia. Brazil, 2004.

79.   IBRAC. Instituto Brasileiro da Cachaça. IBRAC. 03/08/2015. http://www.ibrac.net/

80.   Lobo P, Jaguaribe E, Rodrigues J, Da Rocha F. Economics of alternative sugar cane milling options. Appl Therm Eng. 2007;27:1405-1413.

81.   Meade GP, Chen JC. Cane Sugar Handbook. New York: John Wiley & Sons; 1977.

82.   Duarte WF, Amorim JC, Schwan RF. The effects of co-culturing non-Saccharomyces yeast with S. cerevisiae on the sugar cane spirit (cachaça) fermentation process. Antonie van Leeuwenhoek. 2013;103:175–194.

83.   Aquarone E, Borzani W, Schmidell W, Lima UdA. Biotecnologia Industrial. São Paulo: Edgard Blücher Ltda; 2001. 523 p.

84.   Passoth V, Olstorpe M, Schnurer J. Past, present and future research directions with Pichia anomala. Antonie van Leeuwenhoek. 2011;99:121–125. DOI: 10.1007/s10482-010-9508-3

85.   Gomes F, Silva C, Marini M, Oliveira E, Rosa C. Use of selected indigenous Saccharomyces cerevisiae strains for the production of the

traditional cachaça in Brazil. J Appl Microbiol. 2007;103:2438-2447.

86. Vicente MA, Fietto LG, Castro IM, dos Santos AN, Coutrim MX, Brandao RL. Isolation of Saccharomyces cerevisiae strains producing higher levels of flavoring compounds for production of "cachaça" the Brazilian sugarcane spirit. Int J Food Microbiol. 2006;108:51–59. DOI: 10.1016/j.ijfoodmicro.2005.10.018

87. Oliveira VA, Vicente MA, Fietto LG, Castro IM, Coutrim MX, Schuller D, Alves H, Casal M, Santos JO, Araujo LD, da Silva PH, Brandao RL. Biochemical and molecular characterization of Saccharomyces cerevisiae strains obtained from sugar-cane juice fermentations and their impact incachaça production. Appl Environ Microbiology. 2008; 74:693–701. DOI: 10.1128/AEM.01729-07

88. Campos C, Silva C, Dias D, Basso L, Amorim H, Schwan R. Features of Saccharomyces cerevisiae as a culture starter for the production of the distilled sugar cane beverage, cachaça in Brazil. J Appl Microbiol. 2010; 108:1871–1879.

89. de Souza APG, Vicente MdA, Klein RC, Fietto LG, Coutrim MX, de Cássia Franco Afonso RJ, Araújo LD, da Silva PHA, Bouillet LÉM, Castro IM, Brandão RL. Strategies to select yeast starters cultures for production of flavor compounds in cachaça fermentations. Antonie van Leeuwenhoek. 2012;101:379–392. DOI: 10.1007/s10482-011-9643-5

90. Schwan RF, Mendonca AT, da Silva Jr JJ, Rodrigues V, Wheals AE. Microbiology and physiology of Cachaça (Aguardente) fermentations. Antonie van Leeuwenhoek. 2001; 79:89–96.

91. Oliveira Cd, Garíglio H, Ribeiro M, Alvarenga M, Maia F. Cachaça de Alambique–Manual de Boas Práticas Ambientais e de Produção. Belo Horizonte. SEAPA/SEMAD/AMPAQ/FEAM/IMA, 2005.

92. Martino DBd. Aguardente: O destilado do século 21. Revista Engarrafador Moderno. 1998;84–88.

93. Cardeal ZL, de Souza PP, da Silva MD, Marriott PJ. Comprehensive two-dimensional gas chromatography for fingerprint pattern recognition incachaça production. Talanta. 2008; 74:793–799. DOI: 10.1016/j.talanta.2007.07.021

94. Bortoletto AM, Alcarde AR. Assessment of chemical quality of Brazilian sugar cane spirits and cachaças. Food Control. 2015; 54:1–6.

95. da Silva AA, do Nascimento ES, Cardoso DR, and Franco DW. Coumarins and phenolic fingerprints of oak and Brazilian woods extracted by sugarcane spirit. J Sep Sci. 2009; 32:3681–3691. DOI: 10.1002/jssc.200900306

96. Bortoletto AM, Alcarde AR. Congeners in sugar cane spirits aged in casks of different woods. Food Chem. 2013; 139:695–701. DOI: 10.1016/j.foodchem.2012.12.053

97. de Souza PP, Siebald HG, Augusti DV, Neto WB, Amorim VM, Catharino RR, Eberlin MN, Augusti R. Electrospray ionization mass spectrometry fingerprinting of Brazilian artisan cachaça aged in different wood casks. J Agric Food Chem. 2007; 55:2094–2102. DOI: 10.1021/jf062920s

98. Lei H, Zhao H, Yu Z, Zhao M. Effects of wort gravity and nitrogen level on fermentation performance of brewer's yeast and the formation of flavor volatiles. Appl Biochem Biotechnol. 2012; 166:1562–1574.

99. Araújo TM. Caracterização bioquímico-molecular de cepas de Saccharomyces cerevisiae isoladas de dornas de fermentação de cachaça para produção de cervejas [thesis]. Universidade Federal de Ouro Preto; 2013.

100. Campos ACS. Caracterização de bactérias lácticas para serem utilizadas em processos fermentativos consorciados entre leveduras e bactérias na produção de cerveja. [Thesis]. Universidade Federal de Ouro Preto; 2014.

101. Sakamoto K, Konings WN. Beer spoilage bacteria and hop resistance. Int J Food Microbiol. 2003; 89:105–124.

102. Carvalho FP, Duarte WF, Dias DR, Piccoli RH, Schwan RF. Interaction of Saccharomyces cerevisiae and Lactococcus lactis in the fermentation and quality of artisanal cachaça. Acta Sci Agron. 2014; 37:51–60.

103. Dias DR, Schwan RF, Lima LCO. Metodologia para elaboração de fermentado de cajá (Spondias mombin L.). Ciênc Tecnol Aliment. 2003; 23:342–350.

104. Vogt E, Jakob L, Lemperle E, Weiss E. El vino: obtención, elaboración y análisis. Zaragoza: Acribia; 1986.

105. da Silva MC, de Azevedo LC, de Carvalho MM, de Sá AGB, dos Santos Lima M. Elaboração e avaliação da qualidade de aguardentes de frutas submetidas a diferentes tratamentos. Revista Semiárido De Visu. 2011; 1:92–106.

106. Dragone G, Mussatto SI, Vilanova M, Oliveira J, Teixeira J, Silva JBA. Obtenção e caracterização de bebida destilada a partir da fermentação do soro de queijo. Braz J Food Tech. 2009; 120–124.

Chapter 6

LACTIC FERMENTATION AND BIOACTIVE PEPTIDES

Anne Pihlanto

MTT Agrifood Finland, Biotechnology and Food Research, Jokioinen, Finland

INTRODUCTION

Fermented milk products have naturally high nutritional value, and as an extra benefit many health-promoting effects, such as improvement of lactose metabolism, reduction of serum cholesterol and reduction of cancer risk [1]. The beneficial health effects associated with some fermented dairy products may, in part, be attributed to the release of bioactive peptide sequences during the fermentation process. Numerous peptides and peptide fractions, having bioactive properties have been isolated from fermented dairy products. These activities include immunomodulatory, cytomodulatory, hypocholesterolemic, antioxidative, antimicrobial, mineral binding, opioid and bone formation activities. Many recent articles and book chapters have reviewed the release of various bioactive peptides from milk proteins through microbial proteolysis [2-5].

Many industrially utilized dairy starter cultures are highly proteolytic. The use of bioactive peptides producers microbial cultures (starter and non-starter) may allow the development new fermented dairy products. The proteolytic system of lactic acid bacteria e.g. *Lactococcus (L.) lactis, Lactobacillus (Lb.) helveticus* and *Lb. delbrueckii* ssp. *bulgaricus,* is already well characterized. This system consists of a cell wall-bound proteinase and a number of distinct intracellular peptidases, including endopeptidases, aminopeptidases, tripeptidases and dipeptidases [6]. *Lb. helveticus* are known to have high proteolytic activities [7], causing the release of oligopeptides from digestion of milk proteins [8]. These oligopeptides can be a direct source of bioactive peptides following hydrolysis by gastrointestinal enzymes. Rapid progress has been made in recent years to elucidate the biochemical and genetic characterization of these enzymes. The fact that the activities of peptidases are affected by growth conditions makes it possible to manipulate the formation of peptides to a certain extent [9].

Cardiovascular disease (CVD) is the single leading cause of death for both males and females in technologically advanced countries in the world. In lesser-developed countries it generally ranks among the top five causes of death. The World Health Organization estimates that by 2020, heart disease and stroke will have surpassed infectious diseases to become the leading cause of death and disability worldwide [10]. Consequently, there has been an increased focus on improving diet and lifestyle as a strategy for CVD risk reduction.

Elevated blood pressure is one of the major independent risk factors for CVD [11]. Angiotensin I-converting enzyme (ACE) plays a crucial role in the regulation of blood pressure as it promotes the conversion of angiotensin I to the potent vasoconstrictor angiotensin II as well as inactivates the vasodilator bradykinin. By inhibiting these processes, synthetic ACE inhibitors (ACEI) have long been used as antihypertensive agents. In recent years, some food proteins have been identified as sources of ACEI peptides and are currently the best-known class of bioactive peptides [12, 13]. These nutritional peptides have received considerable attention for their effectiveness in both the prevention and the treatment of hypertension.

Oxidant stress, the increased production of reactive oxygen species (ROS) in combination with outstripping endogenous antioxidant defense mechanisms, is another significant causative factor for the initiation or progression of several vascular diseases. ROS can cause extensive damage to biological macromolecules like DNA, proteins and lipids. Specifically, the oxidative modification of LDL results in the increased atherogenicity of oxidized LDL. Therefore, prolonged production of ROS is thought to contribute to the development of severe tissue injury [14]. Some peptides derived from hydrolyzed food proteins exert antioxidant activities against enzymatic (lipoxygenase-mediated) and nonenzymatic peroxidation of lipids and essential fatty acids [15]. The antioxidant properties of these peptides have been suggested to be due to metal ion chelation, free radical scavenging and singlet oxygen quenching.

This review centers on liberation during fermentation, of bioactive peptides with properties relevant to cardiovascular health including the effects on blood pressure and oxidative stress. The focus is mainly to those peptides with in vivo blood pressure lowering effects. Moreover, bioavailability of peptides and aspects of necessary further information is given.

RELEASE AND IDENTIFICATION OF PEPTIDES

Peptides In Cheese

Proteolysis in cheese has been linked to its importance for texture, taste and flavour development during ripening. Changes of the cheese texture occur due to breakdown of the protein network. It contributes directly to taste and flavour by the formation of peptides and free amino acids as well as by liberation of substrates for further catabolic changes and thereby formation of volatile flavour compounds. Besides sensory quality aspects of proteolysis, formation of bioactive peptides as a result of proteolysis during cheese ripening has been reported. Cheese contains phosphopeptides as natural constituents [16, 17], and secondary proteolysis during cheese ripening leads to the formation of other bioactive peptides, such as those with ACEI activity. The findings by Meisel et al. [18] showed that inhibitory activity increased as proteolysis developed, however, the bioactivity decreased when proteolysis during ripening exceeded a certain level. Another link between potential antihypertensive peptides and proteolysis was found in Parmesan cheese [19]. A bioactive peptide derived from α_{s1}-casein was isolated from 6-month old cheese, but it was degraded further during maturation and was not detectable after 15 month of ripening. ACEI peptide fractions having different potencies have been isolated from various Italian cheeses, e.g. Crescenza (37% inhibition), mozzarella (59% inhibition), Gorgonzola (80% inhibition) and Italico (82% inhibition) [20]. ACEI peptides have also been found in enzyme-modified cheeses [21], in a low-fat cheese made in Finland [22] and Manchego cheeses manufactured with different starter cultures [23]. Mexican Fresco cheese manufactured with *Enterococcus faecium* or a *L. lactis* ssp.*lactis-Enterococcus faecium* mixture showed the largest number of fractions with ACEI activity among tested lactic acid strains [24]. Pripp et al. [25] investigated the relationship between proteolysis and ACE inhibition in Gamalost, Castello, Brie, Pultost, Norvegia, Port Salut and Kesam. The traditional Norwegian cheese Gamalost had per unit cheese weight higher ACE inhibition potential than Brie, Roquefort and Gouda-type cheese. However, ACE inhibition expressed as IC_{50} per unit peptide concentration from ethanol soluble fraction assessed by the OPA-assay was highest for Kesam, a Quark-type cheese with a low degree of proteolysis.

When β-casomorphins were looked from commercial cheese products, no peptides were found or their concentration in the cheese extract was below 2 μg/ml [26]. They further noted that the enzymatic degradation of β-casomorphins was influenced by a combination of pH and salt concentration at the cheese ripening temperature. Therefore, if formed in cheese, β-casomorphins may be degraded under conditions similar to Cheddar cheese ripening. Precursors of β-casomorphins, on the other hand, have been identified in Parmesan cheese [19]. β-Casomorphins were found at a higher level in the mould cheeses (166–648 mg/100 g), whereas the opioid peptides with antagonistic activity (casoxin-6) were identified at a higher level in the semi-hard cheeses (136–276 mg/100 g) and a low quantity of casomorphins (4–100 mg/100 g) [27]. Immunomodulating properties in water-soluble extracts from traditional French Alps cheeses, Abondance and Tomme de Savioe have been observed [28]. However, no correlation between peptide composition and *in vitro* immunomodulation of T-lymphocyte cells could be established.

A limited number of bioactive peptides have been isolated and identified in Gouda, Manchego, Festivo and Crescenza cheeses (Table 1). Several ACEI peptides have been identified from N-terminal of α_{s1}-casein of Gouda, Festivo, Cheddar and Fresco cheeses [22, 24, 29, 30]. In addition, peptides from β–casein, Tyr-Pro-Phe-Pro-Gly-Pro-Ile-Pro-Asn (β-cn, f(60–68)); and Met-Pro-Phe-Pro-Lys-Tyr-Pro-Val-Gln-Pro-Phe (β-cn, f(109–119)) from Gouda [29] and Tyr-Gln-Glu-Val-Leu-Gly-Pro-Val-Arg-Gly-Pro-Phe-Pro-Ile-Ile-Val (β-cn, f(193–209)) from Cheddar [30] have been identified. Antihypertensive peptides Val-Pro-Pro (VPP) (β-cn, f(84–86)) and Ile-Pro-Pro (IPP) (β-cn, f(74–76) and κ-cn, f(108–110)), have also been identified and quantified in different cheese varieties [31-33]. In some varieties physiologically relevant amounts was observed, however, a large variation exists between samples of the same cheese variety, as well as between different varieties.

Table 1: Examples of identified bioactive peptides in different cheese varieties

Cheese variety	Milk protein fragment	Peptide sequence	ACE-inhibition $IC_{50} \mu M$	Ref
Gouda	α_{s1}-cn f(1-9)	RPKHPIKHQ	13.4	29
	α_{s1}-cn f(1-13)	RPKHPIKHQGLPQ	ND	
	β-cn f(68-66)	YPFPGPIPN	14.8	
	β-cn f(109–119)	MPFPKYPVQPF	ND	
Manchego	ovine α_{s1}-cn f(102-109)	KKYNVPQL	77.2	23
	ovine α_{s1}-cn f(205-208)	VRYL	24.1	

Cheddar (with probiotics)	α_{s1}-cn f(1-9)	RPKHPIKHQ	ND	30
	α_{s1}-cn f(1-7)	RPKHPIK		
	α_{s1}-cn f(1-6)	RPKHPI		
	α_{s1}-cn f(24-32)	FVAPFPEVFGK		
	β-cn f(193-209)	YQEPVLGPVRGP-FPIIV		
Swiss cheese varieties	β-cn, f(84–86)	VPP	9	31-34
	β-cn, f(74–76) and	IPP	5	
	κ-cn, f(108–110)			
Fresco cheese	α_{s1}-cn f(1-15)	RPKH-PIKHQGLPQEV	ND	24
	α_{s1}-cn f(1-22)	RPKH-PIKHQGLPQEVL-NENLLR		
	α_{s1}-cn f(14-23)			
	α_{s1}-cn f(24-34)	EVLNENLLRF		
	β-cn f(193-205)	FVAPFPEVFGK		
	β-cn f(193-207)	YQEPVLGPVRGPF		
	β-cn f(193-209)	YQEPVLGPVRGPFPI		
		YQEPVLGPVRGP-FPIIV		

The concentrations of VPP and IPP were in the range of 0-224 mg/kg and 0-95 mg/kg, respectively, indicating that some cheese varieties contain similar concentrations of VPP and IPP to fermented milk products. Milk pretreatment, cultures, scalding conditions, and ripening time were identified as the key factors influencing the concentration of these two naturally occurring bioactive peptides in cheese. Thus, it is necessary to develop a reproducible cheese-making process with selected cultures to produce higher concentrations of these peptides that could be used for clinical trials.

Fermented Milk

During fermentation process, lactic acid bacteria hydrolyze milk proteins, mainly caseins, into peptides and amino acids which are used as nitrogen sources necessary for their growth. Hence, bioactive peptides can be generated by starter and non-starter bacteria used in the manufacture of fermented dairy products (Table 2). Proteolytic system of *Lb. helveticus, Lb. delbrueckii* ssp *bulgaricus, L. lactis* ssp.*diacetylactis, L. lactis* ssp. *cremoris,* and *Streptococcus (Str.) salivarius* ssp. *thermophilu*s strains have demonstrated to hydrolyze milk proteins and release ACEI peptides. Among lactic acid bacteria, *Lb. helveticus* has high extracellular proteinase activity and the ability to release large amount of peptides in fermented milk. As a result, among various kinds of fermented milk, antihypertensive effect related to ACEI peptides were found in milk produced by *Lb. helveticus*. Two ACEI peptides have been purified from sour milk and identified as VPP and IPP [34].

Pihlanto-Leppälä et al. [35] studied the potential formation of ACEI peptides from cheese whey and caseins during fermentation with various commercial dairy starters used in the manufacture of yogurt, ropy milk and sour milk. No ACEI activity was observed in these hydrolysates. Further digestion of the above samples with pepsin and trypsin resulted in the release of several strong ACEI peptides derived primarily from α_{s1}-casein and β-casein. The formation of ACEI peptides was demonstrated in two dairy strains, *Lb. delbrueckii* ssp. *bulgaricus* and *L. lactis* ssp. *cremoris,* after fermentation of milk separately with each strain for 72 hours [36]. The most inhibitory fractions of the fermented milk mainly contained β-casein-derived peptides with inhibitory concentration (IC_{50}) values ranging from 8.0 to 11.2 μg/ml. Fuglsang et al. [37] tested a total of 26 strains of wild-type lactic acid bacteria, mainly belonging to *L. lactis* and *Lb. helveticus*, for their ability to produce a milk fermentate with ACEI activity. All tested strains produced ACEI substances in varying amounts, and two of the strains exhibited high ACE inhibition and a high OPA index, which correlates well with peptide formation. In another study 25 lactic acid strains of *Lactobacillus, Lactococccus* and *Leuconsotoc* were used [38]. The strains were tested alone or in combination and the highest activities were observed in *Lb. jensenii, Lb. acidophilus*and *Leuc. mesenteroides* strains and all strains showed correlation between ACE inhibition and degree of proteolysis. In a recent study, milk was fermented to defined pH values with 13 strains of lactic acid bacteria. The highest ACEI activity was obtained with two highly proteolytic strains of *Lb. helveticus*and with the *Lactococcus* strains. Fermentation from pH 4.6 to 4.3 with these strains slightly increased the ACEI activity, whilst fermentation to pH 3.5 with *Lb. helveticus* reduced the ACEI activity [39]. Moreover, four different *Enterococcus faecalis* strains, isolated from raw milk, produced fermented milk with potent ACEI activity [40]. In a recent research it was found that *L. lactis* strains isolated from artisanal dairy starters or commercial starter cultures are potential for the production of fermented dairy products with ACEI properties. Especially, a strain isolated from artisanal cheese presented the lowest IC_{50} (13μg/ml) [41].

Table 2: ACE-inhibitory and antihypertensive activity in spontaneously hypertensive rats of peptides produced by fermentation of milk

Organisms	ACE-inhibition IC_{50} mg/ml	Sequence	IC_{50} µM	Dose	Response (ΔSBP mmHg)	Ref.
Lb. helveticus and Str. thermophilus	ND	VPP / IPP	9 / 5	5 ml/kg	-21.8 ±4.2 after 6 h	34
Lb. helveticus		VPP / IPP	9 / 5	27 ml/day	-21 after 4 weeks	67
Lb.helveticus CPN4	ND	YP	720	10 ml/kg	32.1 ±7.4 after 6 h	42
Lb. helveticus CHCC637	0.16			10ml/kg	-12 after 4-8 h	37
Lb. helveticus CHCC641	0.26				-11 after 4-8 h	
Lact. delbrueckii ssp. bulgaricus Str. salivarius ssp thermophilus and L.lactis biovar diacetylactis		SKVYPFPGPI / SKVYP	1.7 mg/ml / 1.5 mg/ml		ND	43

Organism		Peptide		Dose	Effect	Ref
Lb. jensenii	0.52	LVYPFPGPIHNSLPQN LVYPFPGPIH	71 89	0.2 kg/kg	approx -12 after 2 h	38
Enterococcus faecalis CECT 5727	0.053	LHLPLP LVYPFPGPIPNSLPQ-NIPP	5.5 5.2	2 mg/kg 6 mg/kg	-21.87 ±4.51 after 4h[1]) approx -15 after 4 h	44
Lb. delbrueckiisubsp. bulgaricus SS1	ND	NIPPLTQTPV	173.3		ND	36
L. lactis subsp.cremoris FT4		LNVPGEIVE DKIHPF	300.1 256.8			
Mixed lactic acid bacteria (Lb. casei, acidophilus, bulcaricus, Str. themophilus, Bifidobacterium) and protease	0.24	GTW GVW	464.4 240.0	5 mg/ml	SBP -22 after 8 weeks	76

Bioactive peptides isolated from skim milk and whey fermented using a range of organisms are summarized in Table 2. The majority of identified peptides are casein-derived ACEI peptides having IC_{50} values ranging from 5 to 500 μM. The best characterized ACEI and antihypertensive peptides liberated with *Lb. helveticus* alone or in combination with *Saccharomyces cerevisiae* are the tripeptides IPP, and VPP. Yamamoto et al. [42] identified an ACEI dipeptide (Tyr-Pro) from a yogurt-like product fermented with *Lb. helveticus* CPN4 strain. This peptide sequence is present in all major casein fractions, and its concentration was found to increase during fermentation, reaching a maximum concentration of 8.1 μg/ml in the product. Ashar and Chand [43] identified an ACEI peptide from milk fermented with *Lb. delbrueckii* ssp. *bulgaricus*. The peptide showed the sequence Ser-Lys-Val-Tyr-Pro-Phe-Pro-Gly-Pro-Ile from β–casein with an IC_{50} value of 1.7 mg/ml. In combination with *Str. salivarius*ssp. *thermophilus* and *L. lactis* biovar. *diacetylactis*, a peptide structure with a sequence of Ser-Lys-Val-Tyr-Pro was obtained from β-casein with an IC_{50} value of 1.4 mg/ml. Both peptides were markedly stable to digestive enzymes, acidic and alkaline pH, as well as during storage at 5 and 10 °C for four days. Two β-casein-derived peptides were identified from water soluble fraction of milk fermented with*Lb. jensenii*. The identified peptides were Leu-Val-Try-Pro-Phe-Pro-Gly-Pro-Ile-His-Asn-Ser-Leu-Pro-Gln-Asn, and Leu-Val-Try-Pro-Phe-Pro-Gly-Pro-Ile-His [38]. Quirós et al. [44] identified two peptides in fermented milk with *Enterococcus faecalis* that corresponded to β -casein fragments Lys-His-Leu-Pro-Leu-Pro and Lys-Val-Tyr-Pro-Phe-Pro-Gly-Pro-Ile-Pro-ASn-Ser-Leu-Pro-Gln-Asn-Ile-Pro-Pro, with potent ACEI activity.

Many kinds of proteolytic enzymes have been reported from lactic acid bacteria, and have been reviewed extensively [6, 45]. The components of the proteolytic systems of lactic acid bacteria are divided into three groups, including the extracellular proteinase that catalyzes casein breakdown to peptides, peptidases that hydrolyze peptides to amino acids and a peptide transport system. The extracellular proteinase activity was almost correlated with ACEI activity in the fermented milk, suggesting that the proteolysis of casein by the extracellular proteinase is the most important parameter in the processing of active components [46]. The importance of the proteinase was also supported by the fact that a proteinase negative mutant was not able to generate antihypertensive peptides in the fermented milk, whereas the wild-type strain had the ability to release strong antihypertensive peptides in the fermented milk [47]. The enzymatic process generating the antihypertensive peptides VPP and IPP in *Lb. helveticus* has been elucidated. By the proteolytic action of the extracellular proteinase long peptide with amino acid residue including VPP and IPP sequences were generated. Next the long peptide would be hydrolyzed to shorter peptides by intracellular peptidases. A key enzyme that can catalyze C-terminal processing of Val-Pro-Pro-Phe-Leu and Ile-Pro-Pro-Leu-Thr to VPP and IPP has been purified from *Lb. helveticus* CM4. The endopeptidase has sequence homology in amino terminal sequence to a previously reported pepO-gene product [48]. Kilpi et al. [49] found out higher ACE inhibition in milk fermentation using peptidase-deletion mutants compared to the wild-type of *Lb. helveticus* strain. Unlike with the wild type strain, ACEI remained constant during the course of fermentation with the proline-specific peptidase mutant. The mutant strains had also different peptide profiles than the wild-type strain.

Other

Various types of fermented soybean foods are consumed in Asian countries such as Korea, China, Japan, Indonesia and Vietnam. Soybeans are traditionally fermented primarily by *Bacilli* species during the early stage of fermentation followed by *Aspergillus* species, which predominate during the remaining fermentation period [50]. ACEI peptides have been found in many traditional Asian fermented soy foods, such as soybean paste, soy sauce, natto and tempeh. ACEI peptide His-His-Leu was isolated from Korean fermented soybean paste [51]. Rye gluten sourdoughs fermented with *Lb. reuteri* and added protease were found to contain the lactoripeptides VPP, IPP [52]. Moreover, our recent studies showed that fermentation of rapeseed or flaxseed meals with *Bacillus subtils* or *Lb. helveticus* strains produced ACEI activity [53].

Other Activities

It is reasonable to expect that lactic acid bacteria produce scavengers for hydroxyl radical, which can be metabolic compounds produced by bacteria or degradation products of milk proteins. The results have demonstrated that the antioxidant production is commonly higher within the group of obligately homofermentative lactobacilli, than within the facultatively or obligately heterofermentative strain groups. Also heterofermentative *Lactobacillus* sp. have been reported to exhibit antioxidative activity.*Lb. acidophilus*, *Lb. bulgaricus*, *Str. thermophilus* and *Bifidobacterium longum* exhibited antioxidative activity by various mechanisms, like metal ion chelating capacity, scavenging of reactive oxygen species (ROS), reducing activity and superoxide dismutase activity [54, 55]. Peptides liberated during fermentation can be partially responsible for the reported antioxidative properties. An antioxidative peptide derived from κ-casein was detected in milk after fermentation with *Lb. delbrueckii* subs.*bulgaricus* [56]. Moreover, Hernández-Ledesma et al. [57] found a moderate ABTS radical scavenging capacity in commercial fermented milk from Europe. Further studies of this radical scavenging activity in different HPLC fractions showed low TEAC values. Virtanen et al. [58] found that fermentation with*Leuc. mesenteroides* ssp. *cremoris*, *Lb. jensenii* and *Lb. acidophilus* strains produced compounds that showed both radical scavenging activity and inhibition of lipid peroxidation.

Inflammation plays a key role in the development of cardiovascular disease. It often begins with inflammatory changes in the endothelium, which begins to express the adhesion molecule VCAM-1. VCAM-1 attracts monocytes, which then migrate through the endothelial layer under the influence of various proinflammatory chemoattractants [59]. Accordingly, fermentation by lactic acid may be able to release components that possess immunomodulatory properties. Most of the studies have been done with synthetic peptides derived from enzymatic treatment of milk proteins using different *in vitro*models. Leblanc et al. [60] investigated the effect of peptides released during the fermentation of milk by *Lb. helveticus* on the humoral immune system and on the growth of fibrosacromas. The study showed that bioactive components were released during fermentation that contributed to the immunoenhancing and antitumor properties. Antimutagenic compounds were produced during fermentation by *Lb. helveticus*, and release of peptides is one possible explanation [61]. The permeate fraction obtained from milk fermented by *Lb. helveticus* was able to modulate the *in vitro* proliferation of lymphocytes by acting on the production of cytokines [62]. Tompa et al. [63] found that peptide fractions form *Lb. helveticus* BGRA43 fermented milk have anti-inflammatory potential. Matar et al. [64] fed milk fermented with a *Lb. helveticus* strain to mice for three days and detected significantly higher numbers of IgA secreting cells in their intestinal mucosa, compared with control mice fed with similar milk incubated with a non-proteolytic variant of the same strain. The immunostimulatory effect of fermented milk was attributed to peptides released from the casein fraction.

ANTIHYPERTENSIVE EFFECTS IN VIVO

The search for *in vitro* ACEI is the most common strategy followed in the selection of potential antihypertensive peptides derived from food proteins. *In vitro* ACEI activity is generally measured by monitoring the conversion of an appropriate substrate by ACE in the presence and absence of inhibitors. The antihypertensive effects have been assessed by *in vivo* experiments using spontaneously hypertensive rats (SHR) as an animal model to study human essential hypertension [7]. Following a positive response in animal studies human studies may be carried out to ascertain the ACEI potential

Animal Studies

A great number of studies have addressed the effects of both short-term and long-term administration of potential antihypertensive peptides using this animal model. Fermented milks with different IC_{50}-values ranging from from 0.08 to 1.88 mg/ml have been shown to decrease blood pressure in SHR from 10 to 32 mmHg (Table 2).

The first antihypertensive effect of milk casein-derived peptides was first demonstrated by casein hydrolysate formed by purified proteinase from *Lb. helveticus* CP790 and milk fermented with the same bacteria [65]. The authors concluded that peptides deliberated from casein by extracellular proteinases were responsible for the antihypertensive effect. The active substances were liberated during fermentation of milk with *Lb. helveticus* and *Saccharomyces cerevisiae* and were identified to be IPP and VPP. Oral administration of fermented milk or pure tripeptides were shown to produce strong antihypertensive effect in SHR after single-dose [34, 66]. Thereafter, several animal studies have been conducted to characterize the long-term effects of lactotripeptides or fermented milk containing them. These studies were mainly conducted with SHR but also Goto-Kakizaki (GK) rats and double transgenic rats (dTGR) with malignant hypertension have been used. The development of hypertension was attenuated significantly in rats receiving fermented milk product containing lactotripeptides, attenuation in systolic blood pressure was 12-21 mmHg in SHR, 10 mmHg in high salt-fed GK rats and 19 mmHg in dTGR in comparison to control group [67-69]. Pure tripeptides did not produce as strong antihypertensive effect as the milk products containing them. In addition, minerals alone did not attenuate the development of blood pressure as much as the fermented milk products [68]. These studies indicate that the bioavailability of peptides may be better from milk in comparison of water or is improved by other milk components.

After the blood pressure monitoring has been completed the effect of long-term intake of lactotripeptides on vascular function has been assessed [68,70,71]. Jauhiainen et al. [70], showed improved endothelium-dependent relaxation in mesenteric arteries and aortas of rats that had received minerals and lactotripeptide. Endothelial function of mesenteric arteries was strongly impaired in all groups of salt-loaded GK rats, and significantly improved endothelium-dependent relaxations were observed after treatment with different fermented milk products [68]. Protection of endothelial function after incubation with tripeptides IPP and VPP for 24 h was found in a study with isolated SHR mesenteric arteries [71].

Evidence from ACE inhibition was gained by Masuda et al. [72], who found that after receiving a single-dose of Calpis™ sour milk, ACE activity was decreased in SHR aorta. The lactotripeptides were detected in solubilized fraction from the abdominal aorta of SHR but not from WKY given the sour milk. Moreover, in SHR, plasma rennin activity increased after long-term treatment of fermented milk product containing the lactotripeptides [67]. In addition, treatment with fermented milk containing lactotripeptides and plant sterols decreased serum ACE activity [73]. In salt-loaded GK rats, fermented milk with lactotripeptides decreased serum ACE and aldosterone levels [68].

Besides the most extensively studied lactotripeptides, also other fermented milk products and peptides have been found. Different strains of lactic acid bacteria, such as *Lb. helveticus* CPN4, *Lb. bulgaricus, Lb. jensenii* and *Str. thermophilus*, have been also shown to provoke liberation of peptides with antihypertensive activity in SHR [36, 37, 41]. Two peptides, corresponding to β -casein fragments Leu-Val-Tyr-Pro-Phe-Pro-Gly-Pro-Ile-Pro-Asn-Ser-Leu-Pro-Gln-Asn-Ile-Pro-Pro and Leu-His-Leu-Pro-Leu-Pro, have been isolated in fermented milk with *Enterococcus faecalis* and their antihypertensive effect in SHR, after acute and long-term administration has been proved. The administration of 2 mg/kg of peptide Leu-His-Leu-Pro-Leu-Pro resulted in a significant decrease of the SBP in SHR 4 h post-administration [74,75]. Fermentation of milk with one or more lactic acid bacteria strains followed by hydrolysis using food-grade enzymes liberated tripeptides (Gly-Thr-Trp and Gly-Val-Trp). Oral administration of this fermented whey lowered significantly SBP in SHR from 9 to 15 weeks of age. Bioactive substances, tripeptides and γ-aminobutyric acid (GABA), contributed to lowering blood pressure of SHR [76].

Some of ACE-inhibitory peptide fractions from cheese have shown *in vivo* activities. A water-soluble peptide preparation isolated from Gouda ripened for 8 months was found to have the most potent antihypertensive activity (maximum decrease in SBP = 24.7 (± 0.3) mmHg (P ≤ 0.01) after 6 h) when administered to SHR by gastric intubation at doses between 6.1 and 7.5 mg/kg body weight. Three peptide fractions were isolated from water-soluble extract by hydrophobic chromatography using different concentrations of acetonitrile. The fractions eluting between 15% and 30%, 30–45% and 60–75% acetonitrile decreased SBP in SHR by 15.0, 29.3 and 18.8 mmHg (P ≤ 0.01), respectively, 6 h after gastric intubation. The peptide fraction eluting between 30% and 45% acetonitrile was shown to contain the sequences (αs1-cn f(1–9)) Arg-Pro-Lys-His-Pro-Ile-Lys-His-Gln and (β-cn f(60–68)) Tyr-Pro-Phe-Gly-Pro-Ile-Pro-Asn (Table 1), which, respectively, decreased SBP in SHR by 9.3 (± 4.8) and 7.0 (± 3.8) mmHg 6 h after gastric intubation [29].

Several sequences have been proposed as responsible for the antihypertensive activity of soy protein hydrolysates and fermented products, but only the peptide His-His-Leu derived from fermented soy paste was assayed in pure form in SHR, where a decrease of 32 mm Hg of SBP was reached at a dose of 100 mg/kg. Moreover, the synthetic tripeptide His-His-Leu resulted in a significant decrease of ACE activity in the aorta [77]. Soybean-derived products contain isoflavones, which are thought to possess a favourable effect in reducing cardiovascular risk factors as well as vascular function [78]. However, on the basis of *in vitro* results and literature review, Wu and Muir [79] have indicated that the contribution of isoflavones to a blood-pressure-lowering effect in soybean ACEI peptides may be negligible. Similarly, it has been reported that the reduction of hypertension of a fermented product from soy milk was contributed mainly by peptides of 800–900 Da but it could be also attributable to GABA [80]. Moreover, fermented soy product, miso, with added tripeptides (VPP and IPP) from casein was reported to act as antihypertensive agents in SHR [81]. Recently, Nakahara et al. [82] used the Dahl salt-sensitive rats as a model of salt-sensitive hypertension to evaluate the antihypertensive effect of a peptide-enriched soy sauce-like seasoning. The results of these tests have highlighted an important lack of correlation between the *in vitro* ACEI activity and the *in vivo* action. This fact has provided doubts on the use of the *in vitro* ACEI activity as the exclusive criteria for potential antihypertensive substances, since physiological transformations may occur *in vivo*, and because other mechanisms of action than ACE inhibition might be responsible for the antihypertensive effect.

Effects in Clinical Studies

Evidence of the beneficial effects of bioactive peptides has to be based on clinical data. Most research has been focused in lactotripeptides, VPP and IPP, and their antihypertensive properties. About twenty human studies have been published linking the consumption of products containing lactotripeptides with significant reductions in both SBP and DBP. Oral administration of these tri-peptides included in different formulas, fermented milk, dried product, fruit juice, etc., products. However, recent studies have provided some conflicting results. Most clinical trials have assessed BP-lowering effects at multiple points over time. Most of the BP studies with lactotripeptides have been done in Japanese subjects, and several studies have been done in Finnish subjects [83-88]. Generally, maximum duration of treatment was 8 weeks at doses between 3 and 52 mg/day (Table 3). From these data, it becomes apparent that the largest part of the total BP reduction takes place in the first 1–2 weeks of treatment. Thereafter, a further gradual lowering is seen, but to a lesser extent than in the first period [84-86]. The first significant effects of lactotripeptides on BP in hypertensive subjects were observed after 1–2 weeks of treatment with dosages as low as 3.8 mg/d. Maximum BP-lowering effects of lactotripeptides approximate 13 mmHg SBP and 8 mmHg DBP active treatment v. placebo, and are likely reached after 8–12 weeks of treatment. Lactotripeptides exert a gradual effect on BP lowering after start of intake and return of BP after end of treatment as well [85, 86, 89]. The highest effective dosage of lactotripeptides was evaluated in a safety study, and consisted of 52.5 mg/d [88]. After 10 weeks of active treatment, mean SBP in subjects with hypertension decreased by 4.1 mmHg and DBP by 1.8 mmHg. The next highest dose of lactotripeptides that was tested amounted to 13.0 mg/d [89]. After 4 weeks of active treatment, SBP in subjects with mild hypertension decreased by 11.2 mmHg compared to placebo, and DBP tended to decrease by 6.5 mmHg. In none of the trials with normotensives were statistically significant BP changes found [90-92]. Even at the highest dosage of lactotripeptides used in normotensives, which included a total of 29.2 mg/d during a period of 7 d, no BP lowering effects by lactotripeptides were observed [93]. Thus lactotripeptides only seem to be active at elevated BP values. Evidence indicates that effectiveness is positively associated with BP level, which is in line with existing data for BP-lowering medication [94].

The results have been included in two meta-analysis [95, 96], which described decreases around 5 mmHg for SBP and 2.3 mmHg for DBP. In general, the effects described in Japanese studies on lactotripeptides are larger than those reported in Finnish studies. However, it is unlikely that genetic differences can account for these differential effects. Moreover, clinical trials in Dutch and Danish subjects have described controversial results since no effect on blood pressure was found [97, 98]. In a recent meta-analysis with a total of 18 trials, it was found a reduction of 3.73 mm Hg for SBP and 1.97 mm Hg for DBP but it was highlighted that the effect was more evident in Asian subjects that in Caucasian ones [99]. The relevance of these findings in genetics or dietary patterns should be further investigated. Comparative studies on antihypertensive medication in different races/ethnic groups have demonstrated that pharmacokinetic parameters and haemodynamic effects are essentially the same in Chinese and Japanese subjects compared with Caucasian subjects [100].

Table 3: Hypotensive effects of fermented milks with bioactive peptides in humans

Design	Duration (weeks)	Study population	Treatment				BP changes mmHg		Ref.
			IPP mg/d	VPP mg/d	Source of peptides	Formula	SBP	DBP	
R, p-c, s-bld, parallel	8	30 eldery hypertensive patients	1.1	1.5	*Lb. helv* + *Str. cer*	1 x 95 ml milk drink	-14.1	-6.9	83
R, p-c, d-bld, parallel	8	64 subjects with SBP 140-159 and DBP 90-99 mmHg	1.58	2.24	*Lb. helv* + *Str. cer.*	2 x 150 g milk drink	-13	-8.4	84
R, p-c, d-bld, parallel	8	32 subjects with SBP 140 - 180 and DBP 90-105 mmHg	1·60	2·66	*Lb. helv* + *Str. cer.*	1 x 120 g milk drink	-12.1	-5.8	85
R, p-c, d-bld, parallel	8	18 hypertensive and 26 normotensive subjects	1.1	1.5	*Lb. helv* + *Str. cer.*	2 x 100 g milk drink	-7.6	-2	91
R, p-c, d-bld, parallel	8	30 subjects with SBP 140-180 and DBP 90-105 mmHg	1.52	2.53	*Lb. helv* + *Str. cer.*	2 x 160 g milk drink	-13.2	-7.8	92

R, p-c, d-bld, parallel[1)	21	39 subjects with SBP 133-176 and DBP 86-108 mmHg	2.25	3.0	Lb. helv LBK-16H	2 x 150 ml milk drink	-6.7	-3.6	86
R, p-c, d-bld, parallel Cross-ove[2)	10 7	60 Finnish subjects with SBP 140-180 and DBP 90-110 mmHg	2.4-2.7	2.4-2.7	Lb. helv LBK-16H	1 x 150 ml milk drink	-2.3 -12.3	-0.5 -3.7	87
R, p-c, d-bld, parallel	10	94 hypertensive patients	30	22.5	Lb. helv LBK-16H	2 x 150 ml milk drink	4.1	1.8	88
R, p-c, d-bld, parallel	1	20 healthy volunteers normal blood pressure (<130 mmHg SBP and <85 mmHg DPB).	11.5	17.7	Lb. helv CM4	1 x 14 tablets	2.6	2	93
R, p-c, d-bld, parallel	8	135 Dutch subjects with untreated high-normal BP or mild hypertension	4.2	5.8	Fermentation	1 x 200 ml yoghurt drink	-0.5	-1.2	97
R, p-c, d-bld, crossover	4	70 Caucasian subjects with prehypertension or stage 1 hypertension	15	-	Hydrolysis by endopeptidase	2 x 7.5 mg capsules	-3.8	-2.3	102

Hypertension is a complex multifactor disorder that is thought to result from an interaction between environmental factors and genetic background. Subject characteristics such as age and race/ethnicity can affect BP, including the BP response to specific antihypertensive medication. For certain antihypertensive drugs, it has been reported that a polymorphism found in humans can affect the clinical effectiveness, and similarly, these differences could be also affecting clinical trials of functional ingredients [101]. Although ACE inhibition has been postulated as the underlying mechanism of these lactotripeptides, results about the inhibition of this enzyme are not conclusive in humans. Several studies have shown that rennin or ACE activity was not affected by the oral administration of the tripeptides [95, 102]. Therefore, other mechanisms could be implicated in the observed blood pressure reduction. It has been found that the intake of fermented milk containing these peptides may decrease sympathetic activity, leading to a diminished heart rate variability, heart rate and total peripheral resistance, although differences did not reach statistical significance [98].

BIOAVAILABILITY

Bioavilability of bioactive peptides is an important target to establish the relationship between *in vitro* and *in vivo* activities. The likelihood of any bioactive peptide released during fermentation mediating a physiological response is dependent on the ability of that peptide to reach an appropriate target site. Therefore, peptides may need to be resistant to further degradation by proteolytic and peptidolytic enzymes in digestive tract. Thereafter peptides should be absorbed and enter systemic circulation. Resistance to hydrolysis is one of the main factors influencing the bioavailability of bioactive peptides. The effects of digestive enzymes on bioactive peptides, in particular ACEI peptides derived from different food matrices, have been evaluated *in vitro* gastrointestinal simulated systems. The common purpose of these experiments was to assess the effects of the peptidases of the stomach and the pancreas on the preservation of the ACEI activity of different hydrolysates. Studies have shown that the ACEI is low after fermentation but increases during hydrolysis that simulates gastrointestinal digestion [35,103]. The ACEI peptides in rapeseed hydrolysate exhibited good stability in an *in vitro* digestion model using human gastric and duodenal fluids [104]. The digestion of some peptides have been reported. For example, Ile-Val-Tyr was hydrolysed by pepsin, trypsin and chymotrypsin alone or in combination and IC_{50}-value did not change significantly during digestion [105]. Proline- and hydroxyproline-containing peptides are usually resistant to degradation by digestive enzymes. Tripeptides containing C-terminal proline-proline are generally resistant to proline-specific peptidases [106]. In some cases, pancreatic digestion is needed to produce active peptide. For instance, the active form of peptide Lys-Val-Leu-Pro-Val-Pro-Glu is generated by hydrolysis of the glutamine residue at the C-terminal during pancreatic digestion [107]. The results are not completely predictive of the resistance of the bioactive peptides because they do not mimic all the physiological factors affecting food digestion, as pH variations, the relative amounts of the enzymes, the interactions with other molecules, and the ratio peptidase/tested compound. These variations may affect the rate of enzymatic degradation of the bioactive peptides under study, therefore affecting the estimated bioavailability of these bioactive peptides. Moreover, commercial enzymes appear to digest whey proteins more efficiently compared with human digestive juices when used at similar enzyme activities [108]. This could lead to conflicting results when comparing human *in vivo* protein digestion with digestion using purified enzymes of non-human species.

Peptides have been reported to have poor permeation across biological barriers (e.g. intestinal mucosa) [109]. Peptides can be transported by active transcellular transport or by passive processes. Although substantial amino acid absorption occurs in the form of di- and tripeptides at the apical side of enterocytes, efflux of intact peptides via the basolateral membrane into the general circulation seems to be negligible [110]. The intestinal absorption of peptides have been performed using *in vitro* tests with monolayer of intestinal cell lines, simulating intestinal epithelium, as well as analysis of peptides and derivatives in blood samples after animal and clinical studies. Foltz et al. [111] investigated the transport of IPP and VPP by using three different absorption models and demonstrated that these tripeptides are transported in small amounts intact across the barrier of the intestinal epithelium. The major transport mechanisms of IPP and VPP were demonstrated to be paracellular transport and passive diffusion [112]. Another ACEI peptide, Leu-His-Leu-Pro-Leu-Pro resisted gastrointestinal simulation but was degraded to His-Leu-Pro-Leu-Pro by cellular peptidases before crossing Caco-2 cell monolayer. The pentapeptide was rapidly transported through Caco-2 cell monolayers through paracellular route [113].

Vascular endothelial tissue peptidases and soluble plasma peptidases further contribute to peptide hydrolysis. As a consequence, for most peptides, the plasma half-life is limited to minutes as shown for endogenous peptides such as angiotensin II and glucagon-like peptide 1 [114]. In order to exert antihypertensive effect ACEI peptides need to resist different peptidases such as ACE. In this regard ACEI peptides can be classified into three groups: the inhibitor type, of which the IC_{50}-value is not affected by preincubation with ACE; the substrate type, peptides that are hydrolysed by ACE to give peptides with a weaker activity; the pro-drug type inhibitor, peptides that are converted to true inhibitors by ACE or other proteases/peptidases. Only peptides belonging to pro-drug or inhibitor type exert antihypertensive properties after oral administration. There are some examples showing that peptides are absorbed and can exert *in vivo* activities. As regard to casein-derived IPP, Jauhiainen et al. [115] used radiolabelled tripeptide and showed that it absorbed partly intact from the gastrointestinal tract after a single oral dose to rats. Considerable amounts of radioactivity were found from several tissues, e.g., liver, kidney and aorta. The excretion of IPP was slow; even after 48 hours the radiolabelled peptide had not been completely excreted. IPP did not bind to albumin or other plasma proteins *in vitro*. Considering this and the long-lasting retention of the radioactivity in the tissues, accumulation of IPP may occur in sufficient concentrations to cause blood pressure lowering effects e.g., by ACE-inhibition in the vascular wall. In another study the absolute bioavailability of the tripeptides in pigs was below 0.1%, with an extremely short elimination half-life ranging from 5 to 20 min [116]. In humans, maximal plasma concentration did not exceed picomolar concentration [117].

The improvement of limited absorption and stability of peptides has been a goal when evaluating their effectiveness. For example, some carriers interact with the peptide molecule to create an insoluble entity at low pH which later dissolves and facilitates intestinal uptake, by enhancing peptide transport over the non-polar biological membrane [118]. Bioavailability of bioactive tripeptides (VPP, IPP, LPP) was improved by administering them with a meal containing fiber, as compared to a meal containing no fiber. High methylated citrus pectin was used as a fiber [119]. Ko et al. [120] applied emulsification, microencapsulation and lipophilization to enhance the antihypertensive activity of a hydrolysate of tuna cooking juice. Among these treatments, lipophilization was the most effective, followed by microencapsulation and lecithin emulsification, getting for each of them a stronger effect than the obtained with the double untreated dosage. Antihypertensive effect of ovokinin (Phe-Arg-Ala-Asp-Pro-Phe-Leu) increased four-times compared to the untreated dosage after administration with egg yolk [121]. In this case, phospholipids were identified as responsible for enhancing the antihypertensive effect, particularly phosphatidylcholine, that could improve intestinal absorption or by protecting ovokinin of peptidases. Among drug delivery systems, emulsions have been used to enhance oral bioavailability or promoting absorption through mucosal surfaces of peptides and proteins [118]. Individually, various components of emulsions have been considered as candidates for improving bioavailability of peptides.

GENERAL CONCLUSIONS

The interest on foods possessing health-promoting or disease-preventing properties has been increasing. An increasing number of foods sold in developed countries bears nutrition and health claims. Fermented milk with putative antihypertensive effect in humans could be an easy applicable lifestyle intervention against hypertension. In fact, much work has been done with dietary antihypertensive peptides and evidence of their effect in animal and clinical studies. Moreover, there are numerous available patents of products containing antihypertensive bioactive peptides. However, certain aspects, such as identification of the active form in the organism and the different mechanisms of action that contribute in the antihypertensive effect still need to be further investigated. Recent advances on specific analytical techniques able to follow small amounts of the peptides or derivatives from them in complex matrices and biological fluids will allow performing these kinetic studies in model animals and humans. Similarly, advances in new disciplines such as nutrigenomic and nutrigenetic will open new ways to follow bioactivity in the organism by identifying novel and more complex biomarkers of exposure and/or of activity. There is still poor knowledge on the resistance of peptides to gastric degradation, and low bioavailability of peptides has been observed. This reinforces the need of various strategies to improve the oral bioavailability of peptides.

More emphasis has been put on the legal regulation of the health claims attached to the products. Authorities around the world have developed systematic approaches for review and assessment of scientific data. Evidence on the beneficial effects of a functional food product should be enough detailed, extensive and conclusive for the use of a health claim in the product labeling and marketing. Besides being based on generally accepted scientific evidence, the claims should be well understood by the average consumer. First, it is necessary to identify and quantify the active sequences. Antihypertensive peptides are only minor constituents in highly complex food matrices and, therefore, a monitoring of the large-scale production by hydrolytic or fermentative industrial process is mandatory. Second, extensive investigations to prove the antihypertensive effect in humans as well as the minimal dose to show this effect are necessary to fulfill the requirements of the legislation concerning functional foods. Japan was the pioneer with the Foods for Special Health Use (FOSHU) legislation in 1991. Europe adopted a joint Regulation on Nutrition and Health Claims made on Foods in 2006 being the European Food Safety Authority (EFSA). At present, EFSA have concludes that the evidence is insufficient to establish a cause and effect relationship between the consumption of the tripeptides VPP and IPP and the maintenance of normal blood pressure. Bearing in mind that 'essential hypertension' consists of disparate mechanisms that ultimately lead to elevations in systemic BP, it is most probably that that products containing lactotripeptides offer a valuable option as a non-pharmacological, nutritional treatment of elevated blood pressure for some groups of people.

REFERENCES

1.N Shah, 2007Functional cultures and health benefitsInt. dairy j. 1712621277

2.T Takano, 2002Anti-hypertensive activity of fermented dairy products containing biogenic peptides.Anton. leeuw. 82333340

3.Korhonen HJTPihlanto-Leppälä A (2004Milk-derived bioactive peptides : formation and prospects for health promotion. In: Edited by Colette Shortt and John O'Brien. Handbook of functional dairy productsFunctional foods and nutraceuticals series 6.0: 109124

4.FitzGerald RJMurray BA(2006Bioactive peptides and lactic fermentationsInt. j. dairy technology 59118125

5.P Jäkälä, H Vapaatalo, 2010Antihypertensive peptides from milk proteinsPharmaceuticals3251272

6.J. E Christensen, E. G Dudley, J. A Pederson, J. L Steele, 1999Peptidases and amino acid catabolism in lactic acid bacteria.Anton. leeuw. 76217246

7.S Luoma, K Peltoniemi, V Joutsjoki, T Rantanen, M Tamminen, I Heikkinen, A Palva, 2001Expression of six peptidases from Lactobacillus helveticus in Lactococcus lactisAppl. environ. microb. 6712321238

8.C Foucaud, V Juillard, 2000Accumulation of casein-derived peptides during growth of proteinase-positive strains of Lactococcus lactis in milk: their contribution to subsequent bacterial growth is impaired by their internal transport.J dairy res. 67233240

9.A. G Williams, J Noble, J Tammam, D Lloyd, J. M Banks, 2002Factors affecting the activity of enzymes involved in peptide and amino acid catabolism in non starter lactic acid bacteria isolated from Cheddar cheeseInt. dairy j. 12841852

10.A. D Lopez, C. C Murray, 1998The global burden of disease, 1990-2020. Nat.med. 412411243

11.T Harris, E. F Cook, W Kannel, A Schatzkin, L Goldman, 1985Blood pressure experience and risk of cardiovascular disease in the elderly. Hypertension711824

12.A Pihlanto, H Korhonen, 2003Bioactive peptides and proteins.Adv. food res. 47175276

13.B Hernández-ledesma, del Mar Contreras M, Recio I (2011Antihypertensive peptides: production, bioavailability and incorporation into foodsAdv. colloid interface sci. 1652335

14.L. F Van Gaal, I. L Mertens, C. E De Block, 2006Mechanisms linking obesity with cardiovascular disease.Nature444876880

15.A Pihlanto, 2006Antioxidative peptides derived from milk proteinsInt. dairy j. 1613061314

16.F Roudot-algaron, D. L Bars, L Kerhoas, J Einhorn, J. C Gripon, 1994Phosptiopeptides from Comté Cheese: Nature and origin. J. food sci. 59544547

17.T. K Singh, P. F Fox, A Healy, 1997Isolation and identification of further peptides from diafiltration retentate of the water-soluble fraction of Cheddar cheese. Jdairy res. 64433443

18.H Meisel, A Goepfert, S Günter, 1997ACE-inhibitory activities in milk productsMilchwissenschaft52307311

19.Addeo F, Chianes L, Salzano A, Sacchi R, Cappuccio U, Ferranti P, Malorni A (1992) Characterization of the 12% tricholoroacetic acid-insoluble oligopeptides of Parmigiano–Reggiano cheese. J. dairy res. 59: 401–411.

20.E Smacchi, M Gobbetti, 1998Peptides from several Italian cheeses inhibitory to proteolytic enzymes of lactic acid bacteria, Pseudomonas fluorescens ATCC 948 and to the angiotensin I-converting enzyme. Enzyme microb.tech. 22687694

21.S. S Haileselassie, B H Lee, B. F Bibbs, 1999Purification and identification of potentially bioactive peptides from enzyme modified cheese. J. dairy sci. 8216121617

22.E-L Ryhänen, A Pihlanto-leppälä, E Pahkala, 2001A new type of ripened low-fat cheese with bioactive properties. Int. dairy j. 11441447

23.J. A Gomez, M Ramos, I Recio, 2002Angiotensin-converting enzyme-inhibitory peptides in Manchego cheeses manufactured with different starter cultures. Int. dairy j. 12697706

24.M. J Torres-llanez, A. F González-córdova, A Hernandez-mendoza, H. S Garcia, B Vallejo-cordoba, 2011Angiotensin-converting enzyme inhibitory activity in Mexican Fresco cheese. J. dairy sci. 9437943800

25.A. H Pripp, R Sorensen, L Stepaniak, and T Sorhaug, 2006Relationship between proteolysis and angiotensin I-converting enzyme inhibition in different cheeses LWT 39677683

26.M. R Muehlenkamp, J. J Warthesen, 1996casomorphins: Analysis in cheese and susceptibility to proteolytic enzymes from Lactococcus lactis ssp. cremoris. J. dairy sci. 792026

27.E Sienkiewicz- Szlapka, B Jarmolowska, S Krawczuk, E Kostyra, H Kostyra, M Iwana, 2009Contents of agonistic and antagonistic opioid peptides in different cheese varieties. Int. dairy j. 19258263

28.C Durrieu, P Degraeve, S Chappaz, A Martial-gros, 2006Immunomodulating effects of water-soluble extracts of traditional French Alps cheeses on a human T-lymphocyte cell line. Int. dairy j. 1615051514

29.T Saito, T Nakamura, H Kitazawa, Y Kawai, T Itoh, 2000Isolation and structural analysis of antihypertensive peptides that exist naturally in Gouda cheese J. dairy sci. 8314341440

30.L Ong, N. P Shah, 2008Release and identification of angiotensin-converting enzyme-inhibitory peptides as influenced by ripening temperatures and probiotic adjuncts in Cheddar cheeses. LWT- Food sci. technol. 4115551566

31.U Bütikofer, J Meyer, R Sieber, D Wechsler, 2007Quantification of the angiotensin-converting enzyme-inhibiting tripeptides Val-Pro-Pro and Ile-Pro-Pro in hard, semi-hard and soft cheeses. Int. dairy j. 17968975

32.U Bütikofer, J Meyer, R Sieber, B Walther, D Wechsler, 2008Occurrence of the angiotensin-converting enzyme-inhibiting tripeptides Val-Pro-Pro and Ile-Pro-Pro in different cheese varieties of Swiss origin. J. dairy sci. 912938

33.J Meyer, U Bütikofer, B Walther, D Wechsler, R Sieber, 2009Hot topic: Changes in angiotensin-converting enzyme inhibition and concentrations of the tripeptides Val-Pro-Pro and Ile-Pro-Pro during ripening of different Swiss cheese varieties. J. dairy sci. 92826836

34.Y Nakamura, N Yamamoto, K Sakai, A Okubo, S Yamazaki, T Takano, 1995Purification and characterization of angiotensin I-converting enzyme inhibitors from sour milk. J. dairy sci. 78777783

35.A Pihlanto-leppälä, T Rokka, H Korhonen, 1998Angiotensin I converting enzyme inhibitory peptides derived from bovine milk proteins. Int. dairy j. 8325331

36.M Gobbetti, P Ferranti, E Smacchi, F Goffredi, F Addeo, 2000Production of angiotensin-I-converting-enzyme-inhibitory peptides in fermented milks started by Lactobacillus delbrueckii subsp. bulgaricus SS1 and Lactococcus lactis subsp. cremoris FT4. Appl. environ. microb. 6638983904

37.A Fuglsang, F. P Rattray, D Nilsson, C. B Nyborg, 2003Lactic acid bacteria:

inhibition of angiotensin-converting enzyme in vitro and in vivo. Anton. leeuw. 832734

38. A Pihlanto, T Virtanen, H Korhonen, 2010Angiotensin I converting enzyme (ACE) inhibitory activity and antihypertensive effect of fermented milk. Int. dairy j. 20310

39. M. S Nielsen, T Martinussen, B Flambard, K. I Sorensen, J Otte, 2009Peptide profiles and angiotensin-I-converting enzyme inhibitory activity of fermented milk products: Effect of bacterial strain, fermentation pH, and storage time Int. dairy j. 19155165

40. B Muguerza, M Ramos, E Sánchez, M. A Manso, M Miguel, A Aleixander, R Lopez-fandino, 2006Antihypertensive activity of milk fermented by Enterococcus faecalis strains isolated from raw milk. Int. dairy j. 166169

41. J. C Rodríguez-figueroa, R Reyes-díaz, A. F González-córdova, R Troncoso-rojas, I Vargas-arispuro, B Vallejo-cordoba, 2010Angiotensin-converting enzyme inhibitory activity of milk fermented by wild and industrial Lactococcus lactis strains. J. dairy sci. 9350325038

42. N Yamamoto, M Maeno, T Takano, 1999Purification and characterization of an antihypertensive peptide from a yogurt-like product fermented by Lactobacillus helveticus CPN4. J. dairy sci. 8213881393

43. M. N Ashar, R Chand, 2004Antihypertensive peptides purified form milks fermented with Lactobacillus belbrueckii ssp. bulgaricus. Milchwissenschaft 591417

44. A Quiros, M Ramos, B Muguerza, M. A Delgado, M Miguel, A Aleixandre, I Recio, 2007Identification of novel antihypertensive peptides in milk fermented with Enterococcus faecalis. Int. dairy j. 173341

45. E. R Kunji, I Mierau, A Hagting, B Poolman, W. N Koning, 1996The proteolytic systems of lactic acid bacteria. Anton. leeuw. 70187221

46. N Yamamoto, Y Ishida, N Kawakami, H Yada, 1991Lactobacillus helveticus bacterium having high capability of producing tripeptide, fermented milk product, and process for preparing the same. EU Patent, 1016709A1.

47. N Yamamoto, A Akino, T Takano, 1993Purification and specificity of a cell-wall associated proteinase from Lactobacillus helveticus CP790. J. biochem, 114740745

48. K Ueno, S Mizuno, N Yamamoto, 2004Purification and characterization of an endopeptidase has an important role in the carboxyl terminal processing of antihypertensive peptides in Lactobacillus helveticus

CM4. Lett. appl. microbiol. 39313318

49.E Kilpi, M Kahala, J. M Steele, A Pihlanto, V Joutsjoki, 2007Angiotensin I-converting enzyme inhibitory activity in milk fermented by wild-type and peptidase-deletion derivatives of Lactobacillus helveticus CNRZ32. Int. dairy j. 17976984

50.D. Y Kwon, J. W Daily, H. J Kim, S Park, 2010Antidiabetic effects of fermented soybean products on type 2 diabetes. Nutr. res. 30113

51.Z. I Shin, R Yu, S. A Park, D. K Chung, C. W Ahn, H. S Nam, K. S Kim, H. J Lee, 2001His-His-Leu, an angiotensin I converting enzyme inhibitory peptide derived from Korean soybean paste, exerts antihypertensive activity in vivo. J. agric. food chem. 4930043009

52.Y Hu, A Stromeck, J Loponen, D Lopes-lutz, A Schieber, M. G Gänzle, 2011LC-MS/MS quantification of bioactive angiotensin I-converting enzyme inhibitory peptides in rye malt sourdoughs. J. agric. food chem. 591198311989

53.A Pihlanto, T Johansson, S Mäkinen, 2012Inhibition of angiotensin I-converting enzyme and lipid peroxidation by fermented rapeseed and flaxseed meal. Eng. life sci. 12 DOI:elsc.201100137

54.M. Y Lin, C. L Yen, 1999Antioxidative ability of lactic acid bacteria J. agric. food chem. 4714601466

55.C. C Ou, T. M Lu, J. J Tsai, J. H Yen, H. W Chen, M. Y Lin, 2009Antioxidative effect of lactic acid bacteria: Intact cells vs. intracellular extracts. J. food drug anal. 17209216

56.Y Kudoh, S Matsuda, K Igoshi, T Oki, 2001Antioxidative peptide from milk fermented with Lactobacillus delbrueckii subsp. bulgaricus IFO13953. Nippon Shokuhin Kagaku Kaishi 484455

57.B Hernández-ledesma, B Miralles, L Amigo, M Ramos, I Recio, 2005Identification of antioxidant and ACE-inhibitory peptides in fermented milk. J. sci. food agric. 8510411048

58.T Virtanen, A Pihlanto, S Akkanen, H Korhonen, 2007Development of antioxidant activity in milk whey during fermentation with lactic acid bacteria. J. appl. microbial. 102106115

59.P Libby, 2006Inflammation and cardiovascular disease mechanisms. Am. j. clin. nutr. 83: 456S- 460S.

60.LeBlanc AMMatar C, Valdéz JC, LeBlanc N, Perdigón G (2002Immunomodulatory effects of peptidic fractions issued from milk fermented with Lactobacillus helveticus. J. dairy sci. 8527332742

61.C Matar, S. S Nadathur, A. T Bakalinsky, J Goulet, 1997Antimutagenic effects of milk fermented by Lactobacillus helveticus L89 and a protease-deficient derivative. J. dairy sci. 8019651970

62.E Laffineur, N Genetet, J Leonil, 1996Immunomodulatory activity of β-casein permeate medium fermented by lactic acid bacteria. J. dairy sci. 7921122120

63.G Tompa, A Laine, A Pihlanto, H Korhonen, I Rogel, P Marnila, 2011Chemiluminescence of non-differentiated THP-1 promonocytes: developing an assay for screening anti-inflammatory milk proteins and peptides. Luminescence 26: 251-258,

64.C Matar, J. C Valdez, M Medina, M Rachid, G Perdigon, 2001Immunomodulating effects of milks fermented by Lactobacillus helveticus and its non-proteolytic variant. J. dairy res. 68601609

65.N Yamamoto, A Akino, T Takano, 1994Antihypertensive effect of the peptides derived from casein by an extracellular proteinase from Lactobacillus helveticus CP790. J. dairy sci. 77917922

66.Y Nakamura, N Yamamoto, K Sakai, T Takano, 1995Antihypertensive effect of sour milk and peptides isolated from it that are inhibitors to angiotensin I-converting enzyme. J. dairy sci. 7812531257

67.M Sipola, P Finckenberg, R Korpela, H Vapaatalo, M. L Nurminen, 2002Effect of long-term intake of milk products on blood pressure in hypertensive rats. J. dairy res. 69103111

68.P Jäkälä, A Hakala, A Turpeinen, R Korpela, H Vapaatalo, 2009Casein-derived bioactive tripeptides Ile-Pro-Pro and Val-Pro-Pro attenuate the development of hypertension and improve endothelial function in salt-loaded Goto-Kakizaki rats. J. funct. foods 1366374

69.T Jauhiainen, T Pilvi, Z. J Cheng, H Kautiainen, D. N Müller, H Vapaatalo, R Korpela, E Mervaala, 2010Milk products containing bioactive tripeptides have an antihypertensive effect in double transgenic rats (dTGR) harbouring human renin and human angiotensinogen genes. J. nutr. metab. doi:10.1155/2010/287030.

70.T Jauhiainen, M Collin, M Narva, Z. J Cheng, T Poussa, H Vapaatalo, R Korpela, 2005Effect of long-term intake of milk peptides and minerals on blood pressure and arterial function in spontaneously hypertensive rats. Milchwissenschaft 60358362

71.P Jäkälä, T Jauhiainen, R Korpela, H Vapaatalo, 2009Milk protein-derived bioactive tripeptides Ile-Pro-Pro and Val-Pro-Pro protect endothelial function in vitro in hypertensive rats. J. funct. foods 1266273

72.O Masuda, Y Nakamura, T Takano, 1996Antihypertensive peptides are present in aorta after oral administration of sour milk containing these peptides to spontaneously hypertensive rats. J. nutr. 12630633068

73.P Jäkälä, E Pere, R Lehtinen, A Turpeinen, R Korpela, H Vapaatalo, 2009Cardiovascular activity of milk casein-derived tripeptides and plant sterols in spontaneously hypertensive rats. J. physiol. pharmacol. 601120

74.M Miguel, I Recio, M Ramos, M. A Delgado, A Aleixandre, 2006Antihypertensive effect of peptides obtained from Enterococcus faecalis-fermented milk in rats J. dairy sci. 8933523359

75.A Quiros, M Ramos, B Muguerza, M. A Delgado, M Migue, A Aleixandre, I Recio, 2007Identification of novel antihypertensive peptides in milk fermented with Enterococcus faecalis Int. dairy j. 173341

76.G. W Chen, J. S Tsai, B. S Pan, 2007Purification of angiotensin I-converting enzyme inhibitory peptides and antihypertensive effect of milk produced by protease-facilitated lactic fermentation. Int. dairy j. 17641647

77.Z. I Shin, R Yu, S. A Park, D. K Chung, C. W Ahn, H. S Nam, K. S Kim, H. J Lee, 2001His-His-Leu, an angiotensin I converting enzyme inhibitory peptide derived from Korean soybean paste, exerts antihypertensive activity in vivo. J. agric. food chem. 4930043009

78.F. S Sacks, A Lichtenstein, L Van Horn, W Harris, P Kris-etherton, M Winston, 2006Soy protein, isoflavones, and cardiovascular health. Circulation 11310341044

79.J Wu, A. D Muir, 2008Hypotensive and physiological effect of angiotensin converting enzyme inhibitory peptides derived from soy protein on spontaneously hypertensive rats J. agric. food chem. 5698999904

80.J. S Tsai, Y. S Lin, B. S Pan, T. J Chen, 2006Antihypertensive peptides and γaminobutyric acid from prozyme 6 facilitated lactic acid bacteria fermentation of soymilk. Process biochem. 4112821288

81.K Inoue, T Gotou, H Kitajima, S Mizuno, T Nakazawa, N Yamamoto, 2009Release of antihypertensive peptides in miso paste during its fermentation, by the addition of casein. J. biosci. bioeng. 108111115

82.T Nakahara, A Sano, H Yamaguchi, K Sugimoto, H Chikata, E Kinoshita, R Uchida, 2010Antihypertensive effect of peptide-enriched soy sauce-like seasoning and identification of its angiotensin I-converting enzyme inhibitory substances, J. agric. food chem.58821827

83.Y Hata, M Yamamoto, M Ohni, K Nakajima, Y Nakamura, T Takano, 1996A Placebo-controlled, study of the effect of sour milk on blood pressure in hypertensive subjects. Am. j. clin. nutr. 64767771

84.O Kajimoto, T Kurosaki, J Mizutani, N Ikeda, K Kaneko, M Yabune, Y Nakamura, 2002Antihypertensive effects of liquid yogurts containing 'lactotripeptides (VPP, IPP)' in mild hypertensive subjects. J. nutr. food 55566

85.H Hirata, Y Nakamura, H Yada, S Moriguchi, O Kajimoto, T Takahashi, 2002Clinical effects of new sour milk drink on mild or moderate hypertensive subjects. J. new. rem. clin 516169

86.L Seppo, T Jauhiainen, T Poussa, R Korpela, 2003A fermented milk high in bioactive peptides has a blood pressure-lowering effect in hypertensive subjects. Am. j. clin. nutr. 77326330

87.J Tuomilehto, J Lindstrom, J Hyyrynen, R Korpela, M. L Karhunen, L Mikkola, T Jauhiainen, L Seppo, A Nissinen, 2004Effect of ingesting sour milk fermented using Lactobacillus helveticus bacteria producing tripeptides on blood pressure in subjects with mild hypertension. J. hum. hypertens. 18795802

88.T Jauhiainen, H Vapaatalo, T Poussa, S Kyrönpalo, M Rasmussen, R Korpela, 2005Lactobacillus helveticus fermented milk lowers blood pressure in hypertensive subjects in 24-h ambulatory blood pressure measurement. Am. j. hypertens. 1816001605

89.K Arihara, O Kajimoto, H Hirata, R Takahashi, Y Nakamura, 2005Effect of powdered fermented milk with Lactobacillus helveticus on subjects with high-normal blood pressure or mild hypertension. J. am. coll. nutr. 24257265

90.O Kajimoto, K Aihara, H Hirata, R Takahashi, Y Nakamura, 2001Hypotensive effects of the tablets containing lactotripeptides (VPP, IPP). J. nutr. food 45161

91.H Itakura, S Ikemoto, S Terada, K Kondo, 2001The effect of sour milk on blood pressure in untreated hypertensive and normotensive subjects. J. jap. soc. clin. nutr. 232631

92.O Kajimoto, Y Nakamura, H Yada, S Moriguchi, H Hirata, T Takahashi, 2001Hypotensive effects of sour milk in subjects with mild or moderate hypertension. J. jpn. soc. nutr. food sci. 54347354

93.K Yasuda, K Aihara, K Komazaki, M Mochii, Y Nakamura, 2001Effect of large high intake of tablets containing 'lactotripeptides (VPP, IPP)' on blood pressure, pulse rate and clinical parameters in healthy volunteers. J. nutr. food 46372

94.M. R Law, N. J Wald, J. K Morris, J. E Jordan, 2003Value of low dose combination treatment with blood pressure lowering drugs: analysis of 354 randomised trials. Br. med. j. 32614271431

95.A. H Pripp, 2008Effect of peptides derived from food proteins on blood pressure: a meta-analysis of randomized controlled trials. Food nutr. res. 519

96.J. Y Xu, L. Q Qin, P. Y Wang, W Li, C Chang, 2008Effect of milk tripeptides on blood pressure: a meta-analysis of randomized controlled trials. Nutrition 24933940

97.M. F Engberink, E. G Schouten, F. J Kok, L. A Van Mierlo, I. A Brouwer, J. M Geleijnse, 2009Lactotripeptides show no effect on human blood pressure: results from a double-blind randomized controlled trial. Hypertension 51399405

98.L Usinger, L. T Jensen, B Flambard, A Linneberg, H Ibsen, 2010The antihypertensive effect of fermented milk in individuals with prehypertension or borderline hypertension. J.hum. hypertens. 24678683

99.Cicero AFGGerocarni B, Laghi L, Borghi C (2011Blood pressure lowering effect of lactotripeptides assumed as functional foods: a meta-analysis of current available clinical trials. J. hum. hypertens. 25425436

100. S Vaidyanathan, J Jermany, C Yeh, M. N Bizot, R Camisasca, 2006Aliskiren, a novel orally effective renin inhibitor, exhibits similarpharmacokinetics and pharmacodynamics in Japanese and Caucasian subjects. Br. j. clin. pharmacol. 62690698

101. J Arsenault, J Lehoux, Lanthier L Cabana J, Guillemette G, Lavigne P, Leduc R, Escher E (2010A single-nucleotide polymorphism of alanine to threonine at position 163 of the human angiotensin II type 1 receptor impairs losartan affinity. Pharmacogenet Genomics 20377388

102. E Boelsma, J Kloek, 2010IPP-rich milk protein hydrolysate lowers blood pressure in subjects with stage 1 hypertension, a randomized controlled trial. J. nutr. 9:52 doi:10.1186/1475-2891-9-52.

103. B Hernández-ledesma, L Amigo, M Ramos, I Recio, 2004Application of high-performance liquid chromatography-tandem mass spectrometry to the identification of biologically active peptides produced by milk fermentation and simulated gastrointestinal digestion. J. chromatogr. A 1049107114

104. S Mäkinen, T Johansson, G Vegarud, J. M Pihlava, A Pihlanto, 2012Angiotensin I-converting enzyme inhibitory and antioxidant properties of rapeseed hydrolysates. J. funct. foods (in press)

105. T Matsui, C. H Li, Y Osajima, 1999Preparation and characterization of novel bioactive peptides responsible for angiotensin I-converting enzyme inhibition from wheat germ. J. pept. sci. 5289297

106. V Vermeirssen, J Van Camp, W Verstraete, 2004Bioavailability of angiotensin I converting enzyme inhibitory peptides. Br. j. nutr. 92357366

107. M Maeno, N Yamamoto, T Takano, 1996Identification of an antihypertensive peptide from casein hydrolysate produced by a proteinase from Lactobacillus helveticus CP790. J. dairy sci. 7913161321

108. E. K Eriksen, H Holm, E Jensen, R Aaboe, T. G Devold, M Jacobsen, G. E Vegarud, 2010Different digestion of caprine whey proteins by human and porcine gastrointestinal enzymes. Br. j. nutr. 104374381

109. G. M Pauletti, S Gangwar, G. T Knipp, M. M Nerurkar, F. W Okumu, T Tamura, T. J Siahaan, 1996Structural requirements for intestinal absorption of peptide drugs. J. control release. 41317

110. H Daniel, 2004Molecular and integrative physiology of intestinal peptide transport. Annu. rev. physiol. 66361384

111. M Foltz, A Cerstiaens, A Van Meensel, R Mols, P. C Van Der Pijl, Duchateau GSMJE, Augustijns 2008The angiotensin converting enzyme inhibitory tripeptides Ile-Pro-Pro and Val-Pro-Pro show increasing permeabilities with increasing physiological relevance of absorption models. Peptides 29: 1312-1320.

112. M Satake, M Enjoh, Y Nakamura, T Takano, Y Kawamura, S Arai, M Shimizu, 2002Transepithelial transport of the bioactive tripeptide Val-Pro-Pro, in human intestinal Caco-2 cell monolayers. Biosci. biotech. bioch. 66378384

113. A Quiros, A Davalos, M. A Lasuncion, M Ramos, I Recio, 2008Bioavailability of the antihypertensive peptide LHLPLP: Transepithelial flux of HLPLP Int. dairy j. 18279286

114. C. F Deacon, M. A Nauck, M Toft-nielsen, L Pridal, B Willms, J. J Holst, 1995Both subcutaneously and intravenously administered glucagon-like peptide I are rapidly degraded from the NH2-terminus in type II diabetic patients and in healthy subjects. Diabetes. 4411261131

115. T Jauhiainen, K Wuolle, H Vapaatalo, O Kerojoki, K Nurmela, C Lowrie, R Korpela, 2007Oral absorption, tissue distribution and excretion of a radiolabelled analog of a milk-derived antihypertensive peptide, Ile-Pro-Pro, in rats. Int. dairy j. 1712161223

116. P. C Van Der Pijl, A. K Kies, Ten Have GA, Duchateau GS, Deutz NE (2008Pharmacokinetics of proline-rich tripeptides in the pig. Peptides. 2921962202

117. M Foltz, E. E Meynen, V Bianco, C Van Platerink, Koning TMMG,

Kloek J (2007Angiotensin converting enzyme inhibitory peptides from a lactotripeptide-enriched milk beverage are absorbed intact into the circulation. J. nutr. 137953958

118.   J Shaji, V Patole, 2008Protein and peptide drug delivery: Oral approaches. J. pharm. sci. 70269277

119.   A. K Kies, P Van Der Pijl, 2012Peptide availability USA Patent Application 20120040895.

120.   W. C Ko, M. L Cheng, K. C Hsu, Y. S Hwang, 2006Absorption-enhancing treatments for antihypertensive activity of oligopeptides from tuna cooking juice: In vivo evaluation in spontaneously hypertensive rats. J. food sci. 711317

121.   H Fujita, R Sasaki, K Kurahashi, M Yoshikawa, 1995Potentiation of the antihypertensive activity of orally administered ovokinin, a vasorelaxing peptide derived from ovalbumin, by emulsification in egg phosphatidyl-choline. Biosci. biotech. bioch.5923442345

Chapter 7

SIMULTANEOUS DETERMINATION OF METHYLCARBAMATE AND ETHYLCARBAMATE IN FERMENTED FOODS AND BEVERAGES BY DERIVATIZATION AND GC-MS ANALYSIS

Ho-Sang Shin and Eun-Young Yang

[1]Department of Environmental Education, Kongju 314-701, Republic of Korea.

[2]Department of Environmental Science, Kongju National University, Kongju 314-701, Republic of Korea.

ABSTRACT

Background

Methylcarbamate (MC) and ethylcarbamate (EC) are toxic compounds that commonly exist in fermented food and beverages. In order to estimate the risk for their exposure, a sensitive simultaneous analytical method is required

Results

A simultaneous determination of MC and EC was described based on derivatization with 9-xanthydrol and consecutive detection using gas chromatography–mass spectrometry. The derivatization of MC and EC was performed directly in food or beverages and the reaction conditions were established through changing various parameters. The detection and the quantification limits were 0.01-0.03 µg/kg and 0.03-0.1 µg/kg, respectively, and the interday relative standard deviation was less than 12% at concentrations of 2.0 and 50 µg/kg. MC and EC were measured from 0.4 µg/kg to 85.8 µg/kg in sixteen Korean fermented foods and eleven beverages.

Conclusion

A simple, sensitive method to detect MC and EC in several solid foods and liquid foods was developed based on derivatization with 9-xanthydrol for 10 min at an ambient temperature. The method may useful for routine analysis of MC and EC in numerous food samples.

BACKGROUND

Ethylcarbamate (EC, urethane, $C_2H_5OCONH_2$) is a known genotoxic carcinogen that commonly exists in fermented food and beverages due to the natural biochemical processes in the fermentation process [1, 2]. EC was re-classified as a carcinogen (Group 2A) by the International Agency for Research on Cancer (IARC) in 2007 [3] and has already been regulated in several countries such as Germany, USA, Canada, France and the Czech Republic [2]. A report from a commission by the European Food Safety Authority (EFSA) issued in 2010 [4] recommended that special attention should be paid to spirits distilled from stone fruits. Furthermore, EC has been detected in various fermented products such as bread, yoghurt, cheese, soy sauce, vinegar and alcoholic beverages [5, 6].

Methylcarbamate (MC, methylurethane, CH_3OCONH_2) is simplest ester of carbamic acid. MC has a relatively low toxicity, otherwise, there is experimental evidence that MC is mutagenic in Droso phila [7] and carcinogenic in rats [8].

EC and MC can co-exist through natural formation during the fermentation processes [9]. In order to estimate the risk for EC and MC exposure, a sensitive simultaneous analytical method in fermented foods and beverages is required.

Many methods for detecting EC in beverages have been reported, such as high-performance liquid chromatography (HPLC) [10–12], liquid chromatography tandem mass spectrometry (LC-MS/MS) [13], gas chromatography (GC) [14–17], and gas chromatography mass spectrometry (GC–MS) [6, 9, 18–27].

Several assay methods have been based on headspace solid-phase micro extraction (HS-SPME) [14, 15, 22, 28], where the headspace is discriminatory in nature because only the volatile compounds in the injection vials can be transferred to the GC system. Many volatile alcohols and interferences exist in fermented food and beverages, give much interference, and have a short fiber life time. Liquid-liquid extraction (LLE) [16, 19, 21] and solid phase extraction (SPE) [20, 21, 26] are often used to determine the EC content in alcoholic beverages. Although it is a traditional extraction technique, LLE represents a convenient method when it is connected with derivatization. Also, 9-xanthydrol has been used to improve the fluorescence of EC in the HPLC method [10–12] and to improve the sensitivity of EC using the GC-MS [21]. However, until now, analytical target compounds and matrices were limited to EC and liquid phases such as spirits or beverages. Another drawback with the methods is that EC is derivatized using 9-xanthydrol after extraction and concentration, and in this case volatile MC and EC can be lost during the evaporation process.

GC coupled with mass spectrometry (GC-MS) is the most widely used due to its good resolution, sensitivity and selectivity. Although the GC-MS methods are very selective and sensitive, it is difficult to detect to ng/kg levels without concentration and derivatization.

In this study, the derivatization parameters that enable the direct reaction of MC and EC in food or beverages are established. The xanthyl methylcarbamate or xanthyl ethylcarbamate derivatives that were formed were extracted by LLE and detected by GC–MS. Therefore, the experiment reported in this paper aimed to optimize the parameters of the derivatization, extraction and GC-MS detection in order to simultaneously determine the MC and EC in fermented foods and beverages, and in order to apply the modified method in the analysis of seventeen real samples.

EXPERIMENTAL

Materials

All organic solvents used were HPLC grade. Sodium chloride, potassium hydroxide, sodium bicarbonate, potassium carbonate, propanol, ethyl acetate, sodium sulfate, 9-xanthydrol (99%), methylcarbamate (98%), ethylcarbamate (EC, 99%), and butylcarbamate (98%) as internal standard were obtained from Sigma-Aldrich (St. Louis, MO, USA).

Apparatus

All mass spectra were obtained with an Agilent 6891/5973N instrument (Agilent Technologies, Santa Clara, CA, USA). The ion source was operated in the electron ionization mode (EI; 70 eV). Full-scan mass spectra (m/z 45–600) were recorded in order to identify the analytes. An HP-5MS capillary column (60 m × 0.25 mm I.D. × 0.25 μm film thickness) was used. The samples were injected in the splitless mode. The flow rate of helium as a carrier gas was 0.6 mL/min. The injector temperature was set at 260°C. The oven temperature programs were set as follows. The initial temperature of 150°C was not held and increased to the first temperature hold of 210°C (held for 1 min) at 30°C/min, and then increased to the final temperature hold of 260°C (held for 4 min) at 10°C/min. The ions selected by SIM were m/z 222, 240 and 255 for xanthyl methylcarbamate, m/z 222, 240 and 269 for xanthyl ethylcarbamate and m/z 222, 240 and 297 for xanthyl butylcarbamate.

Derivatization and extraction procedures

Fermented foods (soybean paste, red pepper paste and soy sauce) were purchased from several local markets or obtained from several homes. Beverages containing makgeolli (raw rice wine), soju (white distilled liquor), jeongjong (refined rice wine) and fruit liquor were purchased from several local markets.

A 2.0 g portion of each sample was homogenized for 10 min at 18,000 rpm in 5.0 mL of NaCl saturated solution using a homogenizer (PowerGen 125, Fisher Scientific, USA) after adding 80 μL of 0.1 M 9-xanthydrol solution in the propanol, 200 μL of 2.0 M HCl, and 20 μL of BC (2.5 mg/L in methanol). The derivatization reaction was conducted at an ambient temperature for 10 min in the dark, and then the solution was neutralized with 1.0 M KOH and the pH of the solution was controlled to 9.5 with 0.2 g of $NaHCO_3/K_2CO_3$ (2:1, w/w). The solution was extracted twice with 5.0 mL of ethyl acetate. The organic layers were combined and dried by passing them through anhydrous sodium sulfate. The dried organic layer was then concentrated in a rotary evaporator (30°C, 300 mbar). The concentrated residue was dissolved in 100 μL of methanol and a 1.0 μL sample of the solution was injected into the GC-MS system.

The derivatization efficiencies were calculated at various temperatures (20, 30, 40, and 50°C), 9-xanthydrol amounts (20, 40, 60, 80, 100, and 120 μL of 0.1 M solution), heating times (5, 10, 15, 20, 30 and 60 min), and acid moralities (0.1, 0.2, 0.3, 0.4, 0.5 and 1.0 M). The pH of each sample was controlled with 2.0 M HCl. The optimum derivatization conditions of MC, EC and BC with 9-xanthydrol were determined using the amounts of the formed xanthyl methylcarbamate, xanthyl ethylcarbamate and xanthyl butylcarbamate.

Calibration and quantification

The calibration curves for MC and EC were established through derivatizations after 1.0, 5.0, 20, 50, 100 and 200 ng of MC and EC standard solutions were added to 2.0 g of a control food (soybean paste), 5 mL of NaCl saturated solution, 20 μL of BC (2.5 mg/L in methanol), 80 μL of 0.1M 9-xanthydrol solution in propanol and 200 μL of 2.0 M HCl. The corresponding concentrations of the standards were 0.5, 2.5, 10, 25, 50 and 100 μg/kg. The ions selected for quantification were m/z 255 for xanthyl methylcarbamate, and m/z 240 for xanthyl ethylcarbamate and xanthyl butylcarbamate. The ratio of the peak area of the standard solution to that of the internal standard was used to quantify the compound.

RESULTS AND DISCUSSION

Optimization of the derivatization conditions in samples

The amino groups of MC, EC, and BC undertook the substitution reaction with 9-xanthydrol under acidic conditions in order to produce xanthyl methylcarbamate, xanthyl ethylcarbamate, and xanthyl butylcarbamate as shown in Figure 1, and it was possible to directly analyze the product by the GC-MS.

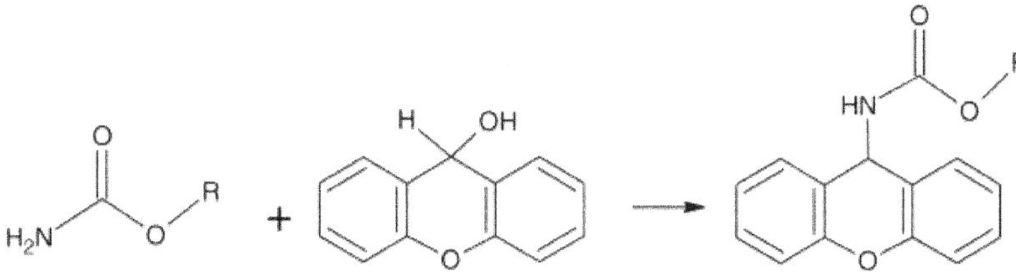

Alkyl carbamate   9-Xanthydrol   Xanthyl alkyl carbamate

$(R=CH_3, C_2H_5$ or $C_4H_9)$

Figure 1: The reaction of alkyl carbamates with 9-xanthydrol.

The optimal reaction conditions for the simultaneous determination of MC and EC in solid fermented foods was also tested. For the first test, the minimum amount of 9-xanthydrol for the derivatization was studied. The derivatization was performed for various 9-xanthydrol concentrations (1.0, 2.0, 3.0, 4.0, 5.0 and 6.0 mM of 9-xanthydrol). The yield stayed continuously beyond 4.0 mM of 9-xanthydrol and the optimal 9-xanthydrol amount was 4.0 mM (Figure 2). The effect of the acid concentration on the reaction of MC, EC and BC with 9-xanthydrol was also studied. The derivative was tested at HCl concentrations of 0.01, 0.05, 0.1, 0.2, 0.3, and 0.5 M. The other reaction conditions were set to have a reaction time of 10 min at a temperature of 20°C. The results showed good recovery at the HCl concentration value of 0.2 M (Figure 3). The reaction rate of MC, EC and BC with 9-xanthydrol was also studied. The reaction rate of the derivative was analyzed at reaction temperatures of 20, 30, 40, and 50°C and the reaction time was analyzed in at 5, 10, 20, 30, and 60 min. From the experiment, the optimal reaction temperature and time was 10 min at 20°C (Figures 4 and 5). The recovery was declined slowly beyond the reaction time of 10 min.

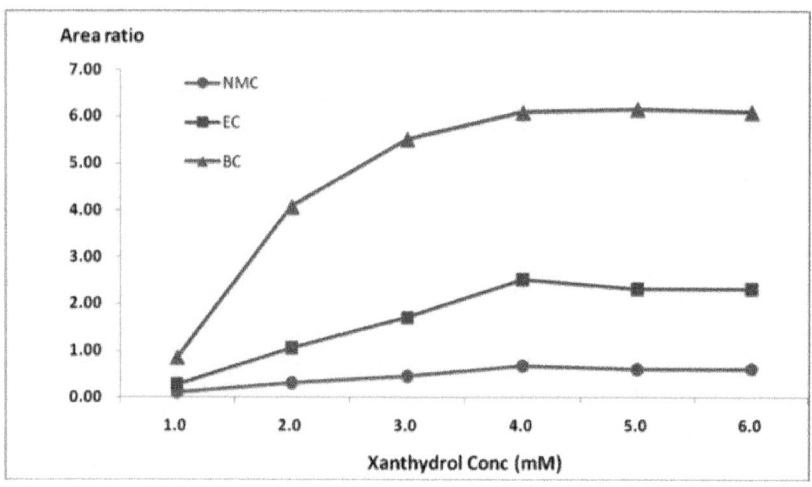

Figure 2: Reaction yield of MC, EC and BC in relation to the amount of 9-xanthydrol. (This experiment was performed at a reaction time of 10 min and a reaction temperature of 20°C).

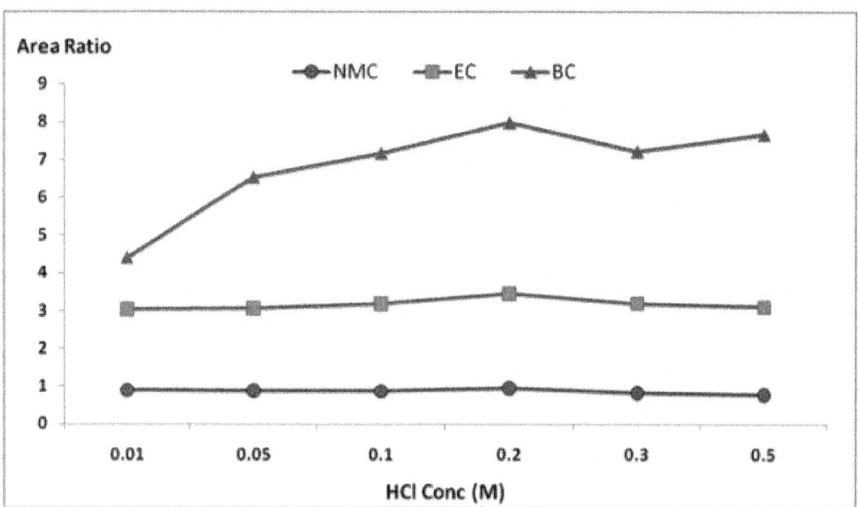

Figure 3: Effect of HCl concentration on the reaction of MC, EC and BC with 9-xanthydrol. (This experiment was performed at a reaction time of 10 min and a reaction temperature of 20°C).

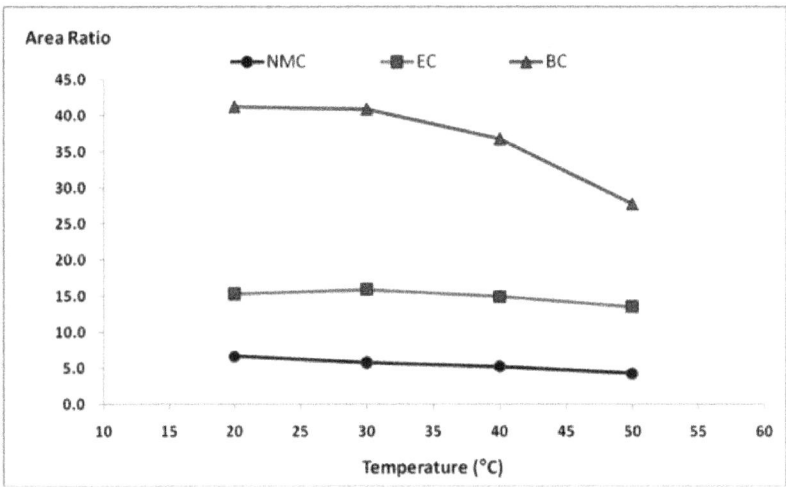

Figure 4: Effect of reaction temperature on the reaction of MC, EC and BC with 9-xanthydrol. (This experiment was performed at a reaction time of 5 min).

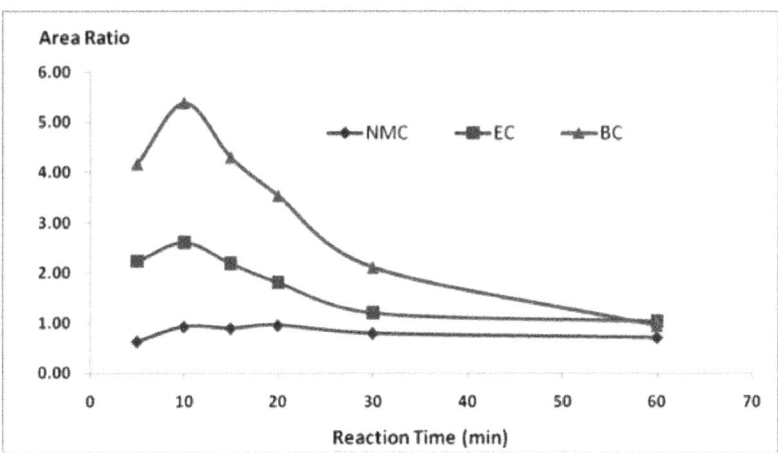

Figure 5: Effect of reaction time on the reaction of MC, EC and BC with 9-xanthydrol. (This experiment was performed at a reaction temperature of 20°C).

As a result, the optimal reaction conditions of MC, EC, and BC with 9-xanthydrol were 4.0 mM 9-xanthydrol, 0.2 M HCl concentration, the reaction time of 10 min at an ambient temperature. The selection of the extraction solvent was of great importance in order to achieve satisfactory extraction efficiency for the target compounds. Based on the consideration for the solvent strength, methylene chloride, ethyl acetate, ethyl ether and hexane were selected as potential extraction solvents for use in this study. As a result, ethyl acetate gave the highest extraction efficiency, and ethyl acetate was selected as an extraction solvent of the analyte derivatives from samples.

Chromatography and mass spectrometry

The optimum derivatization conditions were applied to the analysis of MC, EC, and BC in fermented food and beverages by GC-MS. Figure 6 shows the GC-MS chromatogram after the derivatization of MC, EC, and BC.

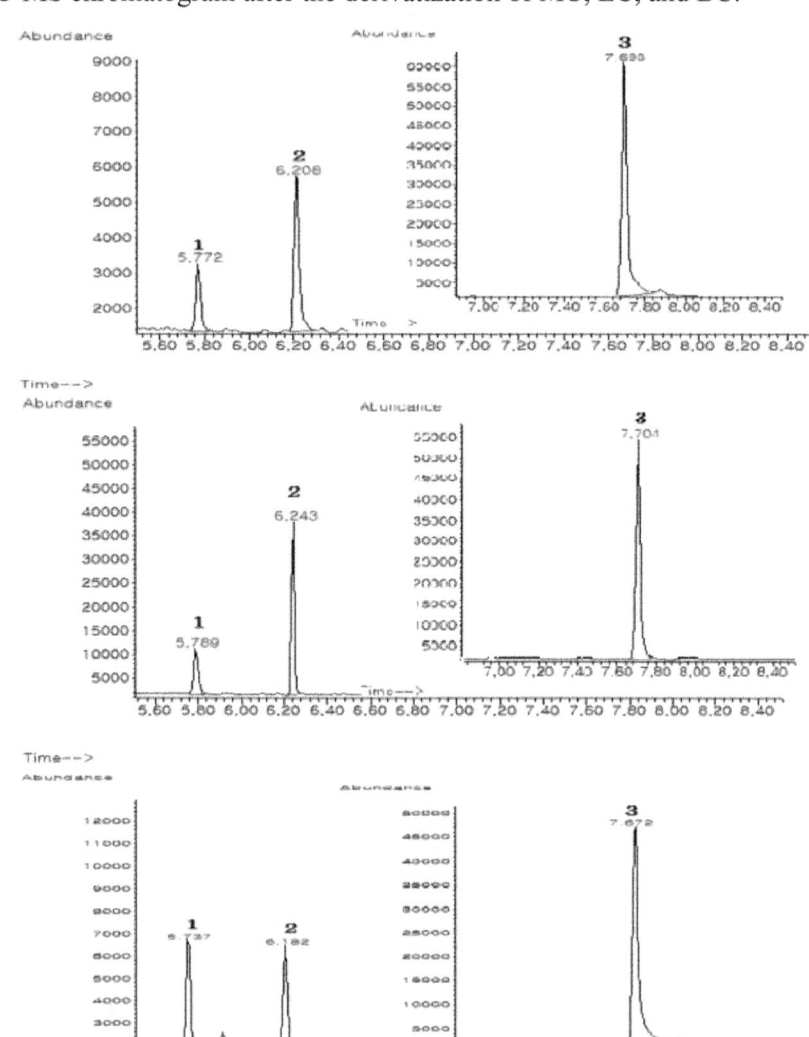

Figure 6: SIM chromatograms after the derivatization of NMC, EC and BC from spiked food (top=25 µg/kg, middle=5 µg/kg) and real sample (bottom). 1=Xanthyl

methylcarbamate, 2=Xanthyl ethylcarbamate, 3=Xanthyl butylcarbamate (ISTD).

For the GC separation of the derivative, the use of a nonpolar stationary phase was found to be efficient. The derivatives of MC, EC, and BC showed a sharp peak, and the compound was quantified as an integration of the peak area. The retention times of xanthyl methylcarbamate, xanthyl ethylcarbamate and xanthyl butylcarbamate are shown in Figure 6. Extraneous peaks were not observed in the chromatograms near the retention times of the analytes.

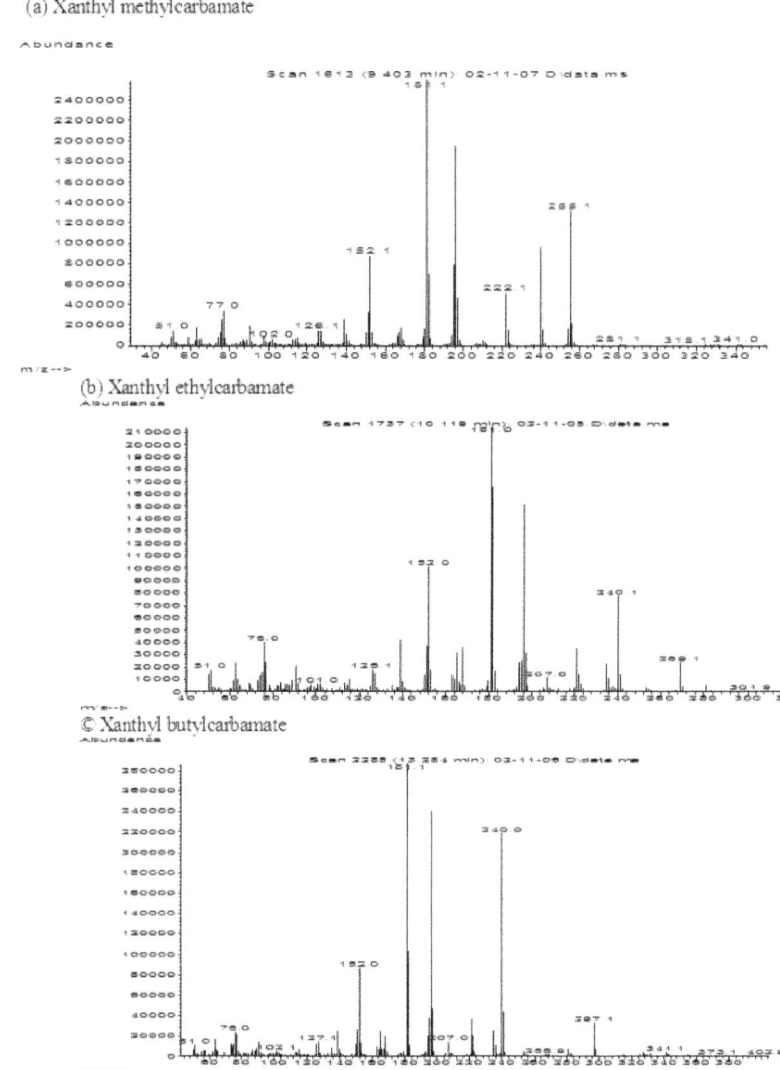

(a) Xanthyl methylcarbamate

(b) Xanthyl ethylcarbamate

© Xanthyl butylcarbamate

Figure 7: Mass spectra of xanthyl methylcarbamate, xanthyl ethylcarbamate and xanthyl butylcarbamate.

The mass spectra of xanthyl methylcarbamate, xanthyl ethylcarbamate and xanthyl butylcarbamate by electron ionization at 70 eV have similar fragmentation pattern as shown in Figure 7. The molecular ions at m/z 255, m/z 269 and m/z 297 were appeared in mass spectra of three compounds. The fragment of m/z 240 was accounted for by the loss of $[CH_3]$, $[C_2H_5]$ and $[C_4H_9]$ from the each molecular ion and that of m/z 196 was accounted for by the loss of $[COOCH_3]$, $[COOC_2H_5]$ and $[COOC_4H_9]$, and m/z 222 were accounted for by the loss of $[H_2OCH_3]$, $[H_2OC_2H_5]$ and $[H_2OC_4H_9]$ from the each molecular ion. The fragment of m/z 181 was a result of the xanthyl group.

Validation of the assay

The combination of a high derivatization yield and the high sensitivity of the derivative by EI-MS (SIM) allowed the detection of MC and EC at concentrations well below those reported previously. The limit of detection (LOD) and the limit of quantification (LOQ) were defined as the analyte concentration corresponding to a signal/noise ratio of 3 and 10 in the control food, in which MC and EC were not detected. The LODs in this study were 0.11 μg/kg for MC, and 0.12 μg/kg for EC, and the LOQs were 0.35 μg/kg for MC and 0.38 μg/kg for EC using a 2.0 g sample. Table 1 compares various analytical methods for determining the MC and EC in fermented food and beverages. The method permits the determination of two analytes below that detected previously using the GC-MS method, which was otherwise slightly higher than GC-HRMS or GC-MS/MS methods.

The calibration curves of the MC and EC were constructed by the reaction and extraction of the spiked food samples. Examination of the standard curve by computing a regression line of the peak area ratios for the MC and EC to the internal standard on concentrations using a least-squares fit demonstrated a linear relationship with correlation coefficients of 0.998 and 0.996, respectively. The line of best fit for the MC was y = 4.191 x - 0.0001 over a range of 1.0-100 μg/kg and that for EC was y = 13.46 x + 0.0051 over a range of 1.0-100 μg/kg, where x is the analyte concentration (mg/kg) and y is the peak area ratio of the analyte to the internal standard.

The accuracy can be assessed by determining the recovery in spiked samples: Intra-day accuracy was evaluated using five spiked samples at concentrations of 0.05 and 0.002 μg/kg for MC and EC, respectively. The inter-day accuracy was determined using the sample recovery on three different days. The accuracy was in range of approximately 90- to 109% and the precision of the assay was less than 12%, as shown in Table 2.

Table 1: Comparison of analytical methods for determining of NMC and EC in fermented food

Reference	Matrix	Preparation method	Derivatization	Measurement	LOD (µg/L or µg/kg) MC	LOD EC	LOQ (µg/L or µg/kg) MC	LOQ EC
[10]	Alcoholic beverage	-	9-xanthydrol	HPLC-FLD		3.0-	-	-
[11]	Wine	-	9-xanthydrol	HPLC-FLD	-	73.2-	-	243.9-
[12]	Cider spirits	-	9-xanthydrol	HPLC-FLD	-	1.64	-	3.56
[13]	Ethanol solution	-	-	LC-MS/MS		2.0		5.1
	Palinka spirits			LC-MS/MS		2.8		8.0
[15]	Alcoholic beverage	HS-SPME	-	GC-NPD	-	34	-	-
[19]	Alcoholic beverage	LLE	-	GC-MS	-	2.3	-	10.4
[20]	Fermented food	SPE	-	GC-HRMS	-	0.03	-	0.05
[22]	Stone-fruit spirits	HS-SPME	-	GC-MS/MS	-	0.03	-	0.11
[21]	Stone-fruit spirits	SPE	-	GC-MS/MS	-	0.01	-	0.04
[27]	Italian aqua vitae	LLE	9-xanthydrol	GC-MS	-	1.0	-	-
This study	Fermented food and beverage	LLE	9-xanthydrol	GC-MS	0.11	0.12	0.33	0.38

Table 2: Intraday and interday laboratory precision and accuracy results for the analysis of NMC and EC in fermented food ($n=5$)

Compound	Spiked Conc. (mg/L)	Intraday measured value			Interday measured value		
		Mean±SD(mg/L)	Accuracy (%)	Precision (%)	Mean±SD(mg/L)	Accuracy (%)	Precision (%)
NMC	0.0500	0.0492 ± 0.0031	98.4	6.30	0.0474 ± 0.0052	94.8	11.0
	0.0020	0.0018 ± 0.0002	90.0	11.1	0.0017 ± 0.0002	106	11.8
EC	0.0500	0.0532 ± 0.0030	106	5.64	0.0546 ± 0.0065	109	11.9
	0.0020	0.0018 ± 0.0002	90.0	11.1	0.0019 ± 0.0002	95.0	10.5

Food analysis

This paper was designed to describe a method to detect MC and EC in solid and liquid state matrices using GC-MS. Generally, many traditional Korean foods are made through fermentation of a mixture of various food materials, and therefore these foods have complicated matrix properties. When the proposed method was applied to the food items, interfering peaks were not observed in the chromatograms near the retention times of the analytes.

Using the proposed method, the levels of MC and EC were analyzed in sixteen traditional fermented Korean foods, including soybean paste, red pepper paste, and soy sauce, and eleven beverages and the results were shown in Table 3. MC was detected in a range from 0.4 to 0.8 μg/L in mainly fruit liquors. Most samples had detectable levels of EC in a range from 0.4 to 85.8 μg/L or μg/kg. The concentration range of the EC of each food or beverage type was found for soybean paste (0.9-2.7 μg/kg), red pepper paste (0.7-2.3 μg/kg), soy sauce (0.4-8.9 μg/L), and beverages (not detected-85.8 μg/L). From the results shown in Table 3, the prolonged mean storage time had no relationship with the detected content of EC.

Table 3: Analytical results of the NMC and EC in fermented food and beverages

Sample	State	Storage time(yr)	Unit	Measured Conc(μg/kg)	
				MC	EC
Red pepper paste-1	Solid	4	μg kg^{-1}	nd	0.7
Red pepper paste-2	Solid	3	μg kg^{-1}	nd	1.3
Red pepper paste-3	Solid	3	μg kg^{-1}	nd	1.9
Red pepper paste-4	Solid	2	μg kg^{-1}	nd	1.8
Red pepper paste-5	Solid	1	μg kg^{-1}	nd	2.3
Red pepper paste-6	Solid	1	μg kg^{-1}	nd	0.9
Soybean paste-1	Solid	4	μg kg^{-1}	nd	0.9
Soybean paste-2	Solid	3	μg kg^{-1}	nd	1.5
Soybean paste-3	Solid	2	μg kg^{-1}	nd	1.2
Soybean paste-4	Solid	4	μg kg^{-1}	nd	1.7
Soy sauce-1	Liquid	2	μg L^{-1}	nd	1.3
Soy sauce-2	Liquid	1	μg L^{-1}	nd	0.4
Soy sauce-3	Liquid	1	μg L^{-1}	0.4	8.9
Soy sauce-4	Liquid	1	μg L^{-1}	nd	1.8
Soy sauce-5	Liquid	1	μg L^{-1}	nd	0.8

Soy sauce-6	Liquid	1	µg L⁻¹	nd	1.3
Beer	Liquid	-	µg L⁻¹	nd	3.9
Soju(white distilled liquor)	Liquid	-	µg L⁻¹	nd	4.8
Jeongjong(refined rice wine)	Liquid	-	µg L⁻¹	0.5	8.3
Soju(distilled liquor)	Liquid	-	µg L⁻¹	nd	nd
Makgeolli(raw rice wine)-1	Liquid	-	µg L⁻¹	nd	nd
Makgeolli(raw rice wine)-2	Liquid	-	µg L⁻¹	0.5	6.9
Makgeolli(raw rice wine)-3	Liquid	-	µg L⁻¹	nd	6.0
Fruit liquor-1	Liquid	-	µg L⁻¹	nd	4.1
Fruit liquor-2	Liquid	-	µg L⁻¹	0.6	78.7
Fruit liquor-3	Liquid	-	µg L⁻¹	0.7	68.6
Fruit liquor-4	Liquid	-	µg L⁻¹	0.8	85.8

The correlations between the levels of EC and MC in beverages also correlated well with each another ($r^2=0.69$, $P=0.001$) due to the similar formation mechanisms. It is suggested that MC is also formed by the reaction of urea with methanol.

CONCLUSIONS

In this paper, a simple, sensitive method to detect MC and EC in several solid foods and liquid foods is presented based on derivatization with 9-xanthydrol for 10 min at an ambient temperature. Using 2.0 g for solid food and liquid food, the LODs of the MC and EC were 0.11 and 0.12 µg/kg, respectively, and the LOQs of the MC and EC were 0.35 and 0.38 µg/kg, respectively. The accuracy and precision of the assay were acceptable: the relative standard deviation was less than 12%. The concentrations of MC and EC in Korean traditional fermented foods were measured to be to 85.8 µg/kg. The natural levels of MC and EC found in these foods are not considered to pose a risk to human health.

AUTHORS' CONTRIBUTIONS

HSS initiated and prepared the draft. EYY conducted the extraction and method developments. All authors designed the study. All authors contributed to data analyses and to finalizing the manuscript. Both authors have read and approved the final version.

REFERENCES

1. Nout MJR: Fermented foods and food safety. *Food Res Int* 1994, 27:291–298.

2. Weber JV, Sharypov VI: 009. Ethyl carbamate in foods and beverages: a review. *Environ Chem Lett* 2009, 7:233–247.

3. IARC (International Agency for Research on Cancer): *Alcoholic beverage consumption and ethyl carbamate (urethane), international agency for research.* 96th edition. Geneva: World Health Organization; 2007. Available from URL: http://monographs.iarc.fr/ENG/Monographs/vol96/index.php. Accessed 15 August 2012

4. EUR-Lex: Commission recommendation of 2 march 2010 on the prevention and reduction of ethyl carbamate contamination in stone fruit spirits and stone fruit marc spirits and on the monitoring of ethyl carbamate levels in these beverages. *Off J Eur Uni*2010, 53:53–57.

5. Funch F, Lisbjerg S: Analysis of ethyl carbamate in alcoholic beverages. *Lebens Wissen Technol* 1988, 186:29–32.

6. Kim YK, Lee KE, Chung HJ: Determination of ethyl carbamate in some fermented Korean foods and beverages. *Food Addit Contam*2000, 17:469–475.

7. Foureman P, Mason JM, Valencia R, Zimmering S: Chemical mutagenesis testing in drosophila. *Environ Mol Mutagen* 1994, 23:51–63.

8. NIH (National Institutes of Health): *U.S. Department of health and human services public health service, toxicology and carcinogenesis studies of methyl carbamate in rats and mice.* 1987. U.S. DEPARTMENT OF HEALTH AND HUMAN SERVICES, Public Health Service National Institutes of Health, NIH Publication No.88–2584, http://ntp.niehs.nih.gov/ntp/htdocs/lt_rpts/tr328.pdf.

9. Sen NP, Seaman SW, Weber D: A method for the determination of methyl carbamate and ethyl carbamate in wines. *Food Addit Contam* 1992, 9:149–160.

10. Herbert P, Santos L, Bastos M, Barros M, Alves A: New HPLC method to determine ethyl carbamate in alcoholic beverages using fluorescence detection. *J Food Sci* 2002, 67:1616–1625.

11. Fu ML, Liu J, Chen QH: Determination of ethyl carbamate in Chinese yellow rice wine using high-performance liquid chromatography with fluorescence detection. *Int J Food Sci Technol* 2010, 45:1297–1302.

12. Madrera RR, Valles BS: Determination of ethyl carbamate in cider spirits by HPLC-FLD. *Food Control* 2009, 20:139–143.

13. Deak E, Gyepes A, Stefanovits-Banyai E: Determination of ethyl carbamate in palinka spirits by liquid chromatography-electrospray tandem mass spectrometry after derivatization. *Food Res Int* 2010, 43:2452–2455.

14. Ye CW, Zhang XN, Gao YL: Multiple headspace solid-phase microextraction after matrix modification for avoiding matrix effect in the determination of ethyl carbamate in bread. *Anal Chim Acta* 2012, 710:75–80.

15. Ye CW, Zhang XN, Huang JY: Multiple headspace solid-phase microextraction of ethyl carbamate from different alcoholic beverages employing drying agent based matrix modification. *J Chromatogr A* 2011, 1218:5063–5070.

16. Ballesteros E, Gallego M, Valcarcel M: Automatic determination of N-methylcarbamate pesticides by using a liquid-liquid extractor derivatization module coupled on-line to a gas chromatograph equipped with a flame ionization detector. *J Chromatogr* 1993,633:169–176.

17. Ya-Ping M, Fu-Quan D, Dai-Zhou C: Determination of ethyl carbamate in alcoholic beverages by capillary multi-dimensional gas chromatography with thermionic specific detection. *J Chromatogr A* 1995, 695:259–265.

18. Baffa Junior JC, Mendonca RC, Pereira JM: Ethyl-carbamate determination by gas chromatography–mass spectrometry at different stages of production of a traditional Brazilian spirit. *Food Chem* 2011, 129:1383–1387.

19. Hong KP, Kang YS, Jung DC: Exposure to ethyl carbamate by consumption of alcoholic beverages imported in Korea. *Food Sci Biotechnol* 2007, 16:975–980.

20. Wai-Cheung CS, Ping KK, Ling-Sze CB: Determination of ethyl carbamate in fermented foods by GC-HRMS. *Chromatographia* 2010,72:571–575.

21. Lachenmeier DW, Frank W, Kuballa T: Application of tandem mass spectrometry combined with gas chromatography to the routine analysis of ethyl carbamate in stone-fruit spirits. *Rapid Commun Mass Sp* 2005, 19:108–112.

22. Lachenmeier DW, Nerlich U, Kuballa T: Automated determination of ethyl carbamate in stone-fruit spirits using headspace solid-phase microextraction and gas chromatography-tandem mass spectrometry. *J Chromatogr A* 2006, 1108:116–120.

23. Whiton RS, Zoecklein BW: Determination of ethyl carbamate in wine by solid-phase microextraction and gas chromatography/mass spectrometry.

Am J Enol Viticult 2002, 53:60–63.

24. Woo IS, Kim IH, Yun UJ: An improved method for determination of ethyl carbamate in Korean traditional rice wine. *J Ind Microbiol Biot* 2001, 26:363–368.

25. Fauhl C, Catsburg R, Wittkowski R: Determination of ethyl carbamate in soy sauces. *Food Chem* 1993, 48:313–316.

26. Giachetti C, Assandri A, Zanolo G: Gas chromatographic-mass spectrometric determination of ethyl carbamate as the xanthylamide derivative in Italian aqua vitae (grappa) samples. *J Chromatogr* 1991, 585:111–115.

27. Lei FF, Zhang XN, Gao YL: Multiple headspace solid-phase microextraction using a new fiber for avoiding matrix interferences in the quantitative determination of ethyl carbamate in pickles. *J Sep Sci* 2012, 35:1152–1159.

28. Ubeda C, Balsera C, Troncoso AM: Validation of an analytical method for the determination of ethyl carbamate in vinegars. *Talanta*2012, 89:178–182.

Chapter 8

SCREENING AND CHARACTERIZATION OF EXTRACELLULAR POLYSACCHARIDES PRODUCED BY LEUCONOSTOC KIMCHII ISOLATED FROM TRADITIONAL FERMENTED PULQUE BEVERAGE

Ingrid Torres-Rodríguez, María Elena Rodríguez-Alegría, Alfonso Miranda-Molina, Martha Giles-Gómez, Rodrigo Conca Morales, Agustín López-Munguía, Francisco Bolívar and Adelfo Escalante

Departamento de Ingeniería Celular y Biocatálisis, Instituto de Biotecnología, Universidad Nacional Autónoma de México (UNAM)

ABSTRACT

We report the screening and characterization of EPS produced by LAB identified as *Leuconostoc kimchii* isolated from *pulque*, a traditional Mexican fermented, non-distilled alcoholic beverage produced by the fermentation of the sap extracted from several (*Agave*) maguey species. EPS-producing LAB constitutes an abundant bacterial group relative to total LAB present in sap and during fermentation, however, only two EPS-producing colony phenotypes (EPSA and EPSB, respectively) were detected and isolated concluding that despite the high number of polymer-producing LAB their phenotypic diversity is low. Scanning electron microcopy analysis during EPS-producing conditions revealed that both types of EPS form a uniform porous structure surrounding the bacterial cells. The structural characterization of the soluble and cell-associated EPS fractions of each polymer by enzymatic and acid hydrolysis, as by 1D- and 2D-NMR, showed that polymers produced by the soluble and cell-associated fractions of EPSA strain are dextrans consisting of a linear backbone of linked α-(1→6) Glc$p$ in the main chain with α-(1→2) and α-(1→3)-linked branches. The polymer produced by the soluble fraction of EPSB strain was identified as a class 1 dextran with a linear backbone containing consecutive α-(1→6)-linked D-glucopyranosyl units with few α-(1→3)-linked branches, whereas the cell-associated EPS is a polymer mixture consisting of a levan composed of linear chains of (2→6)-linked β-D-fructofuranosyl residues with β-(2→6) connections, and a class 1 dextran. According to our knowledge this is the first report of dextrans and a levan including their structural characterization produced by *L. kimchii* isolated from a traditional fermented source.

BACKGROUND

The analysis of the microbial diversity in traditional fermented beverages, cereal doughs and vegetables has revealed the presence of a remarkable diversity of LAB involved in the development of the characteristic sensorial properties of fermented foods (Giraffa 2004). A wide diversity of EPS and genes encoding biosynthetic enzymes from naturally occurring LAB in traditional fermented foods and beverages have been extensively studied for their role in the physicochemical and sensorial characteristics of final fermented products (viscosifying, stabilizing or water-binding agents). This has led to the discovery of a remarkable structural diversity of EPS produced by LAB, particularly of the genus *Leuconostoc* and *Weissella* (Chellapandian et al. 1998; Uzochukwu et al. 2001; Olivares-Illana et al. 2002; Uzochukwu et al. 2002; van Hijum et al. 2006; Eom et al. 2007; Van der Meulen et al. 2007; Bauer et al. 2009; Bounaix et al. 2009; Bounaix et al. 2010; Amari et al.2012; Vasileva et al. 2012). EPS have received additional attention as valuable products because of their potential economic applications that include natural, safe-food additives or natural functional food ingredients for their properties as soluble fiber and prebiotics, and the possibility that they can replace or reduce the use of hydrocolloids (Giraffa 2004; Tieking et al. 2005; Vu et al. 2009; Leemhuis et al. 2013).

Pulque is a traditional Mexican, non-distilled alcoholic fermented beverage currently produced and consumed mainly in the Central Mexico Plateau. It is obtained from the fermentation of a fresh sap, known as *aguamiel*, which is extracted from several maguey (*Agave*) species, such as *Agave atrovensis* and *A. americana*. The production of this traditional beverage requires for the freshly collected *aguamiel* to be deposited in large open containers where previously fermented *pulque* acts as seed for a new batch. Fermentation time varies from a few hours, overnight, or even for several days. Fermented *pulque* is gradually retired from the container but always leaving a residual volume of the fermented beverage to start a new fermentation. The viscosity resultant from EPS synthesis and the alcohol content of the beverage are the main parameters used to determine the extent of fermentation as they define its distinctive sensorial properties (Escalante et al. 2008; Escalante et al. 2012). The LAB *L. mesenteroides* has been traditionally considered one of the most important microorganisms during *pulque* fermentation as a result of its ability to synthesize EPS, primarily dextrans produced by a glucosyltransferase from sucrose present in *aguamiel* and *pulque* (Sanchez-Marroquin and Hope 1953; Chellapandian et al. 1998; Escalante et al. 2012). Although EPS production by *Leuconostoc* species isolated from*pulque* was first reported in 1953, no detailed information was available concerning the properties or structure of these polymers (Sanchez-Marroquin and Hope 1953). Nevertheless, the structure of an EPS produced by *L. mesenteroides* strain IBT-PQ isolated from *pulque*, revealed the presence of a soluble linear dextran with glucose molecules linked primarily by α-(1→6) bonds with branching from α-(1→3) bonds, in a 4:1 ratio, and was produced by a cell-associated dextransucrase displaying a similar biochemical behavior to that reported for this enzyme obtained from the industrial strain *L. mesenteroides* NRRL B512F (Chellapandian et al. 1998).

We have previously reported a great LAB diversity in *aguamiel* and *pulque* samples from different geographical origins, that is composed mainly of *Lactobacillus* and *Leuconostoc* species (Escalante et al. 2004). Among them *L. citreum* and *L. kimchii* were reported for the first time to be the most abundant LAB present in *aguamiel* and during the early stages of *pulque* fermentation (Escalante et al. 2008). This remarkable abundance of *L. citreum* and *L. kimchii* in *aguamiel* and *pulque*, suggests the presence of a possible diversity of EPS produced by these LAB during fermentation contributing to the final sensorial properties of the beverage. The aim of the present work was to screen and characterize the EPS diversity associated to the LAB *L. kimchii* isolated during traditional *pulque* fermentation. Screening of LAB was based in their ability to produce EPS from sucrose; this method resulted in the identification of two unique EPS-producing colony types. Structural characterization of these polymers including SEM during EPS production conditions, enzymatic and acid hydrolysis, as 1D- and 2D- ^{1}H- and ^{13}C-NMR allowed the detection of dextran and levan polymers produced by this LAB.

RESULTS AND DISCUSSION

Isolation and identification of EPS-producing LAB

The total EPS-producing LAB CFU/mL detected by growth on APTS plates was 50% in previously fermented *pulque*, 47% in *aguamiel*, 50.8% at T0, 70.4% at T3, and 50% at the end of the fermentation (T6), compared to the total LAB CFU/mL grown on APT plates. These results indicated a great abundance of EPS-producing LAB in *pulque*, *aguamiel*, and during the fermentation process. The visual analysis of purified EPS-producing colonies for unique morphology grown on APTS, allowed the identification of only two EPS colony phenotypes, leading to the conclusion that despite the high number of polymer-producing LAB, phenotypic diversity is low. The fastest growing colony of each EPS type was selected and designated as EPSA (compact colony morphology) and EPSB (creamy colony morphology) (Figure 1).

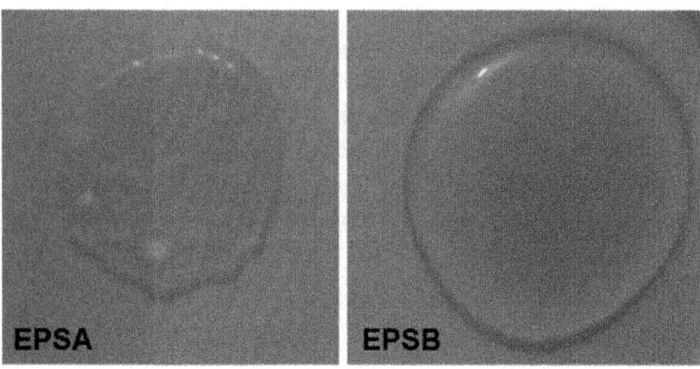

Figure 1: Phenotypic traits of EPS produced by *pulque* isolated *Leuconostoc kimchii* strains. Compact left colony or EPSA and creamy right colony or EPSB grown on APT plates supplemented with 20% sucrose.

Analysis of the 16S rDNA sequence of strains EPSA and EPSB in the non-redundant database of the NCBI archive database showed that the closest matching sequences found, corresponded to *L. kimchii* strain IMSNU 11154 (an isolated LAB from Korean *kimchi*), *L. palmae*strain TMW 2.694, and diverse *L. citreum* 16S rDNA sequences in both cases. To precise identity of EPS isolates, a phylogenetic analysis was performed using several *Leuconostoc* 16S rDNA reference sequences retrieved from the GenBank database, such as *L. citreum, L. kimchii, L. mesenteroides,* and *L. palmae* including type strains. The resulting neighbor-joining tree clearly demonstrated that the 16S rDNA sequences of the EPSA and EPSB isolates were placed in a cluster in which only *L. kimchii* sequences were included. However, the 16S rDNA sequence from *L. palmae* strain: TMW 2.694, isolated from palm wine (Ehrmann et al. 2009), was placed in the closest single terminal node separated from the well-defined clusters of the 16S rDNA sequences from *L. kimchii, L. citreum,* and *L. mesenteroides* (Figure 2). The 16S rDNA sequences corresponding to the EPSA and EPSB *L. kimchii* strains isolated from *pulque* were deposited into the GenBank database and accession numbers KC424437 and KC424438 were assigned to strains EPSA and EPSB, respectively. *L. kimchii* strain IMSNU 11154 was originally isolated from traditional Korean *kimchi,* a traditional fermented vegetable food (Kim et al. 2000). Complete genome sequencing of the EPS-producing *L. kimchii* strain IMSNU 11154 was reported and demonstrated the presence of genes corresponding to GTFs capable to produce dextran from sucrose (Oh et al. 2010).

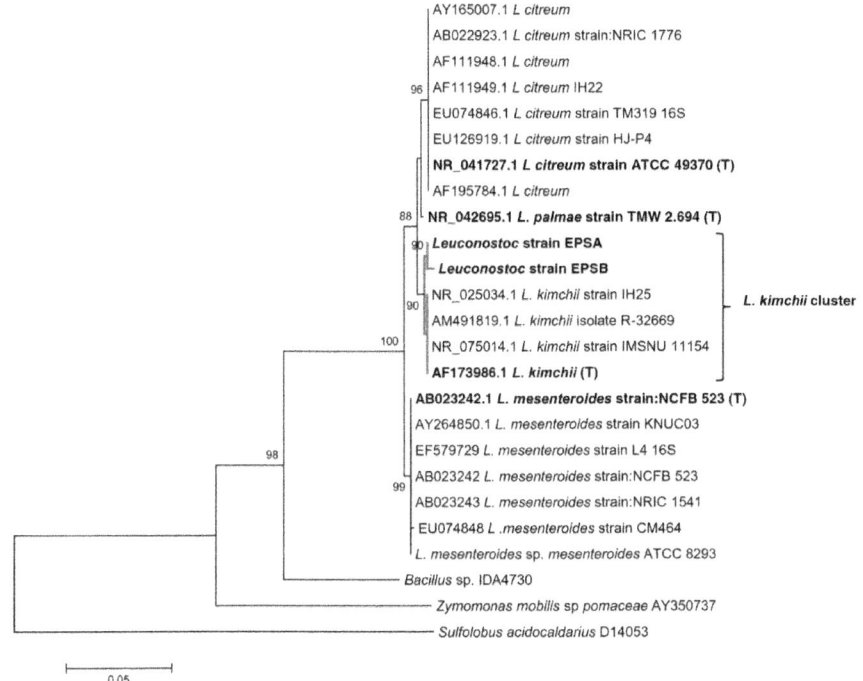

Figure 2: Phylogenetic tree of 16S rDNA sequences of EPSA and EPSB isolates and 16S rDNA reference sequences. Some 16S rDNA sequences of *L. citreum, L. palmae, L.*

kimchii, and *L. mesenteroides* strains deposited in the GenBank database including type strains of each genus (bold) are included as references. Accession numbers of reference sequences are indicated. The 16S rDNA sequences of *Sulfolobus acidocaldarius* and *Zymomonas mobilis* are included as outgroups. The percentage of 1000 bootstrap samplings supporting each topological element in the neighbor-joining analysis is indicated. No values are given for groups with bootstrap values less than 80%. *L. kimchii* cluster including EPSA and EPSB sequences is highlighted.

Scanning electronic microscopy of EPS-producing LAB

The SEM observation of EPS-producing colonies of EPSA and EPSB growing on APTS plates showed the presence of porous -"cocoon-like" structures associated with the cells (cocci) in both strains. These structures surround the bacterial cells and have different sizes depending on the colony, with larger structures observed for EPSA (Figure 3).

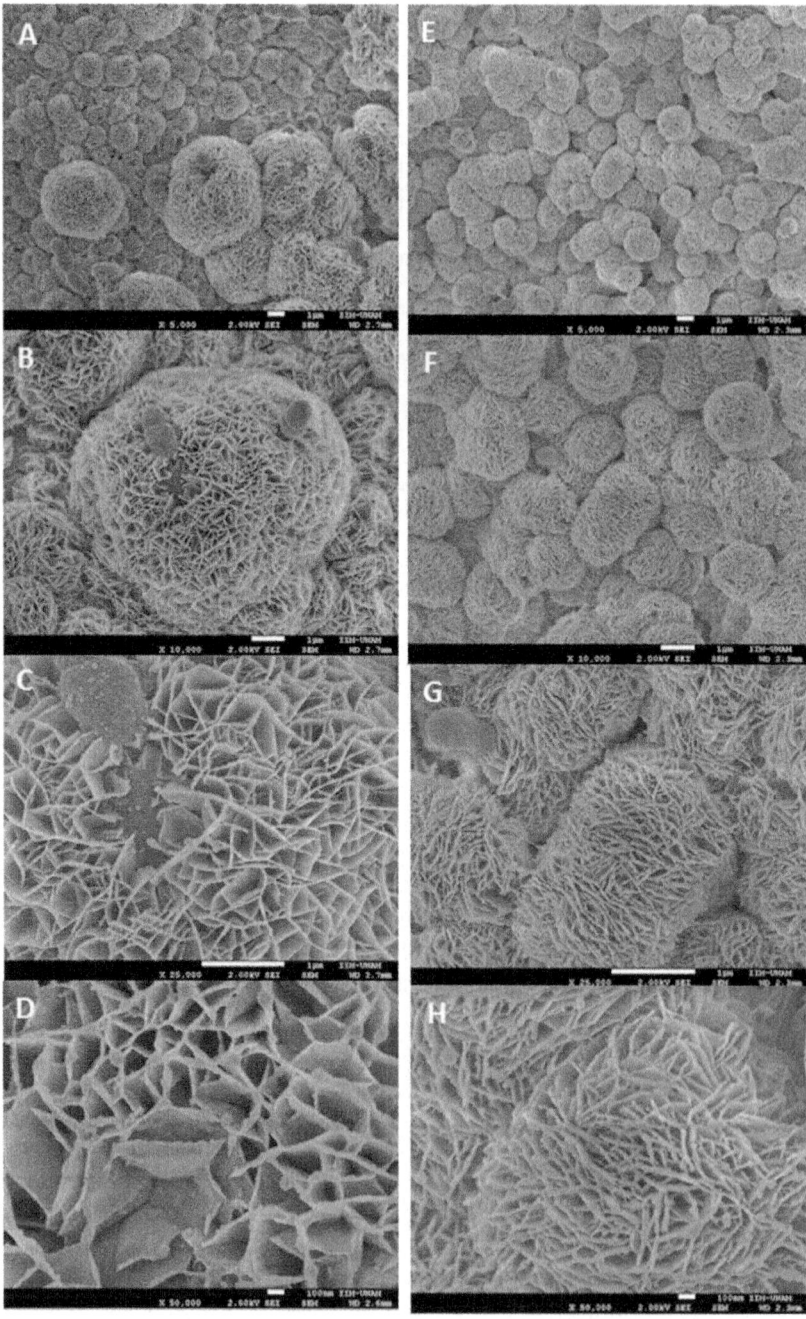

Figure 3: Scanning electron microscopy analysis of EPS-producing *Leuconostoc kimchii*. Images **A-D**, correspond to different fields of EPSA producing strain at 5 000X, 10 000X, 25 000X, and 50 000X, respectively. Images **E-H**,

correspond to different fields of EPSB producing strain at 5 000X, 10 000X, 25 000X, and 50 000X, respectively.

Although for both strains a surrounding hollow porous structure is observed, the polymer matrix produced by the EPSB strain appears more compact when compared to the EPSA polymer matrix. No other dextran produced by LAB has been previously analyzed by SEM, although a rather similar porous structure was previously reported for the dried cell-free insoluble dextran produced by *L. mesenteroides* NRRL B-1149 (Shukla et al. 2011).

Evidence of glycosyltransferase activity

Total GTF activity was determined either in the cell culture supernatant or associated to the cell fraction of cultures for *L. kimchii* EPSA and EPSB isolates on APTS broth. For the EPSA isolate, a maximum of 3.2 U/L was observed in the supernatant after 6 h of fermentation, when the pH decreased to 4.4. GTF activity was also found to be associated with the cultures's cell fraction showing a maximum activity of 1.8 U/L after 4 h. When the culture supernatant reached pH 5.5; the insoluble activity was approximately 36% of the total GTF activity produced by the cells. The total GTF activity was found higher in a similar analysis performed with the EPSB strain. In this case the soluble GTF activity reached 11.21 U/L after 6 h of fermentation when the culture supernatant pH was 4.6, while the cell-associated GTF activity reached a maximum of 5.57 U/L after 6 h, representing 33% of the total GTF activity. These culture conditions were selected to produce soluble and cell-associated polymers for structural characterization. It was also found that the isolated strains EPSA and EPSB showed higher residual total GTF activity in the temperature range of 30°C - 40°C, both in soluble and cell-associated fractions. Regarding the pH, both EPSA and EPSB-soluble fractions showed higher residual total GTF activity at pH 5.4, while both EPSA and EPSB cell-associated fractions showed higher residual total GTF activity at pH 6.0.

The GTF activity level associated only to the soluble or cell-associated fractions, or in both fractions ranging 0.8 to 2 U/mL was previously detected in several *Leuconostoc* and *Weissella* isolated strains from fermented sources (Bounaix et al. 2009; Vasileva et al. 2012). Our results showed that the GTF activity detected in the soluble fractions of *L. kimchii* EPSA and EPSB strains, was higher than the activity found in the cell-associated fractions was but low when compared with previous reports (Bounaix et al. 2010; Bounaix et al. 2009; Vasileva et al.2012). EPS production by LAB from sucrose was reported that depends on diverse factors such as the cultivation conditions (aerobic, anaerobic and temperature) and media composition (liquid or solid media, rich media such as MRS or APT or mineral media supplemented with phosphate sources, tryptone or yeast extract) (Maina et al. 2008; Minervini et al. 2010). While further investigation is required to determine the fine enzymatic properties of GTF responsibles for soluble and cell-associated EPS production in both EPSA and EPSB *L. kimchii* isolates, to our knowledge, these results provide evidence that GTF is involved in EPS production by this LAB isolated from a traditional alcoholic fermented beverage for the first time. Previous studies involving both glucosyltransferase and fructosyltransferase characterization have been performed particularly in diverse EPS-producing *Leuconostoc, Weissella,* and *Lactobacillus* species isolated from fermented beverages, cereal doughs, and vegetables (Table 1).

Table 1: EPS produced from sucrose by LAB isolated from traditional fermented products

EPS type	EPS structure (producing LAB)	Source	Reference
Dextran	Linear backbone linked mainly in α-(1→6) D-Glc*p* with branching in α-(1→3) D-Glc*p*produced by a cell-associated GTF (*L. mesenteroides* IBT-PQ strain)	*Pulque*(fermented alcoholic beverage)	(Chellapandian et al. 1998)
Dextran[a]	Linear backbone linked mainly in α-(1→6) D-Glc*p* with branching in α-(1→2) (*L. mesenteroides*).	Palm wine	(Uzochukwu et al. 2001)
	Linear backbone linked mainly in α-(1→6) D-Glc*p* with branching in α-(1→3) with minor α-(1→4) linked branches (*L. dextranicum*).		
	Linear backbone linked mainly in α-(1→6) D-Glc*p* with branching in α-(1→3) (*Lactobacillus* spp. AW strain)		
Fructan	Inuline-like structure with β-(2→1) glycosidic linkages produced by a cell-associated fructosyltransferase (*L. citreum*)	*Pozol* (maize fermented dough)	(Olivares-Illana et al. 2002)
Dextran[a]	Dextran type I containing α-(1→2) D-Glc*p* and α-(1→3) linked branches (*L. citreum*VTT E-93497 strain).	Malting process	(Maina et al.2008)
	Dextran type I containing few α-(1→3) linked branches (*W. confusa* VTT E-90392 strain)	Soured carrot mash	

Dextran	Dextran type I linked mainly in α-(1→6) D-Glc*p* with few α-(1→3) linked branches produced by soluble GTF (Several *L. mesenteroides* and *Weissella* spp isolates)	Traditional French wheat sourdough	(Bounaix et al.2009)
	Dextran type I linked mainly in α-(1→6) D-Glc*p* with α-(1→2) linked branches produced by soluble GTF (several *L. mesenteroides* and *L. citreum* isolates)		
	Dextran type I linked mainly in α-(1→6) D-Glc*p* with high α-(1→3) linked branches produced by soluble GTF (Several *L. citreum* isolates)		
Dextran/levan mixture	Dextran type I linked mainly in α-(1→6) D-Glc*p* with high α-(1→3) linked branches produced by soluble and cell-associated GTF mixed with a levan (several *L. mesenteroides* isolates)	Traditional French wheat sourdough	(Bounaix et al.2009)
Dextran	Dextran type I linked mainly in α-(1→6) D-Glc*p* with few α-(1→2) linked branches (*W. confusa* LBAE C39-2)	Traditional French wheat sourdough	(Amari et al.2012)
Dextran[a]	Linear backbone linked mainly in α-(1→6) D-Glc*p* (*Lactobacillus curvatus* 69B2 strain and *Leuconostoc lactis* 95A strain)	Wheat sourdough	(Palomba et al.2012)
Dextran	Linear backbone linked mainly in α-(1→6) D-Glc*p* with α-(1→3) linked branches produced by soluble and cell-associated GTF (*L. mesenteroides*)	Bulgarian fermented vegetables	(Vasileva et al.2012)
Dextran/levan mixture	Dextran linked mainly in α-(1→6) D-Glc*p* (56%) and a levan (44%) (*L. mesenteroides*URE and Lm 17 strains)	Bulgarian fermented vegetables	(Vasileva et al.2012)
Dextran[a]	Linear backbone linked mainly in α-(1→6) D-Glc*p* (*L. mesenteroides* and *W. confusa*)	*Kimchi*(traditional fermented vegetable food)	(Park et al.2013)

Dextran	Linear backbone linked mainly in α-(1→6) D-Glc$p$ with α-(1→2) and α-(1→3) linked branches produced by a soluble and cell-associated GTF (*L. kimchii* EPSA strain)	*Pulque*	This work
	Dextran type I containing few α-(1→3) linked branches produced by a soluble GTF (*L. kimchii* EPSB strain)		
Levan/dextran mixture	Polymer mixture composed by linear chains of (2→6)-linked β-D-fructofuranosyl residues with connections β-(2→6) (79%), and a dextran Type I (21%) produced by the cell-associated GTF fraction (*L. kimchii* EPSB strain)	*Pulque*	This work

[a]No information about GTF producing enzymes is provided.

EPSA and EPSB Characterization

Hydrolysis of soluble and cell-associated EPS fractions

Enzymatic hydrolysis assays of the cell-associated and soluble EPSA and EPSB polymer samples showed that polymers produced by the EPSA-soluble and cell-associated GTF were hydrolyzed only by dextranase but not by endolevanase, inulinase or Fructozyme®. These results indicated the presence of a dextran with α-(1→6) D-Glc$p$ linkages (Figure 4), similar to dextrans that result from GTF synthesis. Similarly, the polymer produced by the EPSB-soluble fraction was also only hydrolyzed by dextranase, demonstrating the presence of α-(1→6) D-Glc$p$ linkages as in dextran. Nevertheless, the polymer produced by the EPSB cell-associated fraction was degraded by dextranase, endolevanase and Fructozyme® but not by inulinase, indicating the presence of a polymer mixture composed of a linear dextran and a levan. Furthermore, acid treatment of EPS with H_2SO_4 yielded hydrolysis giving the expected monomers.

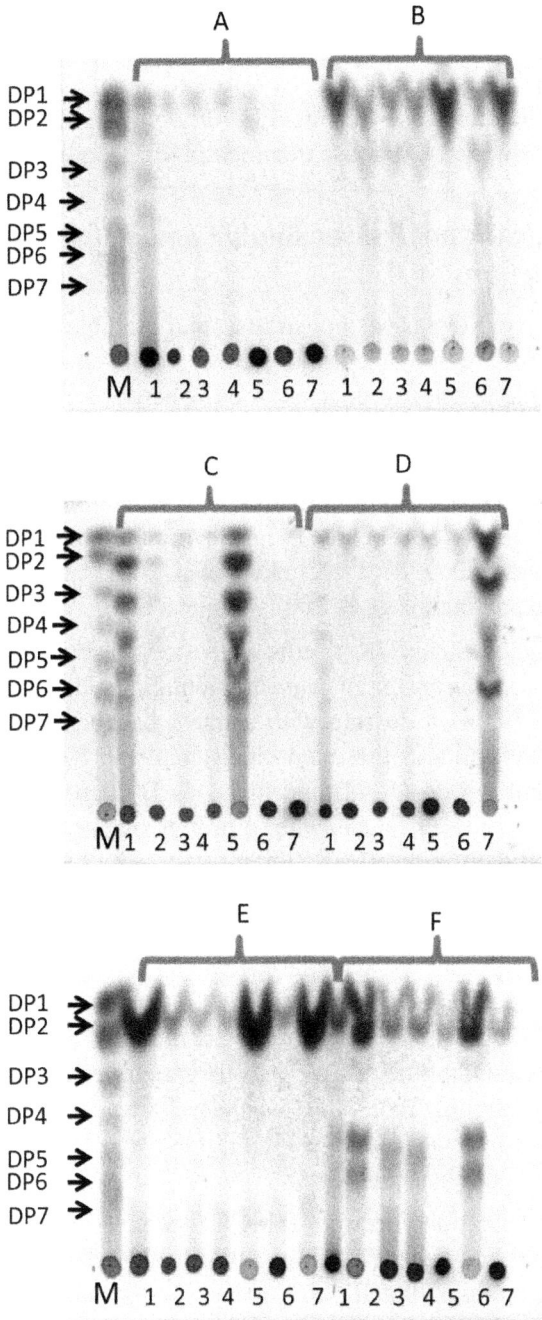

Figure 4: Enzymatic and acid hydrolysis of EPSA and EPSB polymers. 1. EPSB cell-associated fraction; 2. EPSB-soluble fraction; 3. EPSA cell-associ-

ated fraction; 4. EPSA-soluble fraction; 5. Levan from *B. subtilis*; 6. Dextran from *L. mesenteroides* B-512; 7. Inulin from *L. citreum*. M. Glucose + fructose + maltodextrins standards; **A**. Non-treated samples; **B**. Acid hydrolysis treatment; **C**. Endolevanase treatment; **D**. Endoinulinase treatment; **E**. Fructozyme® treatment; **F**. Dextranase treatment. DP = Degree of polymerization.

^{1}H- and ^{13}C-NMR analysis of soluble and cell-associated EPSA and EPSB fractions

As enzymatic hydrolysis demonstrated that soluble and cell-associated fractions of EPSA are the same polymer, only the ^{1}H- and ^{13}C-NMR spectra of the EPSA-soluble fraction was determined, whereas for the soluble and cell-associated polymers of *L. kimchii* EPSB strain both spectra were obtained. Resultant NMR spectra were compared with those previously reported for diverse polymers produced by several *L. mesenteroides*, *L. citreum*, and *Weissella* sp. strains (Colson et al. 1974; Seymour et al. 1976; Seymour et al. 1979; Shimamura et al. 1987; Uzochukwu et al. 2001; Uzochukwu et al. 2002; Maina et al. 2008; Bounaix et al. 2009; Vasileva et al. 2012).

Comparison of the NMR results with previously reported spectra for dextrans produced during palm wine fermentation (Uzochukwu et al.2002) revealed that EPSA is a dextran with a linear backbone of linked α-(1→6) D-glucopyranosyl units in the main chain confirming enzymatic hydrolysis assay results, but with α-(1→2) and α-(1→3) D-Glc*p* linked branches. The ^{1}H-NMR spectra showed four anomeric proton signals at δ 4.92, 5.06, 5.13, and 5.29 labeled A-D, respectively (Figure 5A) corresponding to the α-(1→6) D-Glc*p*, the α-(1→2) branching D-Glc*p*, the 2,6 di-*O*-substituted α-D-Glc*p*, and the α-(1→3) D-Glc*p* units, respectively. The relative intensities of the A-D peaks were 15.30%, 12.41%, 16.86%, and 55.43%, respectively, resulting proportional to the degree of branching. HSQC of EPSA (Additional file 1) showed their attachment to the anomeric carbons at δ 97.64, 96.15, 95.30, and 99.47 corresponding respectively to ^{1}H anomeric signals at δ 4.92, 5.06, 5.13, and 5.29 (Figure 5B). These signals characteristically served as the starting point for the analysis of the ^{1}H–^{1}H COSY and TOCSY experiments for connectivities within the spin system.

The C-H correlation spectrum (HSQC) indicated that ^{13}C resonances at δ 65.48 split into two peaks (Additional file 1). The correlation peak showed the ^{1}H resonance at δ 3.70 and δ 3.93. The other ^{13}C peaks at δ 60.32, 60.47, and 60.65 showed their protons as a multiplet at δ 3.78 and δ 3.66 which corresponded to non-linked C-6. The previous characterization of dextrans from fermented palm wine showed that its backbone contains a bound C-6 with a chemical shift at δ 66.6 (Uzochukwu et al. 2002), whereas for EPSA polymer it was found at δ 65.48, resulting farthest downfield than a free C-6 at δ 61.4 (δ 60.32, 60.47, and 60.65 for EPSA polymer). Finally, the long-range correlation between H-3 (δ 3.81) and the anomeric carbon at δ 99.47 of EPSA polymer in the HMBC spectrum allowed to confirm the presence of α-(1→3)-linked branches (Additional file 1).

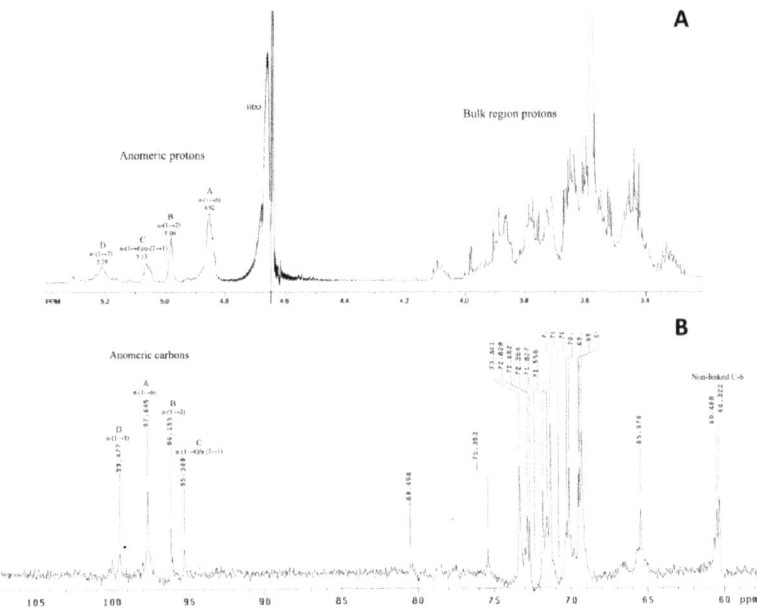

Figure 5: NMR spectra of EPSA soluble fraction. (A) ¹H-NMR spectrum. **(B)** ¹³C-NMR spectrum. Anomeric protons are labeled A–D according to the increasing chemical shifts.

Furthermore, in the δ 68–74 region of the ¹³C-NMR spectrum of EPSA, were found characteristic branching signals for linked C-2 and C-3 at branching points, which split in three peaks found in δ 71.83, 71.56, and 71.31; and in the region at δ 69.53, 69.46, and 69.31, respectively (Figure 5b). These data were consistent with previous results of spectra for the series of dextrans B1254, B1355S, and B1099L (Seymour et al. 1976). The split in the free C-6 around δ 60-61 represents C-6 in different chemical environments with two possible sources of branch-terminating residues: α-(1→2) and α-(1→3) branching. Free C-6 signals in EPSA polymer were detected at δ 60.32, 60.47, and 60.65, respectively and were associated with branching at C-2 and C-3, respectively, indicating that in EPSA polymer the linkage is present both as branch point and intra chain linkages as reported previously for palm wine dextran (Uzochukwu et al. 2002).

NMR spectra results for EPSB polymers corroborated that its soluble fraction is a dextran with a linear backbone containing consecutive α-(1→6)-linked D-glucopyranosyl units with few α-(1→3)-linked branches, structure characteristic of class 1 dextrans (Maina et al. 2008; Amari et al. 2012). The ¹H-NMR spectrum for EPSB-soluble fraction (Figure 6A), revealed anomeric signals characteristic of glucosyl residues linked through α-(1→6) (98.5%) and α-(1→3) (1.5%) linkages appeared at δ 4.97 and δ 5.3, respectively. The other protons appeared from δ 3.4 to 4.1: δ=3.55 (dd, 1H, J=3.2 Hz, J=9.8 Hz, H-2), 3.48 (t, 1H, J=9.2 Hz, J=9.6 Hz, H-3), 3.71 (t, 1H, J=9.2 Hz, J= 9.6 Hz, H-4), 3.90 (d, J=8.4, 1H, H-5), 3.74 (d, J=7.48, H-6a), 3.98 (d, J=9.76, 1H, H-6b). The ¹³C-NMR spectrum (Figure 6B) shows six signals, among them, those appearing at δ 97.56 and δ 65.41, corresponding to C-1 and C-6, which are involved in α-(1 → 6) linkages. The linkage carbon resonance at δ 65.41 showed two correlation peaks (δ 3.74 and δ 3.98) in the HSQC spectrum (Additional file 1). The four remaining signals observed at δ 73.28, 70.06, 69.39, and 71.29 corresponded to C-4, C-5, C-3, and C-2, respectively. Full assignments of the proton and carbon resonances were secured from the TOCSY, COSY, NOESY, and HMBC data.

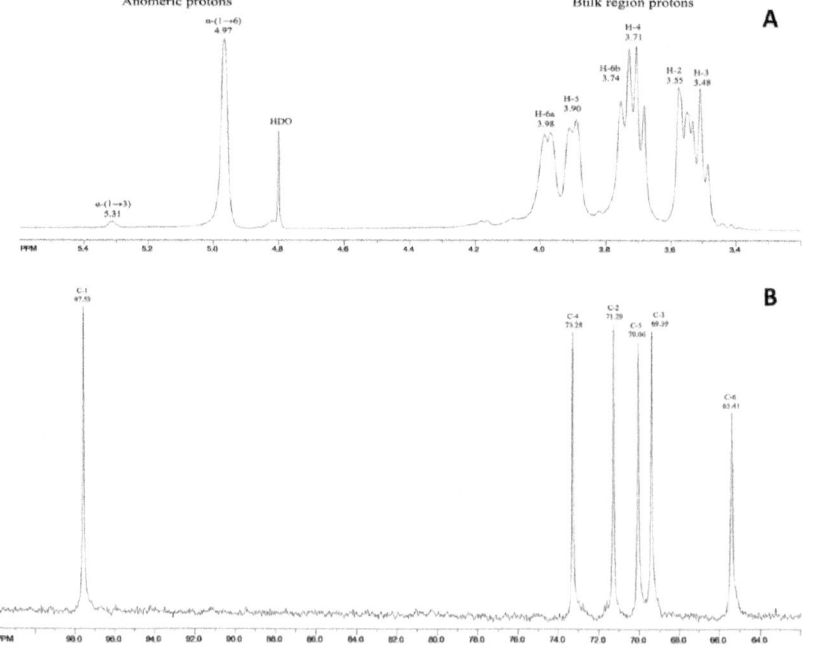

Figure 6: RMN spectra of EPSB-soluble fraction. **(A)** ¹H-NMR spectrum. **(B)** ¹³C-NMR spectrum.

These results are consistent with previous reported ^{1}H- and ^{13}C-NMR data for the dextran produced by *W. confusa* C39-2 strain with 97.6% α-(1 → 6) and 2.4% α-(1 → 3) linkages, respectively (Bounaix et al. 2009). The structure of the soluble EPSB dextran also resembled that of dextrans produced by *W. confusa* VTT E-90392 (0DSM 20194) (Maina et al. 2008) and *W. cibaria* CMGDEX3 (Ahmed et al. 2012).

In other hand, the ^{1}H- and ^{13}C-NMR chemical shifts observed suggest that the polymer from the cell-associated fraction of isolated strain EPSB is a mixture that consist of a levan (79%) composed of linear chains of (2→6)-linked β-D-fructofuranosyl residues with β-(2→6) linkages, and a polymer of class 1 dextran (21%). The assignments of methylene and methine carbons were determined by a DEPT analysis (Additional file 1). The ^{13}C-NMR spectrum gave three upfield resonances which were methylene (δ_c 59.83, 63.32, 65.46) while the remaining resonances were methines and one quaternary carbon (δ_c 104.13).

The ^{1}H-NMR spectrum showed intense signals associated to the fructose moieties in the levan fraction (Figure 7A), which were detected at δ 3.77 (d, J =12.6 Hz, H-6a), 3.44 (d, J =11.2 Hz, H-6b), δ 3.55 (d, J= 11.9 Hz, H-1a), δ 3.64 (d, J= 11.9 Hz, H-1b), δ 3.82 (ddd, J= 2.8 Hz, J= 7.7 Hz, J= 7.7 Hz, H-5), δ 3.96 (t, J= 7.7 Hz, J= 8.4 Hz, H-4), δ 4.06 (d, J= 8.4 Hz, H-3). The COSY spectrum was used to detect cross peaks between H-3/H-4, H-4/H-5, and H-5/H-6. The HMBC spectrum (Additional file 1) showed the presence of cross peaks between H3/C4, H4/C6, H4/C3, H1/C3, and H1/C2 as it was reported previously (Dahech et al. 2013). This result allowed to confirm the presence of β-(2→6) linkage between two fructofuranosyl moieties, but the β-(2→6) linkage was confirmed only by the presence of a downfield shifted signal at δ 63.32 (C-6) in the ^{13}C-NMR spectrum as reported previously (Tomašić et al. 1978). It has been shown that normally glycosylation induces a downfield shift of δ 4-10 (Newbrun and Baker 1968). In the ^{13}C-NMR spectrum (Figure 7B) of the EPSB cell-associated polymer were detected six main resonance shifts at δ 59.83 (C-1), δ 63.32 (C-6), δ 75.12 (C-4), δ 76.22 (C-3), δ 80.22 (C-5), and δ 104.13 (C-2) relating to a quaternary anomeric carbon, corresponding to levan. These results are also similar to those observed previously in levan produced by *L. mesenteroides* B-512 F (Morales-Arrieta et al. 2006) and *Streptococcus mutans* (Shimamura et al. 1987).

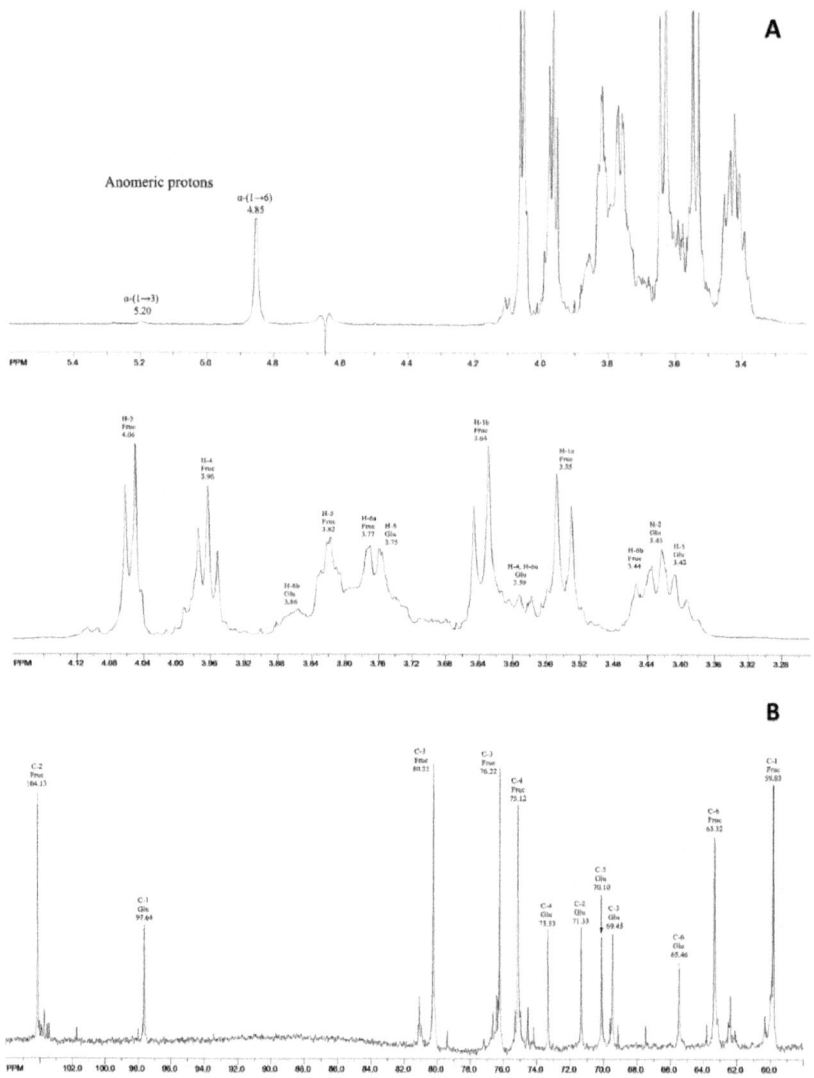

Figure 7: RMN spectra of EPSB-soluble and cell-associated fractions. (A) Full ¹H NMR spectrum (upper graph) and region between 4.2 and 3.1 nm (bottom graph). **(B)** ¹³C-NMR spectrum.

According to 1D- and 2D-NMR analysis, the second polymer in the mixture corresponded to dextran. The chemical shifts for the 6 carbons of the dextran were linked to those of the seven hydrogens as determining by using HSQC (Additional file 1). The characteristically downfield anomeric carbon (δ 97.64) and anomeric hydrogen (d, δ 4.85, $J = 1.4$ Hz) was considered as the starting point for the analysis of the ¹H-¹H COSY (Additional file 1) for connectivity within the spin system. The two diasterotopic H-6 hydrogens at δ 3.86 (m) and at δ 3.59 (m) in the ¹H spectrum were also useful markers and attached to the carbon at δ 65.46 in the ¹³C spectrum. The other protons appeared in the spectrum at: δ = 3.43 (H-2), 3.42 (H-3), 3.59 (H-4 and H-6a), 3.75 (H-5), 3.86 (H-6b). In the ¹³C-NMR spectrum, the peaks corresponding to individual carbons were identified at δ C-4 (73.33), C-2 (71.33), C-5 (70.10), C-3 (69.45), and C-6 (65.46), respectively (Figure 7B).

Polymers synthesized by GTF from sucrose by diverse LAB isolated from different traditional fermented sources include linear and branched dextrans, dextran and levan mixtures, and also inulin like polymers, indicating a high natural diversity of these EPS, particularly interesting considering the wide geographical distribution of these fermented products: palm wine from Africa, cereal sourdoughs and fermented vegetables from Europe, traditional *kimchi* from Korea, and maize sourdough and a fermented *pulque* beverage from Mexico (Table 1). The production of a dextran and levan polymers mixture was previously detected in the cell-associated fraction of cultures on MRS-sucrose of *L. mesenteroides* G15 strain isolated from a French cereal sourdough as revealed by ¹³C-NMR spectroscopy (Bounaix et al. 2009). Interestingly, detection of genes coding for glucosyltransferase and enzymes demonstrated their respective presence in isolated *L. mesenteroides* strains from Bulgarian fermented vegetables, but activity of putative fructosyltransferase was detected only in the presence of raffinose instead sucrose (Vasileva et al. 2012). Additionally, the production of an inulin-like polymer with β-(2→1) glycosidic linkages, was demonstrated by ¹³C-NMR in a cell-associated fructosyltrasferase in cultures containing 20 g/L sucrose of *L. citreum* strain CW isolated from*pozol*, a Mexican fermented maize sourdough (Olivares-Illana et al. 2002) (Table 1).

Several polymers described in Table 1 have relevant potential applications, as it has been proposed that branched dextrans possess prebiotic properties, demonstrating that these polymers and their oligodextrans may be substrates for butyric acid production by intestinal microbiota. Additionally, highly linear dextrans are considered highly soluble and were proposed to confer viscosity properties associated to their molecular weight (Maina et al. 2008). Furthermore, levans were shown to exhibit prebiotic effects, attracting attention for its antitumor properties, cholesterol-lowering properties and application such eco-friendly adhesive and as a promising bio-thickener in food industry (Patel et al. 2011). *Pulque* beverage is consumed directly from the fermentation vessel without further antimicrobial treatment such as pasteurization or filtration, as a consequence, living microorganisms are consumed. Additionally, *pulque* has been traditionally considered has a healthy beverage due to its nutrient content and used as a natural medicine to control several diseases (Escalante et al. 2012). Even though proposed healthy effects have been limited to traditional pharmacopoeia, further evidence on the possible probiotic effect of LAB involved in the fermentation process as the possible prebiotic activity of dextran or dextran and levan mixture produced by LAB such *L. kimchii* EPSA and EPSB strains requires further investigation.

CONCLUSIONS

EPSA and EPSB producing strains of *L. kimchii* isolated from traditional Mexican *pulque* and their polymers described in this contribution are to our knowledge the first study reporting the production and characterization of EPS produced by this LAB isolated from traditional fermented sources. *L. kimchii* was previously identified as the main LAB present in agave sap *aguamiel* and during the first hours of *pulque* fermentation. Although further studies are required to provide additional information concerning the enzymatic properties of the GTF responsibles for the synthesis of dextrans and levan by *L. kimchii* EPSA and EPSB strains, results presented in this contribution are highly relevant for *pulque*microbiology. In effect, while *L. mesenteroides* has traditionally been considered the main bacteria responsible for polymer production during*pulque* fermentation, this study provides new evidence regarding the diversity of EPS produced by LAB involved in the production of this traditional Mexican beverage.

METHODS

Aguamiel and *pulque* sampling

Fresh *aguamiel* and fermented *pulque* were collected from the town of Hiutzilac, State of Morelos, Mexico (Central Mexico Plateau) and transported to the laboratory, where controlled fermentations were performed by the addition of fresh sap to previously fermented *pulque* as described before (Escalante et al. 2008). Samples of *aguamiel*, after inoculation (mixing of sap with previously fermented *pulque*) (sample T0) and after 3 and 6 h of fermentation, were serially diluted in 0.1% tryptone water (DIFCO), and aliquots of 0.1 mL were plated on APT-agar (DIFCO) and APT-agar supplemented with 20% (w/v) sucrose (APTS). Plates were then incubated at 30°C for 24 h to determine the total count of LAB and EPS-producing LAB, respectively.

EPS-producing LAB isolation and identification

An average of 100 colonies of EPS-producing LAB from representative plates of *aguamiel* and each *pulque* sample were isolated, Gram stained, observed with a light microscope to verify colony purity, and tested for catalase activity. The purified isolated colonies were grown on APTS plates (4 colonies per plate), where the colony morphology and size were visually screened. Fast growing colonies were selected and visually screened for unique EPS-producing phenotypes resulting in the identification of two distinct EPS-producing colonies designated as EPSA and EPSB. Finally, selected isolates were conserved in 50% glycerol at -70°C and identified by 16S rDNA sequencing as described previously (Escalante et al. 2004; Escalante et al. 2008) by using fd1 + rd1 primer set resulting in the amplification of the entire 16S rRNA gene (Weisburg et al. 1991). Comparison was performed against the non-redundant nucleotide database using the online nucleotide BLAST application in the NCBI homepage. To corroborate their molecular identity, a phylogenetic tree was constructed using MEGA 6 software (Tamura et al. 2013) including reference and type 16S rDNA sequences retrieved from the NCBI database.

Scanning electron microscopy analysis of EPS-producing LAB

EPS-producing LAB were grown on APTS plates until a viscous morphology was observed. Collodion-coated electron microscope grids were placed on growing colonies, and maintained for four additional days at room temperature to promote LAB adhesion to the grid. The grids were then lifted from the APTS plates and placed on a porcelain spot dish for further processing. Fixation was performed by 1 h of exposure to 1% paraformaldehyde and 2.5% glutaraldehyde in pH 7.4, 0.1 M phosphate buffer, followed by washing with the same buffer, post-fixing with 1% osmium tetroxide in buffer for 30 min and dehydration with graded alcohols. Throughout the procedure, special care was taken when changing solutions to avoid EPS dispersion and damage to the grid surface. The samples were then dried under critical CO_2 (Sandri-780; Tousimis), gold coated (JFC-110 ion sputter; JEOL), and observed with a JEOL JSM-7600 F Field Emission SEM.

Growth kinetics of EPS-producing strains

Flask cultures were performed in 50 mL of APT broth at 30°C and 150 rpm to define optimum growth conditions for selected LAB isolates and to achieve adequate enzyme activities and EPS production. The OD_{600nm} was determined and adjusted to 0.2 before inoculation in a 2.8 L flask with 560 mL of APT broth and incubation for 6 – 8 h at 30°C and 150 rpm. Microbial growth was followed by the OD_{600nm} measurements and the pH was monitored each hour.

GTF activity assay

GTF activity was determined in cell-free supernatants and in harvested cell fractions by measuring the release of reducing sugars using the DNS technique (Sumner and Howell 1935) in the presence of 10% (w/v) sucrose in 100 mM, pH 5.4 acetate buffer. Incubation was performed in an Eppendorf Thermomixer Comfort device (Eppendorf, Hamburg) at 30°C. One activity unit (U) of total GTF activity is defined as the amount of enzyme producing 1 µmol of glucose per minute from GTF activity under the assay reaction conditions (Morales-Arrieta et al.2006). Protein concentration was determined by the Bradford method (Bradford 1976), using the Bio-Rad reagent and BSA (BioRad) as standards.

In vitro EPS production

EPS were produced from the supernatant and cell-associated enzyme fractions. For this purpose, the soluble GTF fraction was precipitated from the supernatant cultures by the addition of one volume of 50% (w/v) polyethylene glycol 5000 and centrifugation (7,000 × g, 5 min at 4°C). The pellet of cells containing the cell-associated GTF fraction, was washed three times and resuspended in a minimal volume of 0.1 mM, pH 5.4 acetate buffer (Quirasco et al. 1999). Polymer was produced in both fractions in the same buffer conditions containing 10% sucrose in final volume of 600 µL and incubated at 30°C and 100 rpm for 14 h (New Brunswick Scientific G24 shaker). The polymers produced from the cell-associated fraction were centrifuged as above to separate them from the cell whereas polymer produced from the supernatant fraction was precipitated by the addition of one volume of absolute ethanol and harvested by centrifugation (2,367 × g, 5 min at 4°C). Both polymers were resuspended in a minimum volume of distilled water and dialyzed against distilled water, eluted in a cellulose membrane (Mw cutoff of 12,400 Da SIGMA-ALDRICH), and dried in a LABCONO FreeZone 5.4 freeze-dryer.

Hydrolysis of EPS

Soluble and cell-associated ESPA and EPSB polymer fractions were subjected to enzyme hydrolysis with Fructozyme L® (kindly provided by Novozymes), a commercial enzymatic preparation obtained from *Aspergillus niger* combining endo- and exoinulinase activities toward β-(2→1) and endo-β-(2→6) D-Fru*p* linkages, in pH 5, 0.05 M, acetate buffer, at 60°C; Dextranase (Enzimas y Productos Químicos S.A. de C.V., Mexico City), which hydrolyses α-(1→6) D-Glc*p* linkages, in 0.05 M acetate buffer, pH 5, at 50°C; recombinant *E. coli* (*levB*) endolevanase from *Bacillus licheniformis*, which degrades endo-β-(2→6)

D-Fru*p* linkages (kindly provided by Dr. A. López-Munguía) in pH 6, 50 mM potassium phosphate buffer at 37°C; and endoinulinase Novozym® 960 (Batch KNN00120 from Novozymes), which degrades endo-β-(2→6) D-Fru*p* linkages, in pH 6, 50 mM potassium phosphate buffer at 60°C. Acid hydrolysis was performed with H_2SO_4 5% (v/v) at 95°C, 1 hr.

Enzymatic and acid hydrolysis assays were performed in a volume of 0.5 mL containing 2% w/v of EPS and were incubated for 12 h. Inulin from *L. citreum*, levan of *B. subtilis* (Ortiz-Soto et al. 2004) and dextran of *L. mesenteroides* B-512 PM 69,000 Da (Sigma) were used as controls. Hydrolyzed samples were analyzed by TLC using pre-coated TLC-sheets of Alugram® Xtra Sil G/UV254 with a mobile phase of 11:3:11:1 acetic acid:chloroform:ethanol:water mixture. The spots of products were detected by spraying with an alcoholic solution of α-naphthol and sulfuric acid, followed by heating at 120°C for 3 min.

^{1}H- and ^{13}C-NMR analysis of EPS

The freeze-dried samples of each polymer were dissolved in 1.0 mL of D_2O (Cambridge Isotope Laboratories, Inc.). NMR spectrum of soluble EPSA polymer was recorded on an Advance 700 MHz spectrometer (Varian) operating at 700 MHz for ^{1}H-NMR and 175 MHz for ^{13}C-NMR. NMR spectra of both EPSB fractions were recorded on an Advance 400 MHz spectrometer (Varian) operating at 400 MHz for ^{1}H-NMR and 100 MHz for ^{13}C-NMR. All measurements were performed at room temperature and were obtained using a 1.0 s relaxation delay. The data were acquired and processed using VNMRJ 2.0 software. Chemical shifts are listed in parts per million (ppm) and were made on the basis of ^{1}H-^{1}H COSY, ^{1}H-^{1}H TOCSY, NOESY, DEPT, HSQC, and HMBC spectral analysis (Additional file 1).

ACKNOWLEDGEMENTS

This project was supported by PAPIIT-UNAM project IN207914. We thank to Mercedes Enzaldo (IBT-UNAM); Alejandro Camacho Cruz and Ma. Antonieta Silva (Facultad de Química, UNAM), for their technical support; Dr. Fernando García Hernández (IFC-UNAM), for biological sample preparation for SEM observation; Dr. Omar Novelo Peralta (IIM-UNAM) for SEM support of EPS samples; and M.C. Silvia Marquina Bahena (CIQ, UAEM) for RMN sample analysis support; Oscar Martínez for the critical review of this document.

AUTHORS' CONTRIBUTIONS

ITR and MRA, characterized the growth profile of *L. kimchii*, preformed

GTF activity assays, produced *in vitro* both EPSA and EPSB polymers, and performed structural characterization of EPS by enzymatic and acid hydrolysis. AMM, performed ^{1}H- and ^{13}C-NMR structural characterization of both EPSA and EPSB polymers. MGG and RCM performed *pulque* fermentation, EPS-producing LAB screening, purification and identification of studied LAB strains. MRA, MGG, ALM, FB, and AE conceived this study and designed experiments. MRA, AMM, ALM, FB, and AE wrote the paper. All the authors read and approved the final manuscript.

REFERENCES

1. Ahmed RZ, Siddiqui K, Arman M, Ahmed N: Characterization of high molecular weight dextran produced by *Weissella cibaria* CMGDEX3. *Carbohydr Polym* 2012 doi:10.1016/j.carbpol.2012.05.063, 90: 441-446.

2. Amari M, Arango LFG, Gabriel V, Robert H, Morel S, Moulis C, Gabriel B, Remaud-Siméon M, Fontagné-Faucher C: Characterization of a novel dextransucrase from *Weissella confusa* isolated from sourdough. *Appl Microbiol Biotechnol* 2012 doi:10.1007/s00253-012-4447-8, 97: 5413-5422.

3. Bauer R, Bekker JP, van Wyk N, du Toit C, Dicks LMT, Kossmann J: Exopolysaccharide production by lactose-hydrolyzing bacteria isolated from traditionally fermented milk. *Int J Food Microbiol* 2009 doi:10.1016/j.ijfoodmicro.2009.02.020, 131: 260-264.

4. Bounaix M-S, Gabriel V, Morel S, Robert H, Rabier P, Remaud-Siméon M, Gabriel B, Fontagné-Faucher C: Biodiversity of exopolysaccharides produced from sucrose by sourdough lactic acid bacteria. *J Agric Food Chem* 2009 doi:10.1021/jf902068t, 57: 10889-10897.

5. Bounaix M-S, Gabriel V, Robert H, Morel S, Remaud-Siméon M, Gabriel B, Fontagné-Faucher C: Characterization of glucan-producing*Leuconostoc* strains isolated from sourdough. *Int J Food Microbiol* 2010 doi:10.1016/j.ijfoodmicro.2010.05.026, 144: 1-9.

6. Bradford MM: A rapid and sensitive method for the quantitation of microgram quantities of protein utilizing the principle of protein-dye binding. *Anal Biochem* 1976, 72: 248-254. 10.1016/0003-2697(76)90527-3

7. Chellapandian M, Larios C, Sanchez-Gonzalez M, Lopez-Munguia A: Production and properties of a dextransucrase from *Leuconostoc mesenteroides* IBT-PQ isolated from "pulque", a traditional Aztec alcoholic beverage. *J Ind Microbiol Biotechnol* 1998, 21: 51-56. 10.1038/sj.jim.2900560

8.   Colson P, Jennings HJ, Smith IC: Composition, sequence, and conformation of polymers and oligomers of glucose as revealed by carbon-13 nuclear magentic resonance. *J Am Chem Soc* 1974, 96: 8081-8087. 10.1021/ja00833a038

9.   Dahech I, Fakhfakh J, Damak M, Belghith H, Mejdoub H, Belghith KS: Structural determination and NMR characterization of a bacterial exopolysaccharide. *Int J Biol Macromol* 2013 doi:10.1016/j.ijbiomac.2013.04.036, 59: 417-422.

10. Ehrmann MA, Freiding S, Vogel RF: *Leuconostoc palmae* sp. nov., a novel lactic acid bacterium isolated from palm wine. *Int J Syst Evol Microbiol* 2009 doi:10.1099/ijs.0.005983-0, 59: 943-947.

11. Eom H-J, Seo DM, Han NS: Selection of psychrotrophic *Leuconostoc* spp. producing highly active dextransucrase from lactate fermented vegetables. *Int J Food Microbiol* 2007 doi:10.1016/j.ijfoodmicro.2007.02.027, 117: 61-67.

12. Escalante A, Rodríguez ME, Martínez A, López-Munguía A, Bolivar F, Gosset G: Characterization of bacterial diversity in Pulque, a traditional Mexican alcoholic fermented beverage, as determined by 16S rDNA analysis. *FEMS Microbiol Lett* 2004 doi:10.1016/j.femsle.2004.04.045, 235: 273-279.

13. Escalante A, Giles-Gómez M, Hernandez G, Cordova Aguilar M, Lopez-Munguia A, Gosset G, Bolivar F: Analysis of bacterial community during the fermentation of pulque, a traditional Mexican alcoholic beverage, using a polyphasic approach. *Int J Food Microbiol* 2008 doi:10.1016/j.ijfoodmicro.2008.03.003, 124: 126-134.

14. Escalante A, Giles-Gómez M, Esquivel Flores G, Matus Acuña V, Moreno-Terrazas R, López-Munguía A, Lappe-Oliveras P: Pulque Fermentation. In *Handb. Plant-Based Fermented Food Beverage Technol., Second Edition*. Edited by: Hui YH. CRC Press, Boca Raton, FL; 2012:691-706.

15. Giraffa G: Studying the dynamics of microbial populations during food fermentation. *FEMS Microbiol Rev* 2004 doi:10.1016/j.femsre.2003.10.005, 28: 251-260.

16. Kim J, Chun J, Han H-U: *Leuconostoc kimchii* sp. nov., a new species from kimchi. *Int J Syst Evol Microbiol* 2000, 50: 1915-1919.

17. Leemhuis H, Pijning T, Dobruchowska JM, van Leeuwen SS, Kralj S, Dijkstra BW, Dijkhuizen L: Glucansucrases: Three-dimensional structures, reactions, mechanism, α-glucan analysis and their implications in biotechnology and food applications. *J Biotechnol* 2013 doi:10.1016/j.jbiotec.2012.06.037, 163: 250-272.

18. Maina NH, Tenkanen M, Maaheimo H, Juvonen R, Virkki L: NMR spectroscopic analysis of exopolysaccharides produced by *Leuconostoc citreum* and *Weissella confusa* . *Carbohydr Res* 2008 doi:10.1016/j. carres.2008.04.012, 343: 1446-1455.

19. Minervini F, De Angelis M, Surico RF, Di Cagno R, Gänzle M, Gobbetti M: Highly efficient synthesis of exopolysaccharides by *Lactobacillus curvatus* DPPMA10 during growth in hydrolyzed wheat flour agar. *Int J Food Microbiol* 2010 doi:10.1016/j.ijfoodmicro.2010.03.014, 141: 130-135.

20. Morales-Arrieta S, Rodríguez ME, Segovia L, López-Munguía A, Olvera-Carranza C: Identification and functional characterization of *levS* , a gene encoding for a levansucrase from *Leuconostoc mesenteroides* NRRL B-512 F. *Gene* 2006 doi:10.1016/j.gene.2006.02.007, 376: 59-67.

21. Newbrun E, Baker S: Physico-chemical characteristics of the levan produced by *Streptococcus salivarius* . *Carbohydr Res* 1968 doi:10.1016/ S0008-6215(00)81506-2, 6: 165-170.

22. Oh H-M, Cho Y-J, Kim BK, Roe J-H, Kang S-O, Nahm BH, Jeong G, Han H-U, Chun J: Complete genome sequence analysis of*Leuconostoc kimchii* IMSNU 11154. *J Bacteriol* 2010 doi:10.1128/JB.00508-10, 192: 3844-3845.

23. Olivares-Illana V, Wacher-Rodarte C, Le Borgne S, López-Munguía A: Characterization of a cell-associated inulosucrase from a novel source: A *Leuconostoc citreum* strain isolated from Pozol, a fermented corn beverage of Mayan origin. *J Ind Microbiol Biotechnol* 2002, 28: 112-117. 10.1038/sj/jim/7000224

24. Ortiz-Soto ME, Olivares-Illana V, López-Munguía A: Biochemical properties of inulosucrase from *Leuconostoc citreum* CW28 used for inulin synthesis. *Biocatal Biotransformation* 2004 doi:10.1080/10242420400014251, 22: 275-281.

25. Palomba S, Cavella S, Torrieri E, Piccolo A, Mazzei P, Blaiotta G, Ventorino V, Pepe O: Polyphasic screening, homopolysaccharide composition, and viscoelastic behavior of wheat sourdough from a *Leuconostoc lactis* and *Lactobacillus curvatus* exopolysaccharide-producing starter culture. *Appl Environ Microbiol* 2012 doi:10.1128/ AEM.07302-11, 78: 2737-2747.

26. Park J-H, Ahn H-J, Kim S, Chung C-H: Dextran-like exopolysaccharide-producing *Leuconostoc* and *Weissella* from kimchi and its ingredients. *Food Sci Biotechnol* 2013 doi:10.1007/s10068-013-0182-x, 22: 1047-1053.

27. Patel S, Majumder A, Goyal A: Potentials of exopolysaccharides from lactic acid bacteria. *Indian J Microbiol* 2011 doi:10.1007/s12088-011-0148-8, 52: 3-12.

28. Quirasco M, Lopez-Munguia A, Remaud-Simeon M, Monsan P, Farres A: Induction and transcription studies of the dextransucrase gene in*Leuconostoc mesenteroides* NRRL B-512 F. *Appl Environ Microbiol* 1999, 65: 5504-5509.

29. Sanchez-Marroquin A, Hope PH: Agave juice, fermentation and chemical composition studies of some species. *J Agric Food Chem* 1953, 1: 246-249. 10.1021/jf60003a007

30. Seymour FR, Knapp RD, Bishop SH: Determination of the structure of dextran by ^{13}C-nuclear magnetic resonance spectroscopy.*Carbohydr Res* 1976 doi:10.1016/S0008-6215(00)83325-X, 51: 179-194.

31. Seymour FR, Knapp RD, Jeanes A: Structural analysis of levans by use of ^{13}C-n.m.r. spectroscopy. *Carbohydr Res* 1979 doi:10.1016/S0008-6215(00)83940-3, 72: 222-228.

32. Shimamura A, Tsuboi K, Nagase T, Ito M, Tsumori H, Mukasa H: Structural determination of D-fructans from *Streptococcus mutans* , serotype b, c, e, and f strains, by ^{13}C-n.m.r. spectroscopy. *Carbohydr Res* 1987, 165: 150-154. 10.1016/0008-6215(87)80091-5

33. Shukla R, Shukla S, Bivolarski V, Iliev I, Ivanova I, Goyal A: Structural characterization of insoluble dextran produced by *Leuconostoc mesenteroides* NRRL B-1149 in the presence of maltose. *Food Technol Biotechnol* 2011, 49: 291-296.

34. Sumner JB, Howell SF: A method for determination of saccharase activity. *J Biol Chem* 1935, 108: 51-54.

35. Tamura K, Stecher G, Peterson D, Filipski A, Kumar S: MEGA6: Molecular evolutionary genetics analysis version 6.0. *Mol Biol Evol* 2013 doi:10.1093/molbev/mst197, 30: 2725-2729.

36. Tieking M, Kaditzky S, Valcheva R, Korakli M, Vogel RF, Ganzle MG: Extracellular homopolysaccharides and oligosaccharides from intestinal lactobacilli. *J Appl Microbiol* 2005 doi:10.1111/j.1365-2672.2005.02638.x, 99: 692-702.

37. Tomašić J, Jennings HJ, Glaudemans CPJ: Evidence for a single type of linkage in a fructofuranan from *Lolium perenne* . *Carbohydr Res*1978 doi:10.1016/S0008-6215(00)83384-4, 62: 127-133.

38. Uzochukwu S, Balogh E, Loefler RT, Ngoddy PO: Structural analysis by [13] C-nuclear magnetic resonance spectroscopy of glucans elaborated by

gum-producing bacteria isolated from palm wine. *Food Chem* 2001, 73: 225-233. 10.1016/S0308-8146(00)00291-0

39. Uzochukwu S, Balogh E, Loefler RT, Ngoddy PO: Structural analysis by ^{13}C-nuclear magnetic resonance spectroscopy of glucan extracted from natural palm wine. *Food Chem* 2002, 76: 287-291. 10.1016/S0308-8146(01)00274-6

40. Van der Meulen R, Grosu-Tudor S, Mozzi F, Vaningelgem F, Zamfir M, Font de Valdez G, De Vuyst L: Screening of lactic acid bacteria isolates from dairy and cereal products for exopolysaccharide production and genes involved. *Int J Food Microbiol* 2007 doi:10.1016/j.ijfoodmicro.2007.07.014, 118: 250-258.

41. Van Hijum SAFT, Kralj S, Ozimek LK, Dijkhuizen L, van Geel-Schutten IGH: Structure-function relationships of glucansucrase and fructansucrase enzymes from lactic acid bacteria. *Microbiol Mol Biol Rev* 2006 doi:10.1128/MMBR.70.1.157-176.2006, 70: 157-176.

42. Vasileva T, Iliev I, Amari M, Bivolarski V, Bounaix M-S, Robert H, Morel S, Rabier P, Ivanova I, Gabriel B, Fontagné-Faucher C, Gabriel V: Characterization of glycosyltransferase activity of wild-type *Leuconostoc mesenteroides* strains from bulgarian fermented vegetables. *Appl Biochem Biotechnol* 2012 doi:10.1007/s12010-012-9812-7, 168: 718-730.

43. Vu B, Chen M, Crawford RJ, Ivanova EP: Bacterial extracellular polysaccharides involved in biofilm formation. *Molecules* 2009 doi:10.3390/molecules14072535, 14: 2535-2554.

44. Weisburg WG, Barns SM, Pelletier DA, Lane DJ: 16S ribosomal DNA amplification for phylogenetic study. *J Bacteriol* 1991, 173: 697-703.

Chapter 9

STRUCTURAL ANALYSIS OF THREE NOVEL TRISACCHARIDES ISOLATED FROM THE FERMENTED BEVERAGE OF PLANT EXTRACTS

Hideki Okada, Eri Fukushi, Akira Yamamori, Naoki Kawazoe, Shuichi Onodera, Jun Kawabata and Norio Shiomi

Department of Food and Nutrition Sciences, Graduate School of Dairy Science Research, Rakuno Gakuen University

ABSTRACT

Background

A fermented beverage of plant extracts was prepared from about fifty kinds of vegetables and fruits. Natural fermentation was carried out mainly by lactic acid bacteria (*Leuconostoc* spp.) and yeast (*Zygosaccharomyces* spp. and *Pichia* spp.). We have previously examined the preparation of novel four trisaccharides from the beverage: O-β-D-fructopyranosyl-(2->6)-O-β-D-glucopyranosyl-(1->3)-D-glucopyranose, O-β-D-fructopyranosyl-(2->6)-O-[β-D-glucopyranosyl-(1->3)]-D-glucopyranose, O-β-D-glucopyranosyl-(1->1)- O-β-D-fructofuranosyl-(2<->1)-α-D-glucopyranoside and O-β-D-galactopyranosyl-(1->1)-O-β-D-fructofuranosyl-(2<->1)- α-D-glucopyranoside.

Results

Three further novel oligosaccharides have been found from this beverage and isolated from the beverage using carbon-Celite column chromatography and preparative high performance liquid chromatography. Structural confirmation of the saccharides was provided by methylation analysis, MALDI-TOF-MS and NMR measurements.

Conclusion

The following novel trisaccharides were identified: O-β-D-fructofuranosyl-(2->1)-O-[β-D-glucopyranosyl-(1->3)]-β-D-glucopyranoside (named "3^G-β-D-glucopyranosyl β, β-isosucrose»), O-β-D-glucopyranosyl-(1->2)-O-[β-D-glucopyranosyl-(1->4)]-D-glucopyranose (4^1-β-D-glucopyranosyl sophorose) and O-β-D-fructofuranosyl-(2->6)-O-β-D-glucopyranosyl-(1->3)-D-glucopyranose (6^2-β-D-fructofuranosyl laminaribiose).

BACKGROUND

A beverage was produced by fermentation of an extract from 50 kinds of fruits and vegetables (see Additional file 1) [1, 2]. The extract was obtained using sucrose-osmotic pressure in a cedar barrel for seven days and was fermented by lactic acid bacteria (*Leuconostoc* spp.) and yeast (*Zygosaccharomyces* spp. and *Pichia* spp.) for 180 days. The fermented beverage showed scavenging activity against 1,1'-diphenyl-2-picrylhydrazyl (DPPH) radicals, and significantly reduced the ethanol-induced damage of gastric mucosa in rats [1]. Analysis by high performance anion exchange chromatography (HPAEC) showed that this beverage contained high levels of saccharides, estimated between 550 and 590 g/L; mainly glucose and fructose, and a small amount of undetermined oligosaccharides. Recently, it was reported that different positions of glycosidic linkage of oligosaccharide isomers affected physiological properties as well as physical properties [3–5]. Development of HPLC analysis with high sensitivity and separation ability enables the detection and isolation of oligosaccharides in the fermented beverage.

We have previously examined the preparation of saccharides of the fructopyranoside series from the fermented beverage of plant extracts, such as O-β-D-fructopyranosyl-(2->6)-D-glucopyranose [2], O-β-D-fructopyranosyl-(2->6)-O-β-D-glucopyranosyl-(1->3)-D-glucopyranose and O-β-D-fructopyranosyl-(2->6)-O-[β-D-glucopyranosyl-(1->3)]-D-glucopyranose [6]. The characteristics of O-β-D-fructopyranosyl-(2->6)-D-glucopyranose were non-cariogenicity and low digestibility, and the unfavorable bacteria that produce mutagenic substances did not use the saccharide [7, 8]. Recently, we have studied isolation and identification of novel non-reducing trisaccharides, such as O-β-D-glucopyranosyl-(1->1)-O-β-D-fructofuranosyl-(2<->1)-α-D-glucopyranoside and O-β-D-galactopyranosyl-(1->1)-O-β-D-fructofuranosyl-(2<->1)- α-D-glucopyranoside from the beverage [9], and those saccharides were confirmed to be produced by fermentation.

In this paper, we have confirmed structures of the novel trisaccharides (Fig. 1): O-β-D-fructofuranosyl-(2->1)-O-[β-D-glucopyranosyl-(1->3)]-β-

D-glucopyranoside (named "3^G-β-D-glucopyranosyl β, β-isosucrose»), *O*-β-D-glucopyranosyl-(1->2)-*O*-[β-D-glucopyranosyl-(1->4)]-D-glucopyranose (4^1-β-D-glucopyranosyl sophorose) and *O*-β-D-fructofuranosyl-(2->6)-*O*-β-D-glucopyranosyl-(1->3)-D-glucopyranose   (6^2-β-D-fructofuranosyl laminaribiose), isolated from the fermented beverage using methylation analysis, MALDI-TOF-MS and NMR measurements.

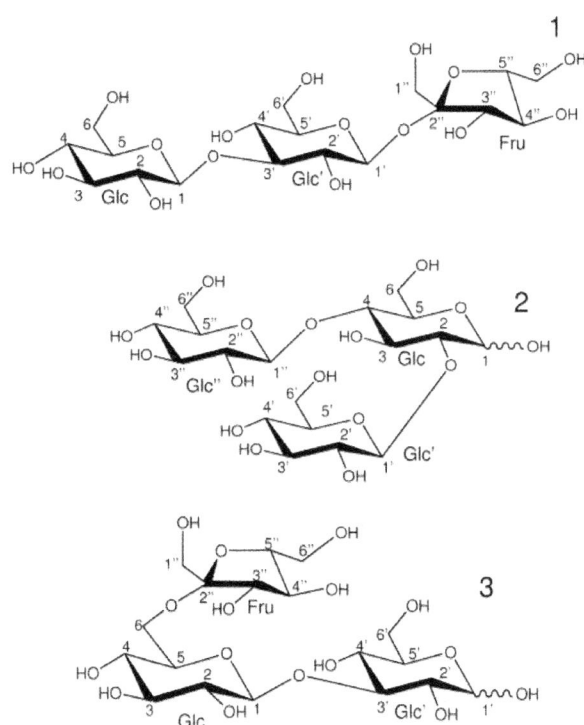

Figure 1: Structures of*O*-β-D-fructofuranosyl-(2->1)-*O*-[β-D-glucopyranosyl-(1->3)]-β-D-glucopyranoside   (1),*O*-β-D-glucopyranosyl-(1->2)-*O*-[β-D-glucopyranosyl-(1->4)]-D-glucopyranose (2) and*O*-β-D-fructofuranosyl-(2->6)-*O*-β-D-glucopyranosyl-(1->3)-D-glucopyranose (3).

RESULTS AND DISCUSSION

Saccharides **1, 2** and **3** were isolated from the fermented beverage of plant extracts using carbon-Celite column chromatography, and were shown to be homogeneous using anion exchange HPLC [$t_{R, \text{sucrose}}$ (relative retention time; retention time of sucrose = 1.0): 1.89, 2.23 and 2.40 respectively]. The retention time of saccharides **1, 2** and **3** did not correspond to that of any authentic saccharides [glucose (0.62), fructose (0.68), sucrose (1.00), maltose (1.43), trehalose (0.58), laminaribiose (1.33), raffinose (1.23), 1-kestose (1.47), 6-kestose (1.75), neokestose (1.90), maltotriose (2.59), panose (1.87), nystose (2.06), fructosylnystose (3.81), O-β-D-fructopyranosyl-(2->6)-D-glucopyranose (0.83) [2], O-β-D-fructopyranosyl-(2->6)-O-β-D-glucopyranosyl-(1->3)-D-glucopyranose (1.74) [6], O-β-D-fructopyranosyl-(2->6)-O-[β-D-glucopyranosyl-(1->3)]-D-glucopyranose (1.72) [6], O-β-D-glucopyranosyl-(1->1)-O-β-D-fructofuranosyl-(2<->1)-α-D-glucopyranoside (1.24) [9], O-β-D-galactopyranosyl-(1->1)-O-β-D-fructofuranosyl-(2<->1)-α-D-glucopyranoside (0.84) [9], 2(2-α-D-glucopyranosyl)isokestose (1.57) [10], 2(2-α-D-glucopyranosyl)$_2$isokestose (1.79) [10], 2(2-α-D-glucopyranosyl)$_3$isokestose (2.09) [10], 2(2-α-D-glucopyranosyl)nystose (2.17) [10], 2(2-α-D-glucopyranosyl)$_2$nystose (2.63) [10], O-α-D-glucopyranosyl-(1->2)-O-α-D-xylopyranosyl-(1->2)-β-D-fructofuranoside (1.51) [11], O-α-D-glucopyranosyl-(1->2)-O-α-D-glucopyranosyl-(1->2)-O-α-D-xylopyranosyl-(1->2)-β-D-fructofuranoside (1.80) [11].

The degree of polymerization of saccharides **1**, **2** and **3** was established as 3 by measurements of [M+Na] ions (m/z: 527) using TOF-MS (see Fig. 2), and analysis of the molar ratios of D-glucose to D-fructose in the acid hydrolysates. Acid hydrolysates of saccharides **1** and **3**were liberated to glucose and fructose, and saccharide **2** was liberated to glucose. From the GC analysis, relative retention times of the methanolysate of the permethylated saccharides were investigated [t_R (relative retention time; retention time of methyl 2, 3, 4, 6-tetra-O-methyl-β-D-glucoside = 1.0; retention time, 9.60 min)]. The methanolysate of permethylated saccharide **1** exhibited six peaks (see Additional file 2) corresponding to methyl 2,3,4,6-tetra-O-methyl-D-glucoside (t_R, 0.94 and 1.48), methyl 2,4,6-tri-O-methyl-D-glucoside (t_R, 3.27 and 4.81) and methyl 1,3,4,6- tetra-O-methyl-D-fructoside (t_R, 1.06 and 1.32). The methanolysate of permethylated saccharide **3** also exhibited six peaks (see Additional file 2) corresponding to methyl 2,3,4-tri-O-methyl-D-glucoside (t_R, 2.58 and 3.59), methyl 2,4,6-tri-O-methyl-D-glucoside (t_R, 3.22 and 4.73), and methyl 1,3,4,6-tetra-O-methyl-D-fructoside (t_R, 1.07 and 1.29). On the other hand, the methanolysate of permethylated saccharide **2** exhibited two peaks (see Additional file 2) corresponding to methyl 2,3,4,6-tetra-O-methyl-D-glucoside (t_R, 0.97 and 1.47). GC-MS analysis on the retention times and fragmentation patterns of the methyl glucosides [12] showed the two peaks (10.08 min and 10.21 min) from the methanolysate of permethylated saccharide **2** to be methyl 3,6-di-O-methyl-D-glucoside. From these findings above, saccharides **1, 2** and **3** were proved to be, O-D-fructofuranosyl-(2->1)-O-[D-glucopyranosyl-(1->3)]-D-glucopyranoside, O-D-glucopyranosyl-(1->2)-O-[D-glucopyranosyl-

(1->4)]-D-glucose and *O*-D-fructofuranosyl-(2->6)-*O*-D-glucopyranosyl-(1->3)-D-glucose, respectively.

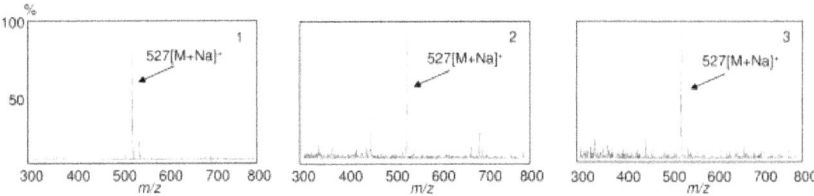

Figure 2: MALDI-TOF-MS spectra of saccharides 1, 2, and 3. 1: saccharide 1, 2: saccharide 2, 3: saccharide 3.

The structural confirmations of saccharides **1**, **2** and **3** according to ¹H and ¹³C NMR analyses and the subsequent complete assignment of ¹H and ¹³C NMR signals of the three saccharides were carried out using 2D-NMR techniques.

First, the NMR spectra of saccharide **1** were analyzed. The HSQC-TOCSY spectrum revealed the ¹H and ¹³C signals of each Glc, Glc' and Fru. The isolated methylene was assigned as H-1 and C-1 in Fru. The other three methylene carbons were assigned as C-6 in these residues. The COSY spectrum assigned the spin systems of these residues; from H-1 to H-3 and H-1' to H-3' (Fig. 3(a)), and from H-3" to H-6". The corresponding ¹³C signals were assigned by HSQC spectrum (Fig. 3(b)). These results clarified the assignment of ¹H and ¹³C NMR signals of each residue. The position of the glucosidic linkage and fructosidic linkage was analyzed as follows. The C-3' showed the HMBC [13, 14] correlations between H-1 (Fig. 3(c)). The *J* (H-1/H-2) value was 7.9 Hz. These results indicated the Glc 1β ->3' Glc linkage, namely the laminaribiose moiety. The C-2" showed the HMBC correlations to H-1'. The *J* (H-1'/H-2') value was 7.4 Hz. These results indicated the Glc' 1β ->2 β Fru linkage, and all ¹H and ¹³C NMR signals were assigned as shown in Additional file 3.

The coupling patterns of overlapped ¹H were analyzed by the SPT method [15, 16]. Due to strong coupling between H-4' and H-5', these couplings could not be analyzed in first order.

The NMR spectra of saccharide **2** showed that it was an anomeric mixture at the Glc. The α anomer was predominant. The COSY spectrum was assigned from H-1 to H-6. The C-4 showed the HMBC correlations between H-1» (Fig. 4(a) and 4(b)). The *J* (H-1"/H-2") value was 7.6–7.8 Hz. These results indicated the Glc" 1β ->4 Glc linkage, namely the cellobiose moiety. The C-2 showed the HMBC correlations to H-1'. The *J* (H-1'/H-2') value was 7.6 Hz. These results indicated the Glc' 1β ->2 Glc linkage, and all ¹H and ¹³C NMR signals were assigned as shown in Additional file 3.

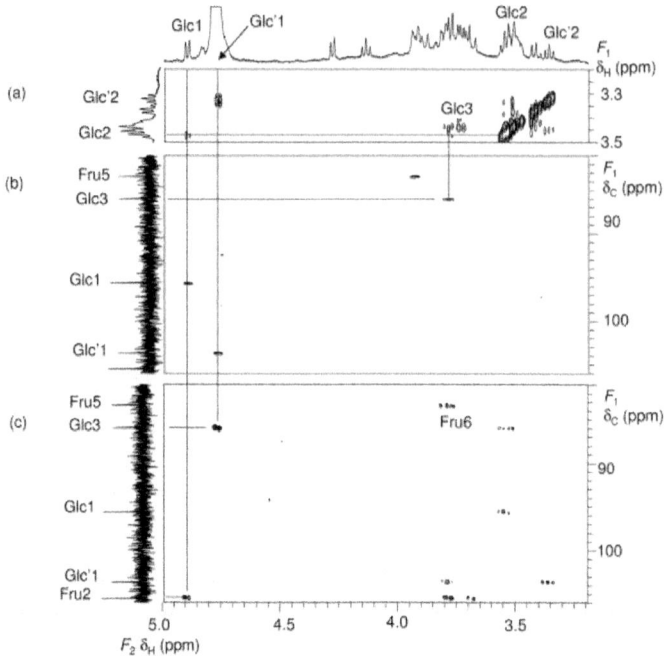

Figure 3: Part of COSY (a), HSQC (b) and HMBC (c) spectra of saccharide 1.

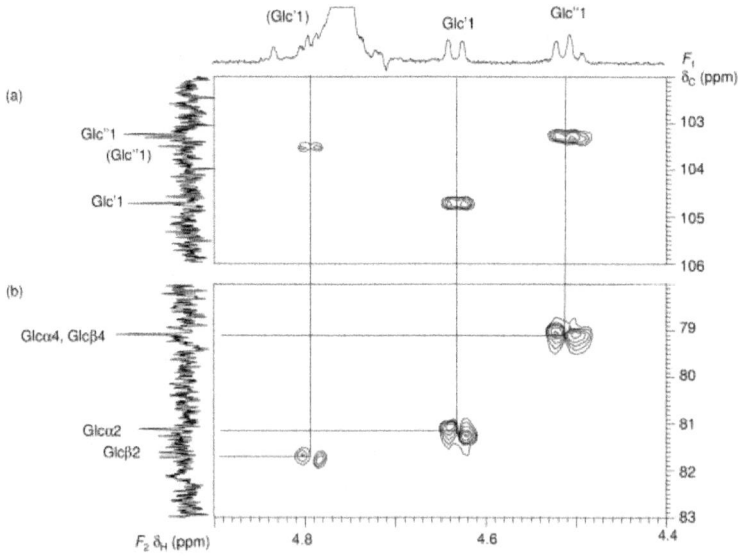

Figure 4: Part of HSQC (a) and HMBC (b) spectra of saccharide 2. ( ) = minor anomer.

The NMR spectra of saccharide **3** were analyzed in the same manner as those of saccharide **2**. Saccharide **3** was also an anomeric mixture at the Glc'. The β anomer was predominant. The HSQC-TOCSY spectrum revealed the[1]H and ^{13}C signals of each Glc, Glc' and Fru. The isolated methylene was assigned as H-1" and C-1". The other three methylene carbons were assigned as C-6 in these residues (Fig. 5(a)). The position of the glucosidic linkage and fructosidic linkage was analyzed as follows. The C-3' showed the HMBC correlations between H-1 (Fig. 5(b)). The J (H-1/H-2) value was 7.9 Hz. These results indicated the Glc 1β ->3 Glc' linkage, namely the laminaribiose moiety. The C-2 showed the HMBC correlations to H-6. These results indicated the Glc 6 <-2 β Fru linkage, and all ^{1}H and ^{13}C NMR signals were assigned as shown in Additional file 3.

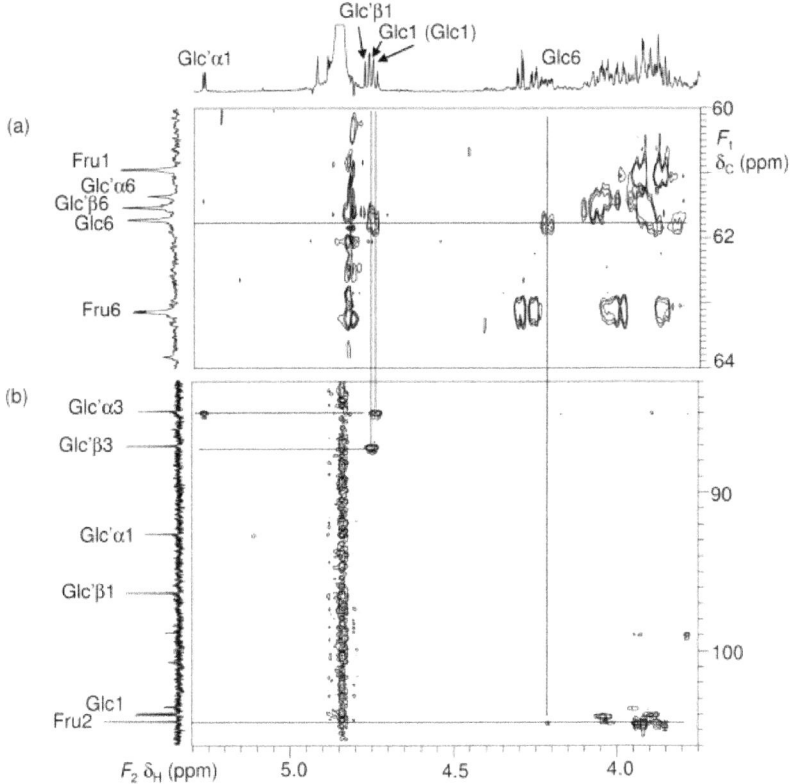

Figure 5: Part of HSQC-TOCSY (a) and HMBC (b) spectra of saccharide 3. ( ) = minor anomer.

From all of these findings, saccharides **1**, **2**, and **3** from the fermented beverage of plant extracts were confirmed to be new oligosaccharides (Fig.

1):O-β-D-fructofuranosyl-(2->1)-O-[β-D-glucopyranosyl-(1->3)]-β-D-glucopyranoside (named "3^G-β-D-glucopyranosyl β, β-isosucrose»), O-β-D-glucopyranosyl-(1->2)-O-[β-D-glucopyranosyl-(1->4)]-D-glucopyranose (4^1-β-D-glucopyranosyl sophorose) and O-β-D-fructofuranosyl-(2->6)-O-β-D-glucopyranosyl-(1->3)-D-glucopyranose (6^2-β-D-fructofuranosyl laminaribiose).

Synthesis of the saccharides by fermentation of plant extracts was investigated using HPAEC. Almost all of the monosaccharides were removed from the fermented and unfermented beverages of plant extracts by the batch method with Charcoal. The saccharides **1**, **2**, and **3**were observed in the fermented beverage, but were not present in the unfermented one. Therefore, saccharides **1**, **2**, and **3** were confirmed to have been produced during fermentation of the beverage of plant extracts (Fig. 6).

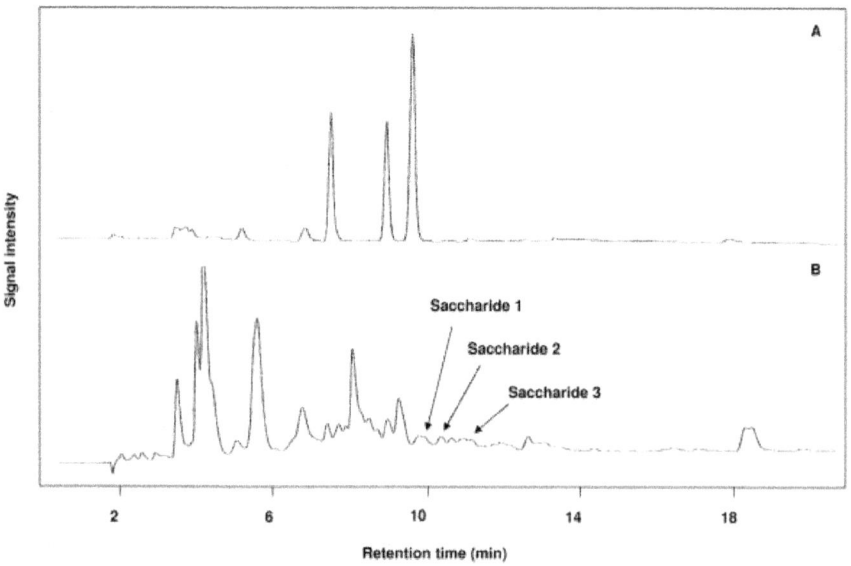

Figure 6: High performance liquid chromatogram of fermentation products. A: The beverage of plant extract was fermented for 0 days. B: The beverage of plant extract was fermented for 180 days. The beverage (100 mL) fermented for 0 or 180 days was mixed with charcoal (10 g), stirred for 3 h. and filtered. The charcoal was extracted with 30% ethanol (500 mL) three times. The ethanol extracts were combined, concentrated to dryness and solubilized with one mL of distilled water. The sugar solution was analyzed by HPAEC.

CONCLUSION

We have previously found that the fermented beverage contained the novel saccharide, O-β-D-fructopyranosyl-(2->6)-D-glucopyranose, which is produced by fermentation. The saccharide showed low digestibility. The saccharide was selectively used by beneficial bacteria, *Bifidobacterium adolescentis* and *B. longum*, but was not used by unfavorable bacteria, *Clostridium perfringens*, *Escherichia coli* and *Enterococcus faecalis* that produce mutagenic substances [8]. It is interesting to study the biological functions of other oligosaccharides existing in the beverage. In this report, three novel oligosaccharides have been found from this beverage, and isolated from the beverage using carbon-Celite column chromatography and preparative high performance liquid chromatography. Structural confirmation of the saccharides was provided by methylation analysis, MALDI-TOF-MS and NMR measurements. These saccharides were identified as new trisaccharides: O-β-D-fructofuranosyl-(2->1)-O-[β-D-glucopyranosyl-(1->3)]-β-D-glucopyranoside   (named "3^G-β-D-glucopyranosyl β, β-isosucrose»), O-β-D-glucopyranosyl-(1->2)-O-[β-D-glucopyranosyl-(1->4)]-D-glucopyranose   (4^1-β-D-glucopyranosyl sophorose) and O-β-D-fructofuranosyl-(2->6)-O-β-D-glucopyranosyl-(1->3)-D-glucopyranose (6^2-β-D-fructofuranosyl laminaribiose). These saccharides were confirmed to be produced during fermentation.

EXPERIMENTAL

Preparation of Fermented Beverage of Plant Extract

For preparation of the initial juice, 50 kinds of fruits and vegetables were used to produce the final extract as shown in a previous paper [1, 2]. The 50 fruits and vegetables were cut, sliced or diced into small pieces, mixed and put in cedar barrels. Afterwards, an equivalent weight of sucrose was added to the samples, mixed well to allow high contact between the samples and sucrose, and then the barrels were left for one week at room temperature. The juice exudate was then separated without compression from solids and used for fermentation. The fermented beverage was obtained by incubation of the juice at 37°C in the dark by natural fermentation using yeast (*Zygosaccharomyces* spp. and *Pichia* spp.) and lactic acid bacteria (*Leuconostoc* spp.). After 7 days, the fermented beverage was kept in a closed enameled tank at 37°C for 180 days for additional maturation and ageing, finally obtaining a brown and slightly sticky liquid.

High performance anion-exchange chromatography (HPAEC)

The oligosaccharides were analyzed using a Dionex Bio LC Series apparatus equipped with an HPLC carbohydrate column (Carbo Pack PA1, inert styrene divinyl benzene polymer) and pulsed amperometric detection (PAD) [17, 18]. The mobile phase consisted of eluent A (150 mM NaOH) and eluent B (500 mM sodium acetate in 150 mM NaOH) with a sodium acetate gradient as follows: 0–1 min, 25 mM; 1–2 min, 25–50 mM; 2–20 min, 50–200 mM; 20–22 min, 500 mM; 22–30 min, 25 mM; using a flow rate through the column of 1.0 mL/min. The applied PAD potentials for E1 (500 ms), E2 (100 ms), and E3 (50 ms) were 0.1, 0.6, and -0.6 V respectively, and the output range was 1 μC.

Isolation of saccharides

The fermented beverage of plant extracts (1000 g) was loaded onto a carbon-Celite [1:1; charcoal (Wako Pure Chemical Industries, Ltd; Osaka, Japan) and Celite-535 (Nacalai Tesque Inc, Osaka, Japan)] column (4.5 × 35 cm), and was successively eluted with water (14 L), 5% ethanol (30 L) and 30% ethanol (10 L). Almost all of the glucose and fructose were eluted with water (4 L), and then saccharides 1, 2 and 3 were eluted with 30% ethanol (1–2 L). The 30% ethanol fraction containing saccharides 1, 2 and 3 was concentrated *in vacuo* and freeze-dried to give 894 mg of sample. Subsequently, the 30% ethanol fraction was successfully repeatedly purified using an HPLC system (Tosoh, Tokyo, Japan) equipped with an Amide-80 column (7.8 mm × 30 cm, Tosoh, Tokyo, Japan) at 80°C, and eluted with 80% acetonitrile at 2.0 mL/min, and using refractive index detection. Furthermore, the saccharides were purified by HPLC with the ODS-100 V column (4.6 mm × 25 cm, Tosoh, Tokyo, Japan) at room temperature, and eluted with water at 0.5 mL/min. Purified saccharides 1 (2.5 mg), 2 (2.2 mg) and 3 (2.0 mg) were obtained as white powders.

Methylation and methanolysis

Methylation of the oligosaccharides was carried out by the method of Hakomori [19]. The permethylated saccharides were methanolyzed by heating with 1.5% methanolic hydrochloric acid at 96°C for 10 or 180 min. The reaction mixture was treated with Amberlite IRA-410 (OH⁻) to remove hydrochloric acid, and evaporated *in vacuo* to dryness. The resulting methanolysate was dissolved in a small volume of methanol and analyzed using gas liquid chromatography.

Gas liquid chromatography (GLC)

For the analysis of the methanolysate, GLC was carried out using a Shimadzu GC-8A gas chromatograph equipped with a glass column (2.6 mm × 2 m) packed with 15% butane 1,4-diol succinate polyester on acid-washed Celite at 175°C. Flow rate of the nitrogen gas carrier was 40 mL/min.

GC-MS analysis

GC-MS analysis was performed using a JMS-AX500 mass spectrometer (JEOL, Japan) using a DB-17HT capillary column (30 m × 0.25 mm I.D., J & W Scientific, USA). Injection temperature was 200°C. The column temperature was kept at 50°C for 2 min after sample injection, increased to 150°C at 50°C/min, kept at 150°C for 1 min, and then increased to 250°C at 4°C/min. The mass spectra were recorded in the electron ionization (EI) mode.

MALDI-TOF-MS

MALDI-TOF-MS spectra were measured using a Shimadzu-Kratos mass spectrometer (KOMPACT Probe) in positive ion mode with 2.5%-dihydroxybenzoic acid as a matrix. Ions were formed by a pulsed UV laser beam (nitrogen laser, 337 nm). Calibration was done using 1-kestose as an external standard.

NMR measurement

The saccharide (ca. 2 mg) was dissolved in 0.06 mL (saccharide **1**) and 0.4 mL (saccharide **2** and **3**) D_2O. NMR spectra were recorded at 27°C with a Bruker AMX 500 spectrometer (1 H 500 MHz, 13C 125 MHz) equipped with a 2.5 mm C/H dual probe (saccharide **1**), a 5 mm diameter C/H dual probe (1D spectra of saccharide **2** and **3**), and a 5 mm diameter TXI probe (2D spectra of saccharide **2** and **3**). Chemical shifts of ^1H (δ_H) and ^{13}C (δ_C) in ppm were determined relative to an external standard of sodium [2, 2, 3, 3-$^2H_4$]-3-(trimethylsilyl)-propanoate in D_2O (δ_H 0.00 ppm) and 1, 4-dioxane (δ_C 67.40 ppm) in D_2O, respectively. ^1H-^1H COSY [20, 21], HSQC [22], HSQC-TOCSY [22, 23] CH_2-selected E-HSQC-TOCSY [24], HMBC [13, 14] and CT-HMBC [13, 14] spectra were obtained using gradient selected pulse sequences. The TOCSY mixing time (0.15 s) was composed of DIPSI-2 composite pulses.

AUTHORS' CONTRIBUTIONS

HO, AY and NK performed data analysis, and contributed to drafting the manuscript. EF and JK collected the NMR data. NS and SO conceived of the study, participated in its design and contributed to drafting the manuscript. All authors read and approved the final manuscript.

REFERENCES

1. Okada H, Kudoh K, Fukushi E, Onodera S, Kawabata J, Shiomi N: Antioxidative activity and protective effect of fermented plant extract on ethanol-induced damage to rat gastric mucosa. *J Jap Soc Nutr Food Sci* 2005, 58:209–215.

2. Okada H, Fukushi E, Yamamori A, Kawazoe N, Onodera S, Kawabata J, Shiomi N: Structural analysis of a novel saccharide isolated from fermented beverage of plant extract. *Carbohydr Res* 2006, 341:925–929.

3. Kohmoto T, Fukui F, Takaku H, Machida Y, Arai M, Mitsuoka T: Effect of isomalto-oligosaccharides on human fecal flora.*Bifidobacteria Microflora* 1988, 7:61–69.

4. Murosaki S, Muroyama K, Yamamoto Y, Kusaka H, Liu T, Yoshikai Y: Immunopotentiating activity of nigerooligosaccharides for the T helper 1-like immune response in mice. *Biosci Biotechnol Biochem* 1999, 63:373–378.

5. Murosaki S, Muroyama K, Yamamoto Y, Liu T, Yoshikai Y: Nigerooligosaccharides augments natural killer activity of hepatic mononuclear cells in mice. *Int Immunopharmacol* 2002, 2:151–159.

6. Kawazoe N, Okada H, Fukushi E, Yamamori A, Onodera S, Kawabata J, Shiomi N: Two novel oligosaccharides isolated from a beverage produced by fermentation of a plant extract. *Carbohydr Res* 2008, 343:549–554.

7. Okada H, Kawazoe N, Yamamori A, Onodera S, Shiomi N: Structural analysis and synthesis of oligosaccharides isolated from fermented beverage of plant extract. *J Appl Glycosci* 2008, 55:143–148.

8. Okada H, Kawazoe N, Yamamori A, Onodera S, Kikuchi M, Shiomi N: Characteristics of *O* -β-D-fructopyranosyl-(2->6)-D-glucopyranose isolated from fermented beverage of plant extract. *J Appl Glycosci* 2008, 55:179–182.

9. Kawazoe N, Okada H, Fukushi E, Yamamori A, Onodera S, Kawabata J, Shiomi N: Structural analysis of two trisaccharides isolated from fermented beverage of plant extract. *Open Glycosci* 2008, 1:25–30.

10. Okada H, Fukushi E, Onodera S, Nishimato T, Kawabata J, Kikuchi M, Shiomi N: Synthesis and structural analysis of five novel oligosaccharides prepared by glucosyltransfer from **β-D-glucose** 1-phosphate to isokestose and nystose using*Thermoanaerobacter brockii* kojibiose phosphorylase. *Carbohydr Res* 2003, 338:879–885.

11. Takahashi N, Fukushi E, Onodera S, Benkeblia N, Nishimato T, Kawabata J, Shiomi N: Three novel oligosaccharides synthesized using

Thermoanaerobacter brockii kojibiose phosphorylase. *Chem Cent J* 2007, 1:18.

12. Funakoshi I: Mass spectrum. In *Seikagaku data book*. Edited by: Yamashina I. Tokyo: Tokyokagakudojin; 1979:606–668.

13. Bax A, Summers MF: [1]H and ^{13}C assignments from sensitivity-enhanced detection of heteronuclear multiple-bond connectivity by 2D multiple quantum NMR. *J Am Chem Soc* 1986, 108:2093–2094.

14. Hurd RE, John BK: Gradient-enhanced proton-detected heteronuclear multiple-quantum coherence spectroscopy. *J Magn Reson* 1991, 91:648–653.

15. Pachler KGR, Wessels PL: Selective Population Inversion (SPI). A pulsed double resonance method in FT NMR spectroscopy equivalent to INDOR. *J Magn Reson* 1973, 12:337–339.

16. Uzawa J, Yoshida S: A new selective population transfer experiment using a double pulsed field gradient spin-echo. *Magn Reson Chem* 2004, 42:1046–1048.

17. Rocklin RD, Pohl CA: Determination of carbohydrate by anion exchange chromatography with pulse amperometric detection. *J Liq Chromatogr* 1983, 6:1577–1590.

18. Johnson DC: Carbohydrate detection gains potential. *Nature* 1986, 321:451–452.

19. Hakomori S: A rapid permethylation of glycolipid and polysaccharide catalyzed by methylsulfinyl carbanion in dimethylsulfoxide. *J Biochem* 1964, 55:205–208.

20. Aue WP, Batholdi E, Ernst RR: Two-dimensional spectroscopy. Application to nuclear magnetic resonance. *J Chem Phys* 1976,64:2229–2246.

21. von Kienlin M, Moonen CTW, Toorn A, van Zijl PCM: Rapid recording of solvent-suppressed 2D COSY spectra with inherent quadrupole detection using pulsed field gradients. *J Magn Reson* 1991, 93:423–429.

22. Willker W, Leibfritz D, Kerssebaum R, Bermel W: Gradient selection in inverse heteronuclear correlation spectroscopy. *Magn Reson Chem* 1993, 31:287–292.

23. Domke T: A new method to distinguish between direct and remote signals in proton-relayed X, H correlations. *J Magn Reson* 1991, 95:174–177.

24. Yamamori A, Fukushi E, Onodera S, Kawabata J, Shiomi N: NMR analysis of mono- and difructosyllactosucrose synthesized by 1 [F]

-fructosyltransferase purified from roots of asparagus (*Asparagus officinalis* L.). *Magn Reson Chem* 2002, 40:541–544

Chapter 10

MIXES OF CARROT JUICE AND SOME FER-MENTED DAIRY PRODUCTS: POTENTIALITY AS NOVEL FUNCTIONAL BEVERAGES

Amany E. El-Abasy, Hany A. Abou-Gharbia, Hamida M. Mousa, Mohammed M. Youssef

Food Science and Technology Department, Faculty of Agriculture, Alexandria University, Alexandria, Egypt.

ABSTRACT

The present study aimed to formulate and evaluate yoghurt and Rayeb (Traditional Egyptian natural fermented milk) mixes with red and yellow carrot juices. Out of 16 different mixing ratios (Eight for yoghurt and eight for Rayeb), data indicated that the most significantly acceptable mixes as judged by panelists were: yoghurt-red carrot juice (3:1 v/v), yoghurt-yellow carrot juice (2:1 v/v), Rayeb-red carrot juice (2:1 v/v) and Rayeb-yellow carrot juice (4:1 v/v). The aforementioned mixes were further investigated versus their counterpart controls (yoghurt and Rayeb). It was obvious that the formulated mixes contained considerably higher concentrations of the bioactive components mainly, ascorbic acid and anthocyanins (red-carrot mixes) and β-carotene (yellow-carrot mixes) along with antioxidant activity (DMPD Radical Scavenging Activity) and higher counts of probiotic LAB, as compared to the controls.

INTRODUCTION

Although there is no consensus on the exact definition of the "Functional Food" term, according to the American Dietetic Association, functional food is defined as: Any modified food or food ingredient that may provide a health benefit beyond the traditional nutrients that contains [1]. Functional food development has enjoyed heightened interest by commercial, academic and governmental sectors over the past decade [2].

The interest in the role of phytochemicals especially as dietary antioxidants in human health has prompted research in the field of food science. Fruits and vegetables are good sources of antioxidants. Consequently, there are a number of commercial polyphenol rich beverages, which base their marketing strategies on antioxidant potency [3].

Carrot (Daucus carota L.) is an inexpensive and highly nutritious vegetable, since it contains appreciable amounts of vitamins B_1, B_2 and B_6 along with carotenes. Dietary carotenes are associated with lowering risk of many cancers. Meanwhile, vitamin A is an antioxidant which plays a key role in growth and repair of tissues in addition to help the body to fight with infections, keep eyes healthy, nourish epithelial tissues in the lungs and skin as well [4].

Extensively reviewed data about Leuconostoc, relative to their habitat, taxonomy, metabolism and genetics, their implications in health and safety, and their present and potential use in dairy technology and functional foods was done [5]. It is worth to mention that the development of probiotics in the last two decades has signaled an important advance in food industry. The number of scientific publications on probiotics has increased a lot stimulated by factors as exciting scientific and clinical findings using well-documented probiotic organisms [6].

The aim of the present work was to formulate mixes of carrot juice (yellow and red varieties) with two fermented dairy products namely, yoghurt and Rayeb, in order to design new functional beverages. Meanwhile, such beverages were evaluated from sensorial, physicochemical and microbiological points of view.

MATERIALS AND METHODS

Materials

Fresh yellow and red carrots (Daucus carota) were purchased from local farm and immediately transferred to cooling room (0°C and 90% RH) until used.

Yoghurt and Rayeb milk (Traditional Egyptian natural fermented milk) were obtained from pilot plant belonging to Department of Dairy Science and Technology, Faculty of Agriculture, Alexandria University, Egypt.

Preparation of Carrot Juice

Both yellow and red carrots were "topped" and "tailed" using a sharp knife, thoroughly washed with tap water, trimmed and peeled using a hand peeler (Approximately 1.00 mm of the periderm was removed by peeling and 2 cm of both tip and top were also removed by trimming), and were cut into sticks by sharp knife ($1 \times 3 \times 1$ cm). The sticks were then blanched for 15 min in boiling water with a ratio of 1:1 w/v carrot: water. The sticks along with blanching water were blended in a household blendor at a maximum velocity for 6 min to get fine paste. Juices were kept in polyethylene bags and frozen at $-18°C$ until used.

Preparation of Yoghurt-Carrot Juice Mixes

One hundered ml of pasteurized milk were tempered to $40°C$ then inoculated with 2 g of commercial yoghurt starter culture (Streptococcus thermophilus and Lactobacillus delbrueckii ssp. Bulgaricus) obtained from Dairy Pilot Plant, Faculty of Agriculture, Alexandria University. Incubation was conducted at $45°C$ for 8 hr, then cooled to $4°C$. Mixes of milk, starter culture and yellow or red carrot juice were prepared by the same method using ratios at 1:0, 1:1, 2:1, 3:1 and 4:1 of milk to carrot juice, respectively.

Preparation of Rayeb-Carrot Juice Mixes

One hundred ml of non-pasteurized milk were tempered to $40°C$ and fermented with natural flora for 24 hr. Mixes of milk and yellow or red carrot juice were prepared by the same method using ratios at 1:0, 1:1, 2:1, 3:1 and 4:1 of milk to carrot juice, respectively. Fermentation was stopped by lowering the temperature to $4°C$ and samples were held overnight at this degree.

Sensory Evaluation

Samples were subjected to sensory evaluation along with their controls. Ten trained panelists were asked to evaluate the samples according to the method described [7] on hedonic scale consisting of 9 points from 1 (Extremely dislike) to 9 (Extremely like). Colour, taste, consistency and overall acceptability were evaluated.

Microbiological Analysis

Food samples were aseptically removed from the bottles and 10 g from each sample were homogenized in 90 ml of sterile distilled water. Serial dilutions were prepared and 1 ml aliquots were plated in each specific medium and incubated at different temperatures, times and growth media as shown in **Table 1**.

Gross Chemical Composition

Moisture content was determined by drying the sample in vacuum oven at

70˚C to a constant weight [10], method No. 945.43. Ash was determined by incineration the sample at 550˚C in an electrical Muffle furnace [10], method No. 923.03.

Crude protein (N × 6.25) was determined according to AOAC method [10], method No. 2001.11. Fat content was determined according to Folch et al. method [12], using a mixture of chloroform and methanol (2:1 v/v). Carbohydrate content, was calculated by difference.

Determination of β-Caroten

Carotenoids were extracted from samples with 80% acetone (1:50 w/v). Absorbance of the extract was measured at 480 nm using Speckol Spectrocoluorimeter (Spekol 11, Carl Zeiss Jena, Germany). The β-carotene concentration was figured out by using the extinction coefficient ($E_{1cm}^{1\%}$) of 2273 [13,14].

Determination of Anthocyanins

Anthocyanins were extracted with 1% conc. HCl in 95% methanol (v/v). The ratio between sample and solvent was 1:50 (w/v). The micromolar concentration of anthocyanins in the extracts was obtained by multiplying the absorbance at 530 nm by 33.3 based on the molar extinction coefficient (1.0 cm light path) of cyanidine chloride being 30,000 [15]. Anthocyanin content was calculated as mg cyanidine chloride/g sample.

Table 1: Media and incubation conditions used for microbiological analysis

Microbiological analysis	Time (hr)	Temperature (˚C)	Growth medium
Total aerobic mesophilic bacteria	48	32	Nutrient Agar (NA)
Molds and yeasts count	72 - 120	28 - 30	Potato Dextrose Agar (PDA)
Lactic acid bacteria (LAB)	72 - 90	35	MRS Agar
Detection of E. coli	48	37	Lactose Broth (LB)

NA and PDA were prepared according to Difco [8] and Badawy [9]. LB was presented according to AOAC method 988 [10]. MRS Agar was prepared according to Man [11].

DMPD Radical Scavenging Activity

Antioxidant activity was measured using N,N-Dimethylp-phenylenediamine dihydrocholoride (DMPD). A dose response curve was derived for ascorbic acid by plotting the absorbance at 505 nm as percentage of the uninhibited radical cation solution.

Antioxidant activity was expressed as ascorbic acid equivalent antioxidant capacity using the calibration curve plotted with different amounts of ascorbic acid [16].

Statistical Analysis

Data were expressed as mean of triplicates ± SD. Data of sensory evaluation were subjected to analysis of variance (ANOVA) and Duncan's multiple range test to separate the treatment means [17]. The analysis was computed using SAS program.

RESULTS AND DISCUSSION

Sensory Properties of Yoghurt-Carrot Juice Mixes

Data presented in Table 2 reveal that the colours of yoghurt-red carrot juice mixes at different ratios were less scored than the control, with ratio of 3:1 being the only exception. In accordance, the taste of the control was higher scored than all mixes. In contrast, scores given for consistency of mixes were significantly comparable to the control. As for the overall acceptability, significant differences could be traced between the control and mixes. Since the mix at ratio 3:1 possessed the highest score for the overall acceptability, it was chosen to be investigated in details.

Table 2 shows that yoghurt-yellow carrot juice (YYC) mix at ratio of 2:1 exhibited the highest score given by the panelists for the overall acceptability as compared to the control and the other mixes. Accordingly, the aforementioned mix was chosen for the further investigation.

Chemical Composition of Yoghurt-Carrot Juice Mixes

Data given in **Table 3** indicate that the moisture content ranged between 86.04% and 89.30% for the control and mixes. The yoghurt red carrot juice (YRC) mix at 3:1 v/v had the highest crude protein content (25.70%) and the lowest fat content (29.06%) as compared to the control and yoghurt-yellow carrot juice (YYC) mix.

The control exhibited the highest ash content (5.31%), while YYC had the least ash content being 4.16%. Carbohydrate contents (**Table 3**) can be ascendingly ordered as followes: YYC (40.84%), YRC (40.56%) and the yoghurt control (31.05%).

The YRC exhibited the highest iron content being 11.83 mg/100g followed by the control (10.34 mg/100g), while YYC was tailed behind (9.01 mg/100g).

The same trend was also observed for phosphorus content (**Table 3**). Increment of iron and phosphrous in YRC mix can be explained on the basis that the resultant lactic acid due to fermentation improves the mineral availability by degradation of some components in the raw material, decrease in pH and growth of lactic acid bacteria is coupled to production of different organic acid [18]. It has been shown that metal chelates originating from vegetables, such as phytate, tannins and oxalate, may be degraded during fermentation as a result of microbial and/or plant enzymatic activities. In contrast, decline of iron and phosphrous in YYC mix may be due to the translocation of minerals from the solution into lactic acid bacteria cells for various functions in the bacteria [18].

It is worth to mention that YRC mix had almost double vitamin C content as compared to each of the control and YYC mix (**Table 3**). Notwithstanding, the YYC mix exhibited almost 5 folds of β-carotene as compared to YRC mix, while β-carotene could not be traced in the control (**Table 3**).

No anthocyanins could be detected in any of the control or YYC mix, on contrary to YRC mix which possessed considerable amount of anthocyanins being 22.05 mg/100g. **Table 3** indicates that YYC mix had more than double of antioxidants as the control had (32.42 versus 15.06 mg/100g). Meanwhile, YRC mix contained higher content (26.42 mg/100g), than its counterpart of the control (15.06 mg/100g).

Table 2: Sensory evaluation of the control yoghurt and its mixes with red and yellow carrot juices

Products	Ratios v/v	Sensory properties			
		Colour	Taste	Consistency	Overall acceptability
	Control	8.4^a	7.0^a	7.1^a	7.5^a
Yoghurt-red carrot juice mix (YRC)	1: 1	6.1^b	4.9^b	6.8^a	6.0^b
	2: 1	6.7^b	4.8^b	6.6^a	6.0^b
	3: 1	6.9^{ab}	5.6^b	6.9^a	6.4^a
	4: 1	5.5^b	5.2^b	6.8^a	6.0^b
	Control	7.1^{ab}	5.9^a	5.0^{bc}	6.1^{ab}
Yoghurt-yellow carrot juice mix (YRC)	1: 1	6.6^{ab}	5.0^{ab}	6.1^{ab}	5.9^{ab}
	2: 1	7.4^b	4.3^b	6.5^a	6.4^a
	3: 1	5.9^b	4.8^{ab}	4.7^c	5.3^{bc}
	4: 1	6.3^{ab}	5.0^{ab}	4.1^c	4.9^c

Means in a column within the same mix not sharing the same superscript are significantly different at P < 0.05.

Table 3: Chemical composition (on dry weight basis) and microbial counts of yoghurt and its mixes with red and yellow carrot juices

Products / Determinations	Control	Yoghurt-red carrot (YRC) juice mix 3:1 v:v	Yoghurt-yellow carrot juice (YYC) mix 2:1 v:v
Moisture (%)	87.35 ± 0.05	89.30 ± 0.04	86.04 ± 0.02
Crude protein (%)	23.94 ± 0.09	25.70 ± 0.00	18.05 ± 0.01
Total fat (%)	39.7 ± 0.32	29.06 ± 0.23	36.95 ± 0.02
Ash (%)	5.31 ± 0.02	4.68 ± 0.01	4.16 ± 0.01
Carbohydrates (%)	31.05	40.56	40.84
Iron (mg/100g)	10.34 ± 0.07	11.83 ± 0.01	9.01 ± 0.03
Phosphorus (mg/100g)	2300 ± 0.49	2370 ± 0.41	2090 ± 0.06
Ascorbic acid (mg/100g)	13.04 ± 0.02	23.64 ± 0.02	12.39 ± 0.00
β-caroten (mg/100g)	0.00	3.68 ± 0.001	16.55 ± 0.00
Anthocyanins (mg/100g)	0.00	22.05 ± 0.04	0.00
Antioxidants (mg/100g)	15.06 ± 0.59	26.42 ± 0.24	32.42 ± 0.05
Total soluble solids (%)	7.00	6.00	7.00
pH_1	6.40	6.36	6.04
pH_2	4.40	4.10	4.20
Total mesophilic bacteria[*]	2.22 ×10^2	2.22 × 10^2	2.22 × 10^2
Moulds and Yeasts[*]	0.00	0.00	0.00
Escherichia coli[*]	0.00	0.00	0.00
Lactic acid bacteria (LAB)[*]	2.3 × 10^6	2.8 × 10^6	3.99 × 10^6

[*]CFU g; pH_1: Prior to fermentation; pH_2: Immediately after fermentation. Results are expressed as means ± SD.

Total soluble solids were 7.0, 6.0 and 7.0 for the control, YRC and YYC, respectively (**Table 3**). The pH was measured twice (prior and immediately after fermentation). It was obvious that the pH values declined considerably form 6.40 to 4.40 (control), from 6.36 to 4.10 (YRC mix) and from 6.04 to 4.20 (YY mix) as shown in **Table 3**. Such declines in pH values are suspected due to formation of lactic acid during fermentation.

Microbial Counts of Yoghurt-Carrot Juice Mixes

Counts of total mesophilic bacteria were quite identical (2.22×10^2 CFU/g) for the yoghurt control, YRC mix and YYC mix. On the other hand, neither moulds and yeasts nor Escherichia coli could be detected neither in the control nor in the two mixes under investigation (**Table 3**).

The point of interest is that the count of lactic acid bacteria (LAB) increased considerably from 2.3×10^6 CFU/g (control) to 2.8×10^6 CFU/g (YRC mix) and 3.99×10^6 CFU (YYC), as it is shown in**Table 3**. Such increment of LAB counts can be attributed to the presence of carrot juice as a source of nutrients for bacteria. In accordance in other studies synbiotic potential of carrot juice supplemented with Lactobacillus spp. was discussed [19]. Their data revealed that both bacterial strains namely lactobacillus rhamnosus and Lactobacillus bulgaricus were capable of growing in carrot juice, reaching nearly 5×10^9 CFU/g after a 48 h fermentation and the pH was reduced to 3.5 - 3.7 or below [19]. Meanwhile, some biochemical characteristics of the fermented juice, such as β-carotene content and antioxidant activity, were also preserved, indicating that the metabolism of the Lactobacillus spp. did not degrade these components after 4 weeks of storage at 4°C.

Sensory Properties of Rayeb-Carrot Juice Mixes

Data presented in **Table 4** indicate that colours of Rayeb red-carrot juice (RRC) mixes were significantly comparable to the control at 1:1 and 3:1 ratios (v/v). No significant differences in taste scores could be figured out regarding RRC mixes. In contrast, these mixes were significantly different from the control in terms of their consistencies. As for overall acceptability, all mixes with an exception of 1:1 ratio mix were significantly comparable to the control. The 2:1 ratio mix was chosen for the further investigation, since it gained the highest score given for overall acceptability among the four RRC mixes.

Rayeb-yellow carrot juice (RYC) mix at 4:1 ratio was significantly similar to the control in terms of its colour and taste. Meanwhile, RYC mix at 3:1 ratio was significantly similar to the control in terms of consistency. The RYC mix at 4:1 ratio was chosen for further investigation because it had the highest score given by panelists for the overall acceptability as compared to the other RYC mixes.

Chemical Composition of Rayeb-Carrot Juice Mixes

Table 5 shows the chemical composition (on dry weight basis) of Rayeb (Natural fermented milk) and its mixes with red (RRC) and yellow (RYC) carrot juices. The RRC mix (2:1 v/v) possessed the highest moisture content being 89.46%, and the RYC mix (4:1 v/v) had the lowest moisture content (85.35%), while the control exhibited moisture content of 87.81%. Notwithstanding, crude protein content ranged between 29.75% (RYC) and 35.10% (Control). Accordingly, mixing of Rayeb milk with carrot juice resulted in decline of crude protein content. The RYC mix exhibited the highest total fat content (34.28%) as compared to 30.70% (RRC mix) and 33.80% (Control). Ash content varied from 4.30% (RYC mix) to 5.89% (RRC mix). As for carbohydrate contents, a range from 25.69% (Control) to 32.18% (RRC mix) was figured out.

Iron content varied from 13.22 mg/100g (RYC mix) to 17.14 mg/100g (RRC). Increment and decrease of iron content due to fermentation can be explained on the basis discussed previously for yoghurt-carrot juice mix. On the other hand, phosphorus content decreased in the mix as compared to its counterpart in the control (**Table 5**).

The RRC mix explored the highest ascorbic acid content (32.84 mg/100g) as compared to the value of 17.47 mg/100g (RYC) and 20.99 mg/100 g (Control). Meanwhile, it was obvious that RYC mix had almost twice content of β-carotene as compared to RRC mix versus nil for control (**Table 5**).

The RRC mix was the only product which contains appreciable amount (37.16 mg/100 g) of anthocyanins versus nil for each of RYC mix and the control (**Table 5**).

Table 4: Sensory evaluation of the control Rayeb and its mixes with red and yellow carrot juices

Products	Ratios v/v	Sensory properties			
		Colour	Taste	Consistency	Overall acceptability
	Control	7.5^a	7.0^a	7.8^a	7.4^a
	1: 1	6.8^{ab}	6.5^a	6.8^b	6.6^b
Rayeb-red carrot juice mix	2: 1	6.6^b	6.6^a	7.1^{ab}	7.0^{ab}
	3: 1	6.7^{ab}	6.6^a	7.1^{ab}	6.9^{ab}
	4: 1	6.6^b	6.5^a	7.0^{ab}	6.8^{ab}
	Control	8.1^a	7.4^a	6.9^a	7.5^a
	1: 1	6.2^b	5.8^b	4.9^b	5.5^c
Rayeb-yellow carrot juice mix	2: 1	6.4^b	6.4^{ab}	5.5^b	6.3^{bc}
	3: 1	6.2^b	6.6^{ab}	6.0^{ab}	6.4^b
	4: 1	7.5^a	6.5^{ab}	5.7^b	6.5^b

Means in a column within the same mix not sharing the same superscript are significantly different at P < 0.05.

Antioxidants (mg/100g) increased considerably on mixing Rayeb with carrot juices. So, the RRC mix had 24.03 while RYC mix exhibited a value of 23.08 versus 10.55 for the control (**Table 5**).

Total soluble solids were 6.6%, 6.0% and 6.4% for the control, RRC mix and RYC mix, respectively (**Table 5**). Notwithstanding, the pH was measured prior to fermentation and immediately after fermentation. Dramatic declines in pH were figured out. Such declines were as follows: from 6.60 to 4.53 (Control), from 6.74 to 4.20 (RRC mix) and from 7.00 to 4.44 (RYC mix). Such declines in pH can be attributed to producing organic acids (mainly lactic acid) as a result of fermentation by lactic acid bacteria present in Rayeb.

Table 5: Chemical composition (on dry weight basis) and microbial counts of Rayeb and its mixes with red and yellow carrot juices

Products / Determinations	Control	*Rayeb*-red carrot juice mix 2:1 v/v (RRC)	*Rayeb*-yellow carrot juice mix 4:1 v/v (RYC)
Moisture (%)	87.81 ± 0.26	89.46 ± 0.01	85.35 ± 0.04
Crude protein (%)	35.10 ± 0.00	31.23 ± 0.00	29.75 ± 0.12
Total fat (%)	33.80 ± 0.14	30.70 ± 0.19	34.28 ± 0.32
Ash (%)	5.41 ± 0.01	5.89 ± 0.001	4.30 ± 0.01
Carbohydrates (%)	25.69	32.18	31.67
Iron (mg/100g)	15.22 ± 0.15	17.14 ± 0.04	13.22 ± 0.16
Phosphorus (mg/100g)	2290 ± 2.48	2060 ± 0.82	1870 ± 2.32
Ascorbic acid (mg/100g)	20.99 ± 0.01	32.84 ± 0.001	17.47 ± 0.02
β-caroten (mg/100g)	0.00	5.52 ± 0.03	10.34 ± 0.01
Anthocyanins (mg/100g)	0.00	37.16 ± 0.02	0.00
Antioxidants (mg/100g)	10.55 ± 0.33	24.03 ± 0.53	23.08 ± 0.41
Total soluble solids (%)	6.60	6.00	6.40
pH_1	6.60	6.74	7.06
pH_2	4.53	4.20	4.44
Total mesophilic bacteria*	2.59×10^2	2.23×10^2	2.18×10^2
Moulds and Yeasts*	10	0.00	0.00
*Escherichia coli**	0.00	0.00	0.00
Lactic acid bacteria (LAB)*	2.80×10^5	2.42×10^6	1.61×10^6

*CFU/g; pH_1: Prior to fermentation; pH_2: Immediately after fermentation. Results are expressed as means ± SD.

Microbial Counts of Rayeb-Carrot Juice Mixes

Counts of total mesophilic bacteria were found to decline as a result of mixing Rayeb with carrot juices. In contrast, RRC and RYC mixes exhibited higher counts (2.42×10^6 CFU/g) and (1.61×10^6 CFU/g) of lactic acid bacteria than the Rayeb control (2.80×10^5 CFU/g). Such an elevation of lactic acid bacteria can be explained on the basis that presence of carrot juices in the medium of bacteria represents pivotal source of nutrients and thereby increase the growth rate of bacteria.

CONCLUSION

High quality and functional beverage mixes were able to be produced by mixing each of yoghurt and Rayeb (Traditional Egyptian natural fermented milk) with red carrot juice and yellow carrot juice. Such mixes can act as one of the most effective means for overcoming vitamin A deficiency. Moreover, these mixes contain high concentrations of natural antioxidants (vitamin C, β-carotene and anthocyanins) along with lactic acid bacteria (LAB) which act as a very good probiotics. All these components possess health benefits, as it has been extensively reported in numerous literatures. Epidmiological studies have demonstrated or at least suggested numerous health effects related to probiotics, prebiotics and natural antioxidants [20].

REFERENCES

1.   A. Bloch and C. A. Thomson, "Position of the American Dietetic Association: Phytochemicals and Functional Foods," Journal of American Dietetic Association, Vol. 95, No. 4, 1995, pp. 493-496. doi:10.1016/S0002-8223(95)00130-1

2.   P. J. Jones and S. Jew, "Functional Food Development: Concept to Reality," Trends in Food Science & Technology, Vol. 18, No. 7, 2007, pp. 387-390. doi:10.1016/j.tifs.2007.03.008

3.   E. Genzález-Molina, D. A. Moreno and Garcia-Viguera. A New Drink Rich in Healthy Bioactive Combining Lemon and Pomegranate Juices," Food Chemistry, Vol. 115, No. 4, 2009, pp. 1364-1372. doi:10.1016/j.foodchem.2009.01.056

4.   B. Singh, P. S. Panesar and V. Nanda, "Utilization of Carrot Pomace for the Preparation of a Value Added Product," World Journal of Dairy & Food Sciences, Vol. 1, No. 1, 2006, pp. 22-27.

5.   D. Hemme and C. Foucaud-Scheunemonn, "Leuconostoc, Characteristics, Use in Dairy Technology and Prospects in Functional Foods," International Dairy Journal, Vol. 14, No. 6, 2004, pp. 467-494. doi:10.1016/j.idairyj.2003.10.005

6.   F. C. Prado, J. L. Parada, A. Pandey and C. R. Soccoi. Review: Trends in Non-Dairy Probiotic Beverages," Food Research International, Vol. 41, No. 2, 2008, pp. 111-123.doi:10.1016/j.foodres.2007.10.010

7.   S. Hooda and S. Jood, "Organoleptic and Nutritional Evaluation of Wheat Biscuits Supplemented with Untreated and Treated Fenugreek Flour," Food Chemistry, Vol. 90, No. 3, 2005, pp. 427-435. doi:10.1016/j.foodchem.2004.05.006

8.   DIFCO, Difco Manual of Dehydrated Culture Media and Reagents for Microbiological and Clinical Laboratory Procedures," 8th Edition, Difco Laboratories, Detroit, 1948.

9.   D. M. Badawy, "Chemical, Technological and Microbiological Studies on Karkade (Hibiscus abdariffa) Petals Extract," M.Sc Thesis, Cairo University, Cairo, 2007.

10. AOAC, Official Methods of Analysis of AOAC International, 17th Edition, Association of Official Analytical Chemists, Maryloand, 2000.

11. J. C. Man, M. Rogosa and M. E. Sharpe, "A Medium for the Cultivation of Lactobacilli," Journal of Applied Bacterial, Vol. 23, No. 1, 1960, pp. 130-135. doi:10.1111/j.1365-2672.1960.tb00188.x

12. J. Folch, M. Less and S. Stanley, "A Simple Method for the Isolation and Purification of Total Lipids from Animal Tissues," Journal Bio-Chemistry, Vol. 226, No. 1, 1957, pp. 497-509.

13. A. Jensen, "Chlorophylls and Carotenoids," In: J. A. Hellebuts and J. S. Carigie, Eds., Hand Book of Physiological and Biochemical Methods, Cambridge University Press, London, 1978, pp. 59-70.

14. A. Ben-Amotz and M. Avronm, "The Factors Which Determines Massive -Carotene Accumulation in the Halotolerant Alga Dunaliella bardawil," Plant Physiology, Vol. 72, No. 3, 1983, pp. 593-597. doi:10.1104/pp.72.3.593

15. F. T. Halaweish and D. K. Dougall, "Sinapolyl Glucose Synthesis by Extracts of Wild Carrot Cell Cultures," Plant Science, Vol. 71, No. 2, 1990, pp. 179-184. doi:10.1016/0168-9452(90)90007-B

16. V. Fogliano, V. Verde, G. Randazzo and A. Ritieni, "Method for Measuring Antioxidant Activity and Its Application to Monitoring the Antioxidant Capacity of Wines," Journal of Agricultural and Food Chemistry, Vol. 47, No. 3, 1999, pp. 1035-1040.doi:10.1021/jf980496s

17. R. B. D. Steel and T. H. Torrie, "Principles and Procedures of Statistics," McGraw Hill Co., Boston, 1980.

18. S.W. Bergqvist, A.S. Sandberg, N. G. Carlsson and T. Andlid, "Improved Iron Solubility in Carrot Juice Fermented by Homoand Hetero-Fermentative Lactic Acid Bacteria," Food Microbiology, Vol. 22, No. 1, 2005, pp. 53-61. doi:10.1016/j.fm.2004.04.006

19. F. Nazzaro, F. Fratianni, A. Sada and P. Orlando, "Synbiotic Potential of Carrot Juice Supplemented with Lactobacillus spp. and Inulin or Fructooligosaccharides," Journal of the Science of Food and Agriculture, Vol. 88, No. 13, 2008, pp. 2271-2276.doi:10.1002/jsfa.3343

20. W. Grajek, A. Olejnik and A. Sip, "Probiotics, Prebiotics and Antioxidants as Functional Foods: A Review," Acta Biochimica Polonica, Vol. 52, No. 3, 2005, pp. 665-671.

Chapter 11

THE MICROBIAL DIVERSITY OF TRADITIONAL SPONTANEOUSLY FERMENTED LAMBIC BEER

Freek Spitaels[1], Anneleen D. Wieme[1, 2], Maarten Janssens[3], Maarten Aerts[1], Heide-Marie Daniel[4], Anita Van Landschoot[2], Luc De Vuyst[3], Peter Vandamme[1]

[1] Laboratory of Microbiology, Faculty of Sciences, Ghent University, Ghent, Belgium,

[2] Laboratory of Biochemistry and Brewing, Faculty of Bioscience Engineering, Ghent University, Ghent, Belgium,

[3] Research Group of Industrial Microbiology and Food Biotechnology (IMDO), Faculty of Sciences and Bioengineering Sciences, Vrije Universiteit Brussel, Brussels, Belgium,

[4] Mycothe`que de l'Universite´ catholique de Louvain (MUCL), Belgian Coordinated Collection of Microorganisms (BCCM), Earth and Life Institute, Applied Microbiology, Mycology, Universite´ catholique de Louvain, Louvain-la-Neuve, Belgium

ABSTRACT

Lambic sour beers are the products of a spontaneous fermentation that lasts for one to three years before bottling. The present study determined the microbiota involved in the fermentation of lambic beers by sampling two fermentation batches during two years in the most traditional lambic brewery of Belgium, using culture-dependent and culture-independent methods. From 14 samples per fermentation, over 2000 bacterial and yeast isolates were obtained and identified. Although minor variations in the microbiota between casks and batches and a considerable species diversity were found, a characteristic microbial succession was identified. This succession started with a dominance of *Enterobacteriaceae* in the first month, which were replaced at 2 months by *Pediococcus damnosus* and *Saccharomyces* spp., the latter being replaced by *Dekkera bruxellensis* at 6 months fermentation duration.

INTRODUCTION

Lambic sour beers are among the oldest types of beers still brewed and are the products of a spontaneous fermentation process that lasts for one to three years [1]. The fermentation process is not initiated through the inoculation of yeasts or bacteria as starter cultures. Rather, microbial growth starts during the overnight cooling of the cooked wort in a shallow open vessel, called the cooling tun or coolship. Lambic beers are traditionally brewed in or near the Senne river valley, an area near Brussels, Belgium. Brewing for the production of lambic traditionally takes place only during the colder months of the year (October to March), since cold nights are needed to lower the wort temperature to about 20°C in one night. The morning following the wort cooking, the cooled wort is assumed to be inoculated with a specific air microbiota of the Senne river valley and is transferred into wooden casks which are stored at cellar or ambient temperatures, *i.e.*, typically between 15 and 25°C. Subsequently, the wort ferments and the lambic beer matures in these same casks. The end product is a noncarbonated sour beer that mainly serves as a base for gueuze or fruit lambic beers. The sour character of the beer originates from the metabolic activities of various yeasts, lactic acid bacteria (LAB), and acetic acid bacteria (AAB) [2], [3].

Previous studies of the lambic beer fermentation process identified four phases: the*Enterobacteriaceae* phase, the main fermentation phase, the acidification phase, and the maturation phase, each characterized by the isolation of specific micro-organisms [2], [3]. The*Enterobacteriaceae* phase starts after 3 to 7 days of fermentation, proceeds until 30 to 40 days, and is characterized by *Enterobacter* spp., *Klebsiella pneumoniae*, *Escherichia coli* and*Hafnia alvei* as the most frequently isolated bacteria [4], along with the cycloheximide-resistant yeasts *Hanseniaspora uvarum* (asexual form *Kloeckera apiculata* [5]) and *Naumovia (Saccharomyces) dairensis* [6] as well as *Saccharomyces uvarum* (synonym *S. globosus* [7])[2], [3]. The main fermentation starts after 3 to 4 weeks of fermentation and is characterized by the isolation of *S. cerevisiae*, *S. bayanus/pastorianus* and *S. uvarum* [2], [3]. After 3 to 4 months of fermentation, the acidification phase occurs and is characterized by the increasing isolation of *Pediococcus* spp. and occasionally *Lactobacillus* spp., while *Brettanomyces* spp. become prevalent after 4 to 8 months of fermentation [2], [3]. The final maturation phase, during which the wort is gradually attenuated, starts after 10 months of fermentation and is characterized by a decrease of LAB [2], [3]. AAB are isolated throughout the fermentation period [2], [3]. Sour beers are currently attracting interest outside Belgium, especially in the USA. In the American craft-brewing sector,

American coolship ales mimic the lambic beer production method [8], and such beers are a seasonal product from craft breweries, which contrasts to traditional Belgian lambic breweries that exclusively produce lambic beers. It is thus likely that *Saccharomyces* spp., used for the brewing of other types of beers in the American craft-brewing sector, are enriched in these brewery environments [8]. A similar microbial succession as described above was recently revealed using culture-independent and culture-dependent techniques for the American coolship ales, whereby 16S rRNA gene sequence analysis was used to identify some morphologically distinct isolates [8]. Although the latter approach is widely applied as part of bacterial identification studies, it lacks resolution between many of the species belonging to the AAB, LAB, and *Enterobacteriaceae* family, and accurate species level identifications can only be obtained after subsequent sequence analysis of more variable protein-encoding genes [9]–[12]. Except for this American brewery study, previous microbial studies on lambic beers used phenotypic identification techniques only, which are nowadays known to have an inadequate taxonomical resolution for the species-level identification of yeasts, LAB, and AAB [2], [3], [13]–[17]. In addition, the discovery of novel species and of many synonymies in these groups of micro-organisms confounds the interpretation of literature data. For instance, *"Pediococcus cerevisiae"* was reported as a key organism in lambic beer fermentation, but this species name has no standing in bacterial nomenclature and has been used for at least two of the currently known *Pediococcus* species, *i.e.*, *P. damnosus* and *P. pentosaceus* [18], [19]. Such *"P. cerevisiae"* isolates likely represent *P. damnosus*, as suggested by Van Oevelen et al. [2]. Also, Kufferath and Van Laer [20] first isolated and described the yeast recognized to confer the characteristic taste to lambic beer as *Brettanomyces bruxellensis* and *B. lambicus*. After the observation of the sexually reproducing form, the name *Dekkera bruxellensis* was introduced [21]. *B. bruxellensis* and *B. lambicus* were later recognized as synonyms of the same species [22].

The present study aimed at the characterization of the microbial communities in two batches of a traditional lambic beer during the first two years of the fermentation process by means of culture-dependent and culture-independent techniques.

MATERIALS AND METHODS

Brewery

Samples were obtained from the Cantillon brewery (http://www.cantillon.be). This brewery is the most traditional, still active, lambic brewery in Brussels and uses the same infrastructure and most of the equipment since 1900, when the brewery was founded.

Sampling

Mash was prepared and boiled according to the brewer's recipe. After 3 h of boiling, the hot wort was pumped into the cooling tun, which was cleaned using hot water and a 500 mL sample was taken aseptically. Subsequent 500 mL samples were taken after overnight cooling in the cooling tun and 15 min; 1, 2 and 3 weeks; and 1, 2, 3, 6, 9, 12, 18 and 24 months after the transfer of the cooled wort into the multiple wooden casks; all these samples were taken from four casks of each of two batches of brews. The brews started on February 25, 2010 (batch 1), and March 23, 2010 (batch 2). Batch 1 was fermented at cellar temperature (ranging from 12°C in winter to 20°C in summer), batch 2 in a different room at ambient temperature (10–30°C). The wooden casks had a volume of approximately 400 L and had two apertures: a bung hole at the top of the cask, which was inaccessible for sampling due to the piling of the casks, and a sampling hole at the front of the cask. The latter was positioned about 10 cm above the cask bottom, plugged by a cork and was used for sampling. After removal of the cork plug, approximately 100 mL of fermenting wort were discarded before collection of the sample. Homogenization of the samples in the casks was not possible and may have introduced a sampling bias towards microbiota that settled onto the bottom of the cask and those at the wort/air interphase. Samples were transported on ice to the laboratory and were processed the same day. One cask per batch was chosen for culture-dependent sampling throughout the whole fermentation period and the microbiota of all eight casks was studied using denaturing gradient gel electrophoresis (DGGE) of the V3 region of the bacterial 16S rRNA genes and the D1/D2 region of the yeast 26S rRNA genes.

DGGE analysis

Crude beer samples were centrifuged at 8000× g for 10 min (4°C) on the day of sampling and cell pellets were stored at −20°C until further processing. DNA was prepared from the pellets as described by Camu et al. [23]. The DNA concentration, purity, and integrity were determined using 1% (wt/vol) agarose gels stained with ethidium bromide and by optical density (OD) measurements at 234, 260, and 280 nm. Total DNA solutions were diluted to an OD_{260} of 1. Amplification of about 200 bp of the V3 region of the 16S rRNA genes with the F357 and R518 primers (with a GC clamp attached to the F357 primer), followed by DGGE analysis, and processing of the resulting fingerprints was performed, as described previously [24], except that DGGE gels were run for 960 min instead of 990 min. For the amplification of about 200 bp of the D1/D2 region of the 26S rRNA genes, NL1 and LS2 primers (NL1 with GC clamp) were used, as previously reported by Cocolin et al. [25]. Similarities in fingerprint patterns were analyzed by means of Dice coefficient analysis, using the BioNumerics 5.1 software package (Applied Maths, Sint-Martens-Latem, Belgium). Gels were also examined using a moving window analysis, in which the percentage change (expressed as 100% - Dice similarity) between two consecutive sample profiles was plotted as a function of time [26].

All DNA bands were assigned to band classes using the BioNumerics 5.1 software. Dense DNA bands and/or bands that were present in multiple fingerprints were excised from the polyacrylamide gels by inserting a pipette tip into the band and subsequent overnight elution of the DNA from the gel slice in 40 μL 1× TE buffer (10 mM Tris-HCl, 5 mM EDTA, pH 8) at 4°C. The position of each extracted DNA band was confirmed by repeat DGGE experiments using the excised DNA as template. The extracted DNA was subsequently re-amplified and sequenced using the same protocol and primers (but without GC-clamp). EzBioCloud and BLAST[27], [28] analyses were performed to determine the most similar sequences in the public sequence databases.

Culture media, enumeration and isolation

The samples were serially diluted in 0.9% (wt/vol) saline and 50 μL of each dilution was plated in triplicate on multiple agar isolation media. The set of isolation media used was selected based on preliminary testing of samples of lambic beers of different ages by comparing DGGE profiles of the original samples with those of all cells that were harvested from the agar isolation media tested (data not shown). A total of twenty-three combinations of different growth media and incubation conditions [20°C vs. 28°C and aerobic vs. anaerobic atmosphere] were tested and this resulted in a set of 7 isolation conditions (see below), which together yielded a community profile that reflected best the diversity obtained in the DGGE profiles of the original beer samples and excluded isolation conditions that yielded redundant results.

All bacterial agar isolation media were supplemented with 5 ppm amphotericin B (Sigma-Aldrich, Bornem, Belgium) and 200 ppm cycloheximide (Sigma-Aldrich) to inhibit fungal growth and were incubated aerobically at 28°C, unless stated otherwise. Samples were incubated after plating on de Man-Rogosa-Sharpe (MRS) agar (Oxoid, Erembodegem, Belgium) [29] at 28°C aerobically and at 20°C anaerobically for the isolation of LAB. Violet red bile glucose (VRBG) agar [30], [31] was used for the isolation of *Enterobacteriaceae* and acetic acid medium (AAM) agar [32] was used for the isolation of AAB.

Yeast isolation media were first supplemented with 30 ppm ampicillin (Sigma-Aldrich), which proved inefficient to inhibit bacterial growth. All samples starting from 3 weeks in batch 1 were subcultured in the presence of 100 ppm chloramphenicol (Sigma-Aldrich). All yeast isolation media were incubated aerobically at 28°C. DYPAI (2% glucose, 0.5% yeast extract, 1% peptone and 1.5% agar; wt/vol) was used as a general yeast agar isolation medium or was supplemented with an additional 50 ppm cycloheximide (DYPAIX) to favor slow-growing *Dekkera/Brettanomyces* spp. [33]–[35]. Furthermore, universal beer agar (Oxoid) was supplemented with 25% (vol/vol) commercial gueuze (Belle-Vue - AB Inbev, Anderlecht, Belgium) as recommended by the manufacturer and was used as an additional general yeast agar isolation medium (UBAGI).

Colonies on plates comprising 25 to 250 colony forming units (CFU) were counted after 3 to 10 days of incubation and for each of the seven isolation conditions about 20–25 colonies, or all colonies if the counts were lower, were randomly picked up.

Matrix-assisted laser desorption/ionization time-of-flight mass spectrometry (MALDI-TOF MS) dereplication and identification

Isolates were subcultured twice using the respective isolation conditions, and MALDI-TOF MS was performed using the third generation of pure cultures by means of a 4800 Plus MALDI TOF/TOF™ Analyzer (AB SCIEX, Framingham, MA, USA), as described previously [36]. In short, Data Explorer 4.0–software (AB SCIEX) was used to convert the mass spectra into .txt-files to import them into a BioNumerics 5.1 (Applied Maths) database. Spectral profiles were compared using Pearson product moment correlation coefficient and a dendrogram was built using the unweighted pair group method with arithmetic mean (UPGMA) cluster algorithm. Homogeneous clusters consisting of isolates with visually identical and/or virtually identical mass spectra were delineated. From each cluster, isolates were chosen randomly for further identification through sequence analysis of 16S rRNA genes and other molecular markers. Sequence analysis of dnaJ and rpoB genes was performed to identify members of the Enterobacteriaceae [37], [38], of the pheS gene to identify LAB [10]–[12], [39] and of dnaK, groEL and rpoB genes to identify AAB [9]. Yeast isolates were identified through sequence analysis of the D1/D2 region of the 26S rRNA gene [14] and, whenever needed, also by determination of ACT1 and/or ITS sequences [40].

All PCR assays were performed as described by Snauwaert et al. [41]. Bacterial DNA was obtained via the protocol as described by Niemann et al. [42], whereas yeast DNA was obtained using the protocol of Harju et al. [43].

Analysis of the microbiota of the brewery environment

To analyze the microbiota of the brewery environment, samples were taken from the cooling tun, the roof above the cooling tun, the walls and ceiling of the cellar, and the outside of the casks by swabbing about 100 cm^2 using a moist swab that was transferred into 5 mL of saline and transported to the laboratory. The inside of a cask was sampled by rinsing it with 5 L of saline. In the laboratory, 5–10 mL portions of each sample were subsequently filtered over a 0.45-μm filter that was transferred into 30 mL of MRS, VRBG, AAM, DYPAI and DYPAIX broth, each, and incubated as described above. Enrichment cultures that showed growth after 3–10 days of incubation were plated on their respective agar media and different morphotypes were selected for further analysis. Isolates were identified as described above. Additionally, the swabs and water sample were directly streaked or plated on the agar isolation media. Air samples were taken using a MAS-100 air sampler (Merck, Darmstadt, Germany) with a flow rate of 0.1 m^3/min placed about 1 m above the floor, for one or ten minutes using yeast and bacterial agar isolation media, respectively.

RESULTS

DGGE analysis

Bacterial and yeast DNA was successfully extracted from most samples and PCR amplicons were generated subsequently. As expected, none of the cooling tun samples collected directly after boiling the wort yielded DNA (the wort temperature at the time of sampling was about 90°C). The samples of the overnight-cooled wort yielded DNA, but this was of low quality (data not shown) and no amplicons could be obtained. The first amplicons were obtained from the cask samples immediately after the transfer of the wort into the casks. For both batches, bacterial and yeast community fingerprints were generated for each of the four casks. Analysis of these community fingerprints revealed highly similar to identical community fingerprints for each sampling moment (Fig. S1). DGGE banding patterns of both bacterial and yeast communities of the casks that were used in the culture-dependent analysis of batch 1 and 2 (see below) are shown in Fig. 1.

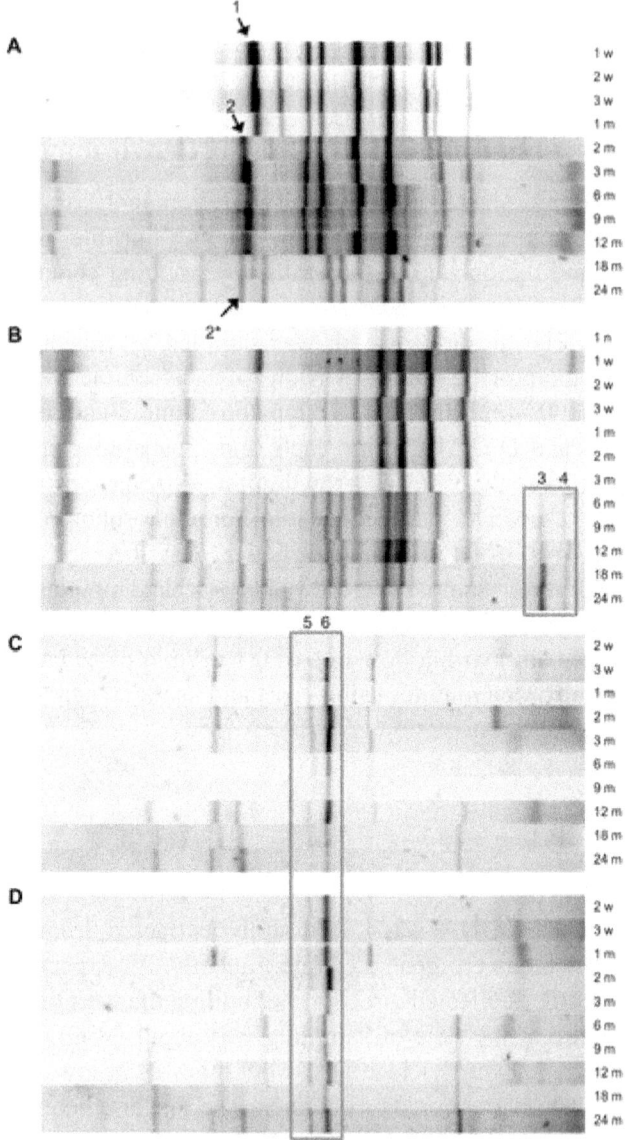

Figure 1: DGGE banding patterns of bacterial and yeast communities of the plated samples. DGGE banding patterns of the bacterial and yeast communities of batch 1, cask 1 (A and C, respectively) and batch 2, cask 2 (B and D, respectively) n, night; w, week(s); m, month(s). Band classes 1–6 are indicated on the figure. Samples after one night in cask 1 of batch 1 did not yield any amplicons with the V3 primer, the other casks yielded banding patterns highly similar to the pattern of the one-week sample (data not shown). Yeast community profiles were obtained from 2 weeks onwards for

all casks. Nevertheless, some samples also yielded amplicons after wort transfer to the casks and after one week; these profiles were comparable to the profiles obtained after 2 weeks for all casks (data not shown). doi:10.1371/journal.pone.0095384.g001

Visual inspection of the bacterial community profiles revealed differences primarily during the first 12 months of the fermentation process, both in terms of presence and intensity of DNA bands. With the exception of two amplicons in the high % G+C region of the fingerprints (Fig. 1, band classes marked 3 and 4), the bacterial community profiles generated after 18 months were virtually identical in both batches. This bacterial community profile was reached in batch 1 after 18 months of fermentation, compared to 6 months in batch 2. The latter may be due to the incubation of batch 2 casks at ambient temperature, which was higher during the summer months compared to batch 1 casks that were incubated at more constant but lower temperatures in the cellar. In batch 1, a very dense band disappeared after 1 month of fermentation (Fig. 1, band class 1), while another band appeared in the subsequent sample taken after 2 months of fermentation time (Fig 1, band class 2).

Visual inspection of the yeast community profiles revealed more simple fingerprints comprising one to six DNA bands throughout the fermentation process. Again, the communities in both batches reached a fairly stable and highly similar composition after 6 months in batch 2 compared to 18 months in batch 1, with two amplicons in the central % G+C region of the fingerprints that were consistently present (Fig. 1, band classes 5 and 6).

The moving window analysis of the Dice similarity values between DGGE profiles (Fig. 2A and 2C) demonstrated that the bacterial community profiles of the four casks of both batches showed a similar evolution in diversity. Consecutive samples displayed few changes. After 2 months, the appearance and disappearance of two dense bands (Fig. 1, band classes 1 and 2) resulted in a higher percentage change. The major transition in bacterial community profile appeared to occur after 18 months in batch 1, whereas the bacterial community profile changed after 6 months in batch 2 (Fig. 2A and 2C).

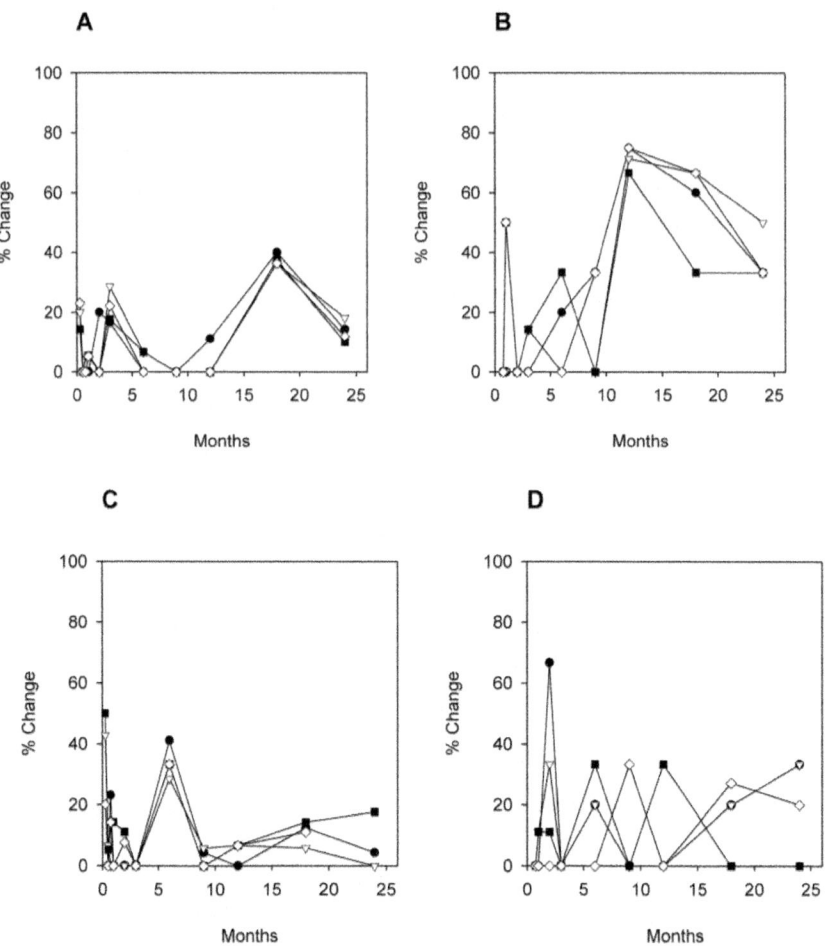

Figure 2: Moving windows analysis of the DGGE bacterial and yeast community profiles. Moving window analysis of the Dice-based similarity values between DGGE analyses of 4 casks from batches 1 and 2. (A) and (C) represent the bacterial diversity in batches 1 and 2, respectively, (B) and (D) visualize the yeast diversity of both batches 1 and 2. The last data point of the bacterial community profile analysis of batch 2, cask 4 was omitted due to the poor quality of the banding patterns. • Cask 1; ∇ Cask 2; ▪ Cask 3; ◊ Cask 4. doi:10.1371/journal.pone.0095384.g002

The moving window analysis of the yeast community profiles (Fig. 2B and 2D) revealed a higher variability. These higher percentages of change are most likely explained by the higher impact of changes in band presence or intensity in these profiles that comprised fewer bands. A total of 64 bands (28 from yeast community fingerprints and 36 from bacterial community fingerprints) were excised (Fig. S2) and sequenced to tentatively assign these band classes to microbial taxa. Because of the short length of the sequences (about 200 bp), EzBioCloud andBLAST analyses resulted in genus or family level identifications only. An overview of these identification data is shown in Table S1 and demonstrates that members of the*Enterobacteriaceae* family could be detected throughout the fermentation process in both batches. Both band class 1 and 2 (Fig. 1) were assigned to members of the*Enterobacteriaceae* family. Band class 2* (Fig. 1) that migrated at nearly the same position as band class 2 was assigned to *Pediococcus*/*Lactobacillus* (which could not be distinguished by using this short rRNA gene fragment). Also, additional band classes in a higher % G+C region of the profile were assigned to LAB, which were rarely found before month 3 in batch 1 samples, but which were nearly consistently present in batch 2 samples (Table S1A and S1B). Band classes 3 and 4 (Fig. 1) were assigned to AAB, which were detected from month six onwards in batch 2 samples and primarily during year 2 in batch 1 (Table S1A and S1B). Several DNA bands of the bacterial community fingerprints were assigned to yeast taxa (Table S1A and S1B), confirming that the V3 primers were not specific for bacteria [44], [45]. The yeast band classes 5 and 6 were assigned to the genus *Saccharomyces* (Table S1C and S1D) and were present throughout the fermentation. Bands originating from other yeast taxa (*Candida*, *Dekkera*/*Brettanomyces*, *Hanseniaspora*, *Kregervanrija*, *Naumovia* and*Wickerhamomyces*) were found frequently, albeit on an irregular basis.

Enumeration and identification of bacteria and yeasts

Table 1 presents an overview of the enumeration analyses and Table S2 presents the identifications of the MALDI-TOF MS clusters. A total of 1304 bacterial and 892 yeast isolates were obtained from the 2 batches. The freshly boiled wort did not allow microbial growth. However, both batches were spontaneously inoculated overnight in the cooling tun, as shown by the colony counts on MRS and VRBG agars, but no colonies were found on AAM agar. All cooling tun isolates (48 from batch 1 [Fig. 3] and 77 from batch 2 [data not shown]) were identified as members of the *Enterobacteriaceae* family. These bacteria were also isolated from MRS agar, which was thus not fully specific for the isolation of LAB. Both MRS and VRBG supported the growth of *Enterobacteriaceae*, but the relative species distribution differed (Fig. 3). Batch 1 isolates were identified as *Escherichia*/*Shigella* (*Escherichia coli* and *Shigella*species are extremely closely related [46] and cannot be distinguished by sequence analysis of conserved genes [47], [48]), *Enterobacter hormaechei* or *Enterobacter kobei*, whereas only the latter two were identified in batch 2 samples (31 and 46 of the 77 isolates, respectively).

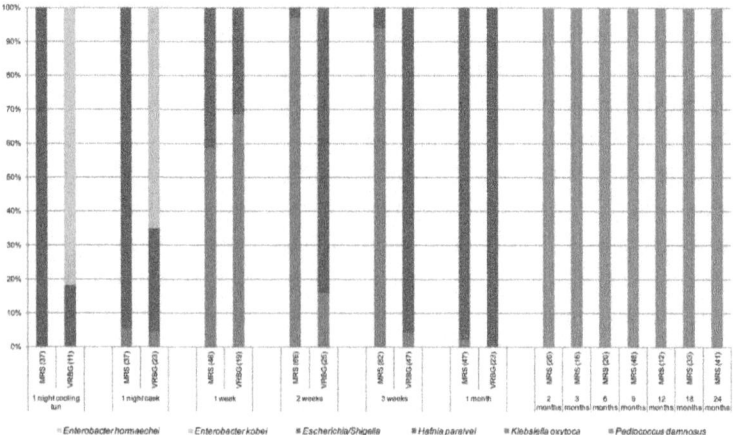

Figure 3: Identification of random isolates from MRS and VRBG agars of batch 1. The identification of isolates belonging to the *Enterobacteriaceae* are reported to the species level, when reliable identification by housekeeping gene sequences could be obtained. The number of isolates is given between brackets. doi:10.1371/journal. pone.0095384.g003

Enterobacteriaceae counts reached up to 10^7–10^8 CFU/mL after one to two weeks of fermentation. A total of 415 isolates from batch 1 samples taken during the first month were identified. *E. hormaechei* was no longer isolated after the transfer of the wort into the cask (performed 15 min after the sampling of the cooling tun), whereas *Klebsiella oxytoca* was then first isolated (Fig. 3). In the following weeks, the number of isolates identified as*Escherichia/ Shigella* and *E. kobei* decreased, while the numbers of *Hafnia paralvei* and*Klebsiella oxytoca* isolates increased until the end of the first month, after which*Enterobacteriaceae* were no longer isolated. In batch 2, from which a total of 398 isolates were identified, a similar evolution was found: the major occurrence of *H. paralvei* from week 1 onwards was confirmed and members of the *Enterobacteriaceae* were again no longer isolated after one month of fermentation (data not shown). However, batch 2 *Enterobacteriaceae* were more diverse and included also *Citrobacter gillenii* and *Raoultella terrigena* (data not shown).

From months 2 until 24, *Pediococcus damnosus* was consistently the only micro-organism isolated from MRS agar (batch 1 [Fig. 3]; batch 2, n=124 [data not shown]). The bacterial counts on MRS agar remained stable at about 10^4 CFU/mL until the end of the fermentation. Colony counts on AAM agar were generally low (below 10^4 CFU/mL; Table 1). AAM counts of the samples up to 3 months of fermentation were influenced by the presence of yeasts, which was due to the apparent loss of activity of amphotericin B under acidic conditions [49]. Amphotericin B was also reported to be unstable in other media with a composition similar to AAM [50]. A combination of amphotericin B and cycloheximide was subsequently found to be more effective in inhibiting yeast growth under all isolation conditions used.

Table 1: Results of plate counts on different agar isolation media. doi:10.1371/journal. pone.0095384.t001

Batch 1	VRBG 28°C	MRS 28°C	MRS 20°C AN	AAM 28°C	DYPAI 28°C	UBAGI 28°C	DYPAIX 28°C
Freshly boiled wort	ULD	ULD	ULD	ULD	ULD	ULD	ULD
1 night cooling tun	6.03	ULD	5.90	ULD	ND	ND	ND
1 night cask	6.51	5.25	6.42	ULD	ND	ND	ND
1 week	6.72	6.98	8.05	ULD	ND	ND	ND
2 weeks	7.73	7.68	7.77	ULD	ND	ND	ND
3 weeks	6.92	7.20	7.41	3.72	6.36	6.31	2.90
1 month	4.63	4.92	4.83	ULQ (466)	6.33	6.47	4.02
2 months	ULQ (40)	ULQ (80)	ULQ (180)	3.39	5.73	5.58	3.29
3 months	ULD	3.23	ULQ (33)	3.28	5.78	5.62	ULQ (273)
6 months	ULD	ULQ (300)	ULQ (447)	ULD	4.56	4.60	4.03
9 months	ULD	3.51	4.38	3.01	3.20	3.24	2.87
12 months	ULD	ULD	2.79	ULQ (26)	4.30	4.35	3.16
18 months	ULD	2.83	2.93	ULD	2.80	ULQ (347)	ULQ (293)
24 months	ULD	3.08	4.19	2.96	3.74	3.83	2.94
Batch 2							
Freshly boiled wort	ULD	ULD	ULD	ULD	ULD	ULD	ULD
1 night cooling tun	5.12	6.71	6.92	ULD	ULQ (50)	ULQ (253)	ULQ (40)
1 night cask	6.13	6.79	6.11	ULD	3.29	3.42	3.00
1 week	7.91	8.41	8.29	ULD	4.36	4.33	ULQ (140)
2 weeks	7.54	7.67	7.50	ULD	6.21	6.18	2.72
3 weeks	6.92	6.78	7.00	ULQ (40)	5.51	5.49	ULQ (120)
1 month	4.88	4.91	4.86	ULQ (270)	5.18	5.17	ULQ (67)
2 months	ULD	4.46	4.58	3.54	5.37	5.31	ULQ (13)
3 months	ULD	6.42	6.35	4.73	4.50	4.46	ULQ (353)
6 months	ULD	4.58	5.02	ULD	4.26	4.30	4.34
9 months	ULD	5.45	5.48	ULQ (40)	4.09	3.02	3.07
12 months	ULD	5.80	5.77	ULD	3.15	2.72	3.51
18 months	ULD	3.81	4.38	ULD	ULQ (173)	ULQ (300)	ULQ (240)
24 months	ULD	4.07	4.26	ULQ (66)	3.08	3.18	3.18

AAB were isolated from batch 1 samples at 9 and 24 months (n=35) and from batch 2 samples at 3, 9 and 24 months (n=17). All but one of the isolates were identified as a novel*Acetobacter* species, for which the name *Acetobacter lambici* has been proposed [51]. One batch 2 isolate represented a novel *Gluconobacter* species, for which the name *Gluconobacter cerevisiae* has been proposed [52]. This erratic isolation of AAB was not in accordance with the consistent presence of AAB-derived DNA bands in the DGGE profiles from 6 months of fermentation onwards in batch 2 (Fig. 1).

An overview of the identified yeast species of batch 1 is graphically represented in Fig. 4 andFig. S3. Isolation and accurate enumeration of yeasts during the first two weeks of fermentation of batch 1 was not possible, due to an insufficient suppression of bacterial growth. In batch 2 samples (data not shown) yeasts could not be detected in the wort after one night in the cooling tun, but increased in numbers directly after the wort was transferred into the casks not more than 15 min after the cooling tun was sampled. Maximal counts (10^6CFU/mL) were reached after 2 weeks to 1 month of fermentation. *Debaryomyces hansenii*(17/18 isolates examined) and *S. cerevisiae* (1/18) were the sole species isolated directly after the transfer of the wort into the cask in batch 2. *S. cerevisiae* (22/44), *S. pastorianus* (21/44) and *Naumovia castellii* (1/44) were isolated after 1 week of fermentation. The relative number of *S. pastorianus* isolates increased further during the first three months of fermentation (a total of 198 isolates examined), until it was the only yeast species isolated on DYPAI and UBAGI agars after 2 months (32 isolates examined). After 3 months, *S. pastorianus* was still the predominant yeast (30/31); one isolate was identified as *N. castellii*.

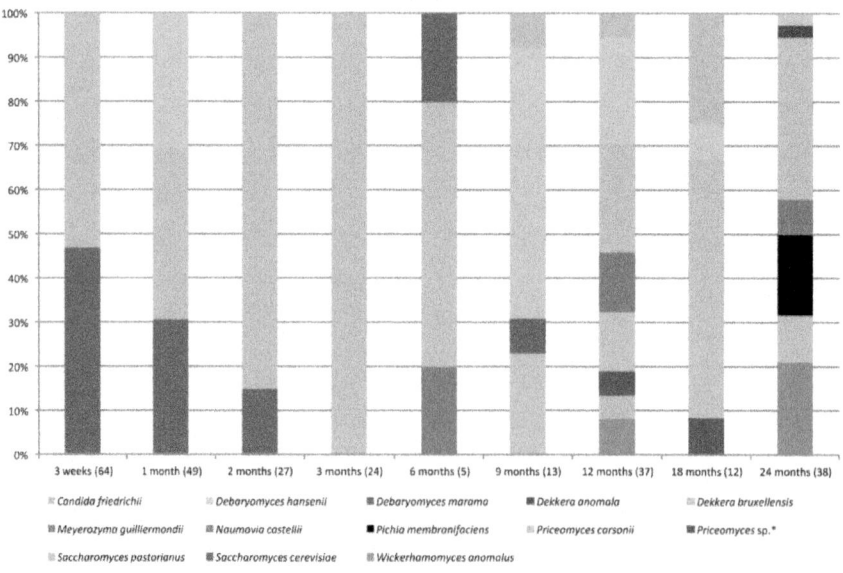

Figure 4: Identification of random isolates from DYPAI and UBAGI agars of batch 1. The number of isolates is given between brackets. *One yeast cluster from MALDI-TOF MS profiles could not be identified unambiguously (Table S2). doi:10.1371/journal.pone.0095384.g004

The same trend occurred during the first three months of fermentation of batch 1 (Fig. 4). *S. cerevisiae* and *S. pastorianus* were the most prevalent species and the latter one was the only yeast species present after three months. Yeast counts on DYPAIX agar were initially lower compared to DYPAI and UBAGI agars, but were comparable from 6 months onwards. The few

DYPAIX isolates that were obtained from samples after 2 months (batch 2) or 3 months (batches 1 and 2) failed to grow on the same growth agar medium upon subculture, indicating that there were no cycloheximide-resistant yeast species present in these samples (Fig. S3). DYPAIX isolates obtained from samples of the first 2 months of batch 1 included *N. castellii*,*Kazachstania servazzii* and *Db. hansenii* (Fig. S3), whereby the former was the only species isolated in the first month of batch 2 (n=58, data not shown).

Saccharomyces spp. were not isolated in large numbers after 6 months of fermentation, while*D. bruxellensis* was isolated at this point for the first time. *D. bruxellensis* was the major yeast species isolated from DYPAI and UBAGI agar media from 6 months until the end of the fermentation of batch 2 (n=102, data not shown) and the only yeast species isolated from DYPAIX agar in the same period (n=82). The cultivated yeast diversity in batch 2 was low compared to batch 1 (see below) and the three yeast media yielded the same species diversity from 6 months onwards.

The yeast species distribution in batch 1 samples after 6 months of fermentation (Fig. 4) was more complex than that of samples of the same age in batch 2. The most frequently cultivated species were *D. bruxellensis, Db. hansenii, Priceomyces carsonii* and *Wickerhamomyces anomalus* along with other species in lower numbers (Fig. 4 and Fig. S3). In contrast to batch 2 where the three yeast agar isolation media yielded the same species diversity from 6 months onwards, the species diversity recovered from different yeast agar isolation media in batch 1 was not comparable. For example, *D. bruxellensis* was not detected on the non-selective yeast agar media in batch 1 after 9 months, but was detected at this sampling point on DYPAIX agar (Fig. 4 and Fig. S3). The use of DYPAIX agar allowed isolating some unusual species from batch 1, such as *Candida patagonica* and *Yarrowia lipolytica* (Fig. S3), of which the latter has never been associated with a beer fermentation process. The total yeast and bacterial counts were similar in both batches after 24 months at about 10^3–10^4 CFU/mL (Table 1).

Air and brewery environment

None of the directly plated samples yielded growth. A total of 139 isolates from the brewery environment were picked up from the bacterial and yeast agar isolation media after enrichment and were identified through MALDI-TOF MS and sequence analysis of 16S rRNA genes or other molecular markers as described above (Table 2). Several species or taxa that were previously isolated during the fermentation process as described above were also found in environmental samples. *E. hormaechei* and *Escherichia*/*Shigella* were isolated from the cellar air.

Table 2: Overview of micro-organisms isolated from the brewery environment and their isolation sources. doi:10.1371/journal.pone.0095384.t002

Identification	Accession number	Accession number highest hit	Similarity	Present in fermentation	Air attic before cooling	Cooling tun	Roof	Air attic after cooling	Air cellar	Cellar ceiling	Cellar wall	Cask outside	Cask inside
Bacteria[a]													
Acetobacter cerevisiae[b]		KF537492	100%										+
Aerococcus urinaeequi		D87677	100%		+								
Bacillus licheniformis		AE017333	100%				·						
Enterobacter hormaechei				+				+					
Enterococcus faecium[c]	KJ186124	AJ843428	97%		+	+							
Escherichia/Shigella				+				·					
Hafnia alvei		M59155	100%										+
Lactobacillus curvatus		AJ621550	100%		+								
Lactobacillus malefermentans		BACN01000105	100%								+		
Lactobacillus nenjiangensis[c]	KJ186125	HF679044	99%		+								
Leuconostoc mesenteroides		CP000414	100%		+			·		+	+		
Leuconostoc pseudomesenteroides		AEOQ01000906	100%		+								
Pediococcus pentosaceus[c]		AM899822	100%		+								
Pseudomonas azotoformans		D84009	100%		+								
Pseudomonas libanensis		AF057645	100%		+								
Pseudomonas psychrotolerans		AJ575816	100%		+								
Rahnella aquatilis		CP003244	100%										+
Raoultella terrigena				+									+
Staphylococcus hominis		X6601	100%		+			+	·				
Yeasts[d]													
Brettanomyces custersianus		DQ406717	100%										+
Candida friedrichii				+	+								
Candida pomicola		AF245400	100%										+
Cryptococcus heveanensis		AF075467	100%										+
Cryptococcus magnus		AF181051	100%		+			+					+
Debaryomyces hansenii				+	+				+	·	-		+
Debaryomyces marama				+									+
Meyerozyma guilliermondii				+	+								
Pichia membranifaciens				+									+
Priceomyces sp.[a]				+					·				
Saccharomyces pastorianus				+					+				
Trichosporon gracile		JN939453	100%		+			·					
Trichosporon cutaneum		AF075483	100%									+	
Wickerhamomyces anomalus				+	+				+			+	+

The bacteria and yeasts present in the fermentation were identified based on their MALDI-TOF MS spectra. *One yeast cluster from MALDI-TOF MS profiles could not be identified unambiguously (Table S2).
[a]Identification is based on 16S rRNA gene sequence.
[b]Identification is based in *rpoB* sequence.
[c]Identification is based in *pheS* sequence.
[d]Identification is based on D1/D2 26S rRNA gene sequence.
doi:10.1371/journal.pone.0095384.t002

Raoultella terrigena, Pichia membranifaciens, Debaryomyces marama and *Db. hansenii* were isolated from the inside of a cask. The latter species was also isolated from the ceiling, the attic and cellar air, along with *S. pastorianus, Meyerozyma guilliermondii, Candida friedrichii* and *Wickerhamomyces anomalus*. The latter species was also found on the outside of a cask. A considerable number of additional micro-organisms that were not detected during the fermentation process were also isolated from environmental samples. These included species previously related to beverage fermentation or spoilage, such as *Brettanomyces custersianus* [53], *Pediococcus pentosaceus* [54], *Lactobacillus malefermentans* [55] and *Acetobacter cerevisiae* [56].

Discussion

Serious limitations of the few available microbiological studies of the lambic beer fermentation process are the rather low numbers of isolates identified using biochemical methods only [2],[3]. Recent polyphasic taxonomic studies revealed that phenotypic identification approaches alone have an inadequate taxonomical resolution for the accurate species level identification of these micro-organisms [10], [14]–[17], [38]. Therefore, the present study revisited the microbiology of the lambic beer fermentation process of the most traditional lambic brewery (Cantillon) in Belgium and identified and monitored the microbiota using MALDI-TOF MS as a high-throughput dereplication technique. This allowed to compare numerous fingerprints and to reduce these isolates to a non-redundant set of different species that were further identified using an array of DNA sequence-based methods [15], [57]. This approach allowed a more in depth analysis of the culturable microbiota of this ecosystem and resulted in the isolation and description of two novel AAB species, *i.e.*, *Acetobacter lambici* and *Gluconobacter cerevisiae*[51], [52]. The former species was even the most frequently isolated AAB species during the lambic fermentation process of Cantillon. The present study also used DGGE profiles of variable prokaryotic and eukaryotic rRNA gene regions to identify and monitor the microbial communities in two batches of lambic beer during a two-year fermentation period at Cantillon.

In both lambic batches, members of the *Enterobacteriaceae* were isolated during the first month, which corresponded to previous studies on Belgian lambic and American coolship ales[2], [3], [8]. The bacteria identified included *E. hormaechei*, *E. kobei*, *Es. coli*, *H. paralvei*, *K. oxytoca*, *Citrobacter gillenii* and *R. terrigena*, from which some of these were already detected in the cooling tun sample, suggesting their origin from the cooling tun environment. Remarkably, DNA from members of the *Enterobacteriaceae* family was detected in the DGGE experiments throughout the two-year fermentation period. This suggests that DNA from these cells persisted for a long time or, alternatively, that these bacteria remained present in a viable but non-culturable form, even under conditions to which *Enterobacteriaceae* are susceptible,*i.e.*, pH<4.0 and ethanol concentrations over 2.0% [58]. This has also been seen during cocoa bean fermentation [59], [60]. Yeast isolations during the first three months yielded*Saccharomyces* spp., but no *Hanseniaspora* spp., as expected from previous studies [2], [3]. However, *Hanseniaspora* spp. were detected by DGGE profiles over several months in both batches. Species of this genus are frequently found in spontaneously fermenting fruit and their preparations, and a positive contribution to wine flavor development is increasingly recognized (*e.g.*, Medina et al. [61]).

After the initial *Enterobacteriaceae* phase, the effects of ethanol production by the main fermentation were reflected in the dominance of *P. damnosus* at two months, along with some AAB (primarily *Acetobacter lambici*) that were occasionally isolated. AAB may survive in the cask due to the diffusion of oxygen through the wood [62], [63] or the short vacuum-releasing opening of the bung hole during sampling. Similarly, AAB seem to survive the anaerobic phase of cocoa bean fermentations [59], [60]. The irregular isolation of AAB may suggest that they are also present in a viable but non-culturable form that could be reversed when oxygen becomes available, as for example in wine production [64]. In both batches, *P. damnosus*remained present throughout the fermentation process and these bacteria were accompanied by *D. bruxellensis* after the decrease of *Saccharomyces* spp. Remarkably, no other LAB were isolated, while *Lactobacillus* spp. and other LAB species have also been isolated from American coolship ales recently [8].

The culture-independent detection of micro-organisms by DGGE was useful to observe the similar succession of microorganisms in each of the four casks of both lambic batches, and to visualize the relative stability of community profiles over time and their homogenization in the two batches at the advanced stage of the fermentation, but it confirmed some of the established pitfalls of this methodology. For instance, some cultivated yeast genera were not detected by DGGE (*Debaryomyces, Kazachstania, Meyerozyma, Pichia, Priceomyces, Yarrowia*), while other genera were detected by DGGE but not cultivated (*Hanseniaspora, Kregervanrija*). Also, some organisms were detected by DGGE before appearing in culture or after having disappeared from cultures, such as *Enterobacteriaceae* which were detected throughout the sampling period. Similar observations using T-RFLP and barcoded amplicon sequencing were made in spontaneous fermentations of American coolship ales [8]. Cultivation experiments too can be strongly biased, for instance, by the presence of VBNC cells, the selection of the culture media in the experiment design and by culture media that favor specific organisms. Therefore, a combination of multiple complementary techniques including both culture-based and culture-independent methods and a cautious interpretation of the results remains the best approach for microbial diversity analyses [65].

The microbial community analyses of the present study did not provide evidence for an extended acidification phase [3], as after six months *P. damnosus* and *D. bruxellensis* were both present and *Saccharomyces* spp. were no longer isolated. In addition, neither lambic batch showed a clear decrease of LAB. Pending a detailed analysis of the microbial metabolites and other biochemical characteristics, the data of the present study suggest that the acidification took place rapidly at the transition from the main fermentation phase to the long maturation phase, as was also found for American coolship ale fermentations [8]. The two nearly simultaneously fermented wort batches were inoculated by micro-organisms present in the brewery air, equipment or casks. As discussed above, members of the *Enterobacteriaceae* family were present in the wort before its transfer into the casks. These rather adventitious bacteria, *S. pastorianus* and some other yeast species, may have at least partially originated from the brewery air, but the present study failed to isolate the key micro-organisms *P. damnosus, S. cerevisiae* and *D. bruxellensis* from environmental samples. These micro-organisms were either missed by the sampling protocol or were concealed in niches that were not sampled. Examples of such niches are biofilms in and the pores of the wooden casks. Micro-organisms may have penetrated and effectively be immobilized and protected from washing steps in the wood of the cask, as demonstrated by Swaffield et al.[66], [67]. All casks had been used for lambic production before, preceded by their use in different fermentations, mostly red wine, so they could have retained specific microbiota in spite of cleaning procedures after previous fermentations [66], [67].

This study generally confirmed and extended the microbial diversity and succession known from previous accounts of lambic beers. The more than 2000 microbial isolates from two fermentation batches of the present study showed diverse members of the *Enterobacteriaceae* family during the first month, and *S. cerevisiae* and *S. pastorianus* from the first week until two and three months, respectively. No LAB were recovered during this first phase, which was previously denoted as the 'mixed acid fermentation'. The main fermentation was characterized by *Saccharomyces* spp. and the completion of the shift from *Enterobacteriaceae* to *P. damnosus*, the latter being isolated from 2 months onwards. The increase of LAB in months 2 and 3 and the concomitant decrease of *Saccharomyces* spp. was followed by the highly acid- and ethanol-resistant *D. bruxellensis,* which dominated from 6 months onwards together with *P. damnosus. Hanseniaspora* spp. that were previously reported in the first fermentation weeks were not isolated, but their presence was evidenced by DGGE analyses. The role of these and other taxa, such as *N. castellii* and *Kazachstania* spp., both also seen in lambic beer fermentations before, is not known.

Despite apparent differences in the microbial diversity, both batches examined reached similar community profiles at the end of the fermentation. The time needed to reach these final community fingerprints differed between the two batches and it is likely that the lower ambient temperature in the localization of batch 1 explains both the longer period needed to reach the characteristic community fingerprints as well as the larger diversity observed in later phases of the fermentation process.

ACKNOWLEDGMENTS

The authors highly appreciate the help and collaboration of Jean Van Roy of the Cantillon brewery and his brewery staff. The Genbank/EMBL accession numbers for the sequences generated in this study are KJ186115-KJ186128.

AUTHOR CONTRIBUTIONS

Conceived and designed the experiments: FS MJ AVL LDV PV. Performed the experiments: FS. Analyzed the data: FS ADW HMD AVL LDV PV. Contributed reagents/materials/analysis tools: ADW MA HMD. Wrote the paper: FS HMD AVL LDV PV.

REFERENCES

1.   De Keersmaecker J (1996) The mystery of lambic beer. Sci Am 275: 74–81. doi: 10.1038/scientificamerican0896-74

2.   Van Oevelen D, Spaepen M, Timmermans P, Verachtert H (1977) Microbiological aspects of spontaneous wort fermentation in the production of lambic and gueuze. J Inst Brew 83: 356–360. doi: 10.1002/j.2050-0416.1977.tb03825.x

3.   Verachtert H, Iserentant D (1995) Properties of Belgian acid beers and their microflora. Part I. The production of gueuze and related refreshing acid beers. Cerevisia, Belgian Journal of Brewing and Biotechnology 20: 37–41.

4.   Martens H, Dawoud E, Verachtert H (1991) Wort enterobacteria and other microbial-populations involved during the 1st month of lambic fermentation. J Inst Brew 97: 435–439. doi: 10.1002/j.2050-0416.1991.tb01082.x

5.   Meyer SA, Smith MT, Simione FP Jr (1978) Systematics of *Hanseniaspora* zikes and *Kloeckera* janke. Antonie Leeuwenhoek 44: 79–96. doi: 10.1007/bf00400078

6.   Kurtzman CP (2003) Phylogenetic circumscription of *Saccharomyces*, *Kluyveromyces* and other members of the Saccharomycetaceae, and the proposal of the new genera *Lachancea, Nakaseomyces, Naumovia, Vanderwaltozyma* and *Zygotorulaspora*. FEMS Yeast Res 4: 233–245. doi: 10.1007/springerreference_17039

7.   Nguyen H-V, Gaillardin C (2005) Evolutionary relationships between the former species *Saccharomyces uvarum* and the hybrids *Saccharomyces bayanus* and *Saccharomyces pastorianus*; reinstatement of *Saccharomyces uvarum* (Beijerinck) as a distinct species. FEMS Yeast Res 5: 471–483.

doi: 10.1016/j.femsyr.2004.12.004

8.   Bokulich NA, Bamforth CW, Mills DA (2012) Brewhouse-resident microbiota are responsible for multi-stage fermentation of American coolship ale. PLoS One 7: e35507. doi: 10.1371/journal.pone.0035507

9.   Cleenwerck I, De Vos P, De Vuyst L (2010) Phylogeny and differentiation of species of the genus *Gluconacetobacter* and related taxa based on multilocus sequence analyses of housekeeping genes and reclassification of *Acetobacter xylinus* subsp.*sucrofermentans* as *Gluconacetobacter sucrofermentans* (Toyosaki et al. 1996) sp. nov., comb. nov. Int J Syst Evol Microbiol 60: 2277–2283. doi: 10.1099/ijs.0.018465-0

10.   De Bruyne K, Franz CM, Vancanneyt M, Schillinger U, Mozzi F, et al. (2008)*Pediococcus argentinicus* sp. nov. from Argentinean fermented wheat flour and identification of *Pediococcus* species by *pheS*, *rpoA* and *atpA* sequence analysis. Int J Syst Evol Microbiol 58: 2909–2916. doi: 10.1099/ijs.0.65833-0

11.   De Bruyne K, Schillinger U, Caroline L, Boehringer B, Cleenwerck I, et al. (2007)*Leuconostoc holzapfelii* sp. nov., isolated from Ethiopian coffee fermentation and assessment of sequence analysis of housekeeping genes for delineation of*Leuconostoc* species. Int J Syst Evol Microbiol 57: 2952–2959. doi: 10.1099/ijs.0.65292-0

12.   Naser SM, Dawyndt P, Hoste B, Gevers D, Vandemeulebroecke K, et al. (2007) Identification of lactobacilli by *pheS* and *rpoA* gene sequence analyses. Int J Syst Evol Microbiol 57: 2777–2789. doi: 10.1099/ijs.0.64711-0

13.   Cleenwerck I, Gonzalez A, Camu N, Engelbeen K, De Vos P, et al. (2008) *Acetobacter fabarum* sp. nov., an acetic acid bacterium from a Ghanaian cocoa bean heap fermentation. Int J Syst Evol Microbiol 58: 2180–2185. doi: 10.1099/ijs.0.65778-0

14.   Kurtzman CP, Robnett CJ (1998) Identification and phylogeny of ascomycetous yeasts from analysis of nuclear large subunit (26S) ribosomal DNA partial sequences. Antonie Leeuwenhoek 73: 331–371.

15.   Vandamme P, Pot B, Gillis M, De Vos P, Kersters K, et al. (1996) Polyphasic taxonomy, a consensus approach to bacterial systematics. Microbiol Rev 60: 407–438.

16.   Cleenwerck I, De Vos P (2008) Polyphasic taxonomy of acetic acid bacteria: An overview of the currently applied methodology. Int J Food Microbiol 125: 2–14. doi: 10.1016/j.ijfoodmicro.2007.04.017

17.   Latouche GN, Daniel HM, Lee OC, Mitchell TG, Sorrell TC, et al. (1997) Comparison of use of phenotypic and genotypic characteristics

for identification of species of the anamorph genus *Candida* and related teleomorph yeast species. J Clin Microbiol 35: 3171–3180.

18. Garvie E (1974) Nomenclatural problems of the pediococci. Request for an opinion. International Journal of Systematic Bacteriology 24: 301–306. doi: 10.1099/00207713-24-2-301

19. Judicial Commission of the International Committee on Systematic Bacteriology (1976) Opinion 52: Conservation of the Generic Name Pediococcus Claussen with the Type Species Pediococcus damnosus Claussen. International Journal of Systematic Bacteriology 26: 292. doi: 10.1099/00207713-26-2-292

20. Kufferath H, Van Laer M (1921) Études sur les levures du Lambic. Bull Soc Chim Belgique 30: 270–276.

21. Van der Walt J (1964) *Dekkera*, a new genus of the *Saccharomycetaceae*. Antonie Leeuwenhoek 30: 273–280. doi: 10.1007/bf02046733

22. Smith MT, Yamazaki M, Poot G (1990) *Dekkera*, *Brettanomyces* and *Eeniella*: Electrophoretic comparison of enzymes and DNA–DNA homology. Yeast 6: 299–310.

23. Camu N, De Winter T, Verbrugghe K, Cleenwerck I, Vandamme P, et al. (2007) Dynamics and biodiversity of populations of lactic acid bacteria and acetic acid bacteria involved in spontaneous heap fermentation of cocoa beans in Ghana. Appl Environ Microbiol 73: 1809–1824. doi: 10.1128/aem.02189-06

24. Duytschaever G, Huys G, Bekaert M, Boulanger L, De Boeck K, et al. (2011) Cross-sectional and longitudinal comparisons of the predominant fecal microbiota compositions of a group of pediatric patients with cystic fibrosis and their healthy siblings. Appl Environ Microbiol 77: 8015–8024. doi: 10.1128/aem.05933-11

25. Cocolin L, Bisson LF, Mills DA (2000) Direct profiling of the yeast dynamics in wine fermentations. FEMS Microbiol Lett 189: 81–87. doi: 10.1111/j.1574-6968.2000.tb09210.x

26. Marzorati M, Wittebolle L, Boon N, Daffonchio D, Verstraete W (2008) How to get more out of molecular fingerprints: practical tools for microbial ecology. Environ Microbiol 10: 1571–1581. doi: 10.1111/j.1462-2920.2008.01572.x

27. Altschul SF, Madden TL, Schaffer AA, Zhang J, Zhang Z, et al. (1997) Gapped BLAST and PSI-BLAST: a new generation of protein database search programs. Nucleic Acids Res 25: 3389–3402. doi: 10.1093/nar/25.17.3389

28. Kim OS, Cho YJ, Lee K, Yoon SH, Kim M, et al. (2012) Introducing EzTaxon-e: a prokaryotic 16S rRNA gene sequence database with phylotypes that represent uncultured species. Int J Syst Evol Microbiol 62: 716–721. doi: 10.1099/ijs.0.038075-0

29. De Man J, Rogosa M, Sharpe ME (1960) A medium for the cultivation of lactobacilli. J Appl Microbiol 23: 130–135. doi: 10.1111/j.1365-2672.1960.tb00188.x

30. Mossel D, Elederink I, Koopmans M, Van Rossem F (1978) Optimalisation of a MacConkey-type medium for the enumeration of *Enterobacteriaceae*. Lab Practice 27: 1049–1050.

31. Mossel D, Mengerink W, Scholts H (1962) Use of a modified MacConkey agar medium for the selective growth and enumeration of *Enterobacteriaceae*. J Bacteriol 84: 381.

32. Lisdiyanti P, Katsura K, Potacharoen W, Navarro RR, Yamada Y, et al. (2003) Diversity of acetic acid bacteria in Indonesia, Thailand, and the Philippines. Microbiol Cult Coll 19: 91–98.

33. Abbott DA, Hynes SH, Ingledew WM (2005) Growth rates of *Dekkera/Brettanomyces* yeasts hinder their ability to compete with *Saccharomyces cerevisiae* in batch corn mash fermentations. Appl Microbiol Biotechnol 66: 641–647. doi: 10.1007/s00253-004-1769-1

34. Licker J, Acree T, Henick-Kling T (1998) What is "brett"(*Brettanomyces*) flavor?: A preliminary investigation. ACS Publications.pp. 96–115.

35. Suárez R, Suárez-Lepe J, Morata A, Calderón F (2007) The production of ethylphenols in wine by yeasts of the genera *Brettanomyces* and *Dekkera*: A review. Food Chem 102: 10–21. doi: 10.1016/j.foodchem.2006.03.030

36. Wieme A, Cleenwerck I, Van Landschoot A, Vandamme P (2012) *Pediococcus lolii* DSM 19927^T and JCM 15055^T are strains of *Pediococcus acidilactici*. Int J Syst Evol Microbiol 62: 3105–3108. doi: 10.1099/ijs.0.046201-0

37. Mollet C, Drancourt M, Raoult D (1997) *rpoB* sequence analysis as a novel basis for bacterial identification. Mol Microbiol 26: 1005–1011. doi: 10.1046/j.1365-2958.1997.6382009.x

38. Nhung PH, Ohkusu K, Mishima N, Noda M, Shah MM, et al. (2007) Phylogeny and species identification of the family *Enterobacteriaceae* based on *dnaJ* sequences. Diagn Microbiol Infect Dis 58: 153–161. doi: 10.1016/j.diagmicrobio.2006.12.019

39. Naser SM, Thompson FL, Hoste B, Gevers D, Dawyndt P, et al. (2005) Application of multilocus sequence analysis (MLSA) for rapid

identification of *Enterococcus* species based on *rpoA* and *pheS* genes. Microbiology 151: 2141–2150. doi: 10.1099/mic.0.27840-0

40. Daniel HM, Meyer W (2003) Evaluation of ribosomal RNA and actin gene sequences for the identification of ascomycetous yeasts. Int J Food Microbiol 86: 61–78. doi: 10.1016/s0168-1605(03)00248-4

41. Snauwaert I, Papalexandratou Z, De Vuyst L, Vandamme P (2013) Characterization of strains of *Weissella fabalis* sp. nov. and *Fructobacillus tropaeoli* from spontaneous cocoa bean fermentations. Int J Syst Evol Microbiol 63: 1709–1716. doi: 10.1099/ijs.0.040311-0

42. Niemann S, Puhler A, Tichy HV, Simon R, Selbitschka W (1997) Evaluation of the resolving power of three different DNA fingerprinting methods to discriminate among isolates of a natural *Rhizobium meliloti* population. J Appl Microbiol 82: 477–484. doi: 10.1046/j.1365-2672.1997.00141.x

43. Harju S, Fedosyuk H, Peterson KR (2004) Rapid isolation of yeast genomic DNA: Bust n'Grab. BMC Biotechnol 4: 8.

44. Scheirlinck I, Van der Meulen R, Van Schoor A, Vancanneyt M, De Vuyst L, et al. (2008) Taxonomic structure and stability of the bacterial community in belgian sourdough ecosystems as assessed by culture and population fingerprinting. Appl Environ Microbiol 74: 2414–2423. doi: 10.1128/aem.02771-07

45. Van der Meulen R, Scheirlinck I, Van Schoor A, Huys G, Vancanneyt M, et al. (2007) Population dynamics and metabolite target analysis of lactic acid bacteria during laboratory fermentations of wheat and spelt sourdoughs. Appl Environ Microbiol 73: 4741–4750. doi: 10.1128/aem.00315-07

46. Brenner D (1984) Family I. *Enterobacteriaceae* Rahn 1937, Nom. Fam. Cons. Opin. 15, Jud. Comm. 1958, 73; Ewing, Farmer and Brenner 1980, 674; Judicial Commission 1981, 104. In: Krieg NR, Holt JG, editors. Bergey's Manual of Systematic Bacteriology. Baltimore: Williams & Wilkins. pp. 408–420.

47. Lan R, Reeves PR (2002) *Escherichia coli* in disguise: molecular origins of *Shigella*. Microbes Infect 4: 1125–1132. doi: 10.1016/s1286-4579(02)01637-4

48. Pupo GM, Lan R, Reeves PR (2000) Multiple independent origins of *Shigella* clones of *Escherichia coli* and convergent evolution of many of their characteristics. Proc Natl Acad Sci U S A 97: 10567–10572. doi: 10.1073/pnas.180094797

49. te Dorsthorst DT, Verweij PE, Meis JF, Mouton JW (2005) Relationship

between in vitro activities of amphotericin B and flucytosine and pH for clinical yeast and mold isolates. Antimicrob Agents Chemother 49: 3341–3346. doi: 10.1128/aac.49.8.3341-3346.2005

50. Cheung SC, Medoff G, Schlessinger D, Kobayashi GS (1975) Stability of amphotericin B in fungal culture media. Antimicrob Agents Chemother 8: 426–428. doi: 10.1128/aac.8.4.426

51. Spitaels F, Li L, Wieme A, Balzarini T, Cleenwerck I, et al. (2013) *Acetobacter lambici*sp. nov. isolated from fermenting lambic beer. Int J Syst Evol Microbiol: published ahead of print December 20, 2013, doi:2010.1099/ijs.2010.057315-057310

52. Spitaels F, Wieme AD, Balzarini T, Cleenwerck I, Van Landschoot A, et al. (2013)*Gluconobacter cerevisiae* sp. nov. isolated from the brewery environment. Int J Syst Evol Microbiol: published ahead of print December 24, 2013, doi:2010.1099/ijs.2010.059311-059310

53. Martens H, Iserentant D, Verachtert H (1997) Microbiological aspects of a mixed yeast-bacterial fermentation in the production of a special Belgian acidic ale. J Inst Brew 103: 85–91. doi: 10.1002/j.2050-0416.1997.tb00939.x

54. Hutzler M, Müller-Auffermann K, Koob J, Riedl R, Jacob F (2013) Beer spoiling microorganisms – a current overview. Brauwelt International 2013/ I: 23–25.

55. Farrow JA, Phillips BA, Collins MD (1988) Nucleic acid studies on some heterofermentative lactobacilli: Description of *Lactobacillus malefermentans* sp. nov. and *Lactobacillus parabuchneri* sp. nov. FEMS Microbiol Lett 55: 163–167. doi: 10.1111/j.1574-6968.1988.tb13927.x

56. Cleenwerck I, Vandemeulebroecke K, Janssens D, Swings J (2002) Re-examination of the genus *Acetobacter*, with descriptions of *Acetobacter cerevisiae* sp. nov. and*Acetobacter malorum* sp. nov. Int J Syst Evol Microbiol 52: 1551–1558. doi: 10.1099/ijs.0.02064-0

57. Dieckmann R, Graeber I, Kaesler I, Szewzyk U, von Dohren H (2005) Rapid screening and dereplication of bacterial isolates from marine sponges of the sula ridge by intact-cell-MALDI-TOF mass spectrometry (ICM-MS). Appl Microbiol Biotechnol 67: 539–548. doi: 10.1007/s00253-004-1812-2

58. Priest FG, Stewart GG (2006) Microbiology and microbiological control in the brewery. Handbook of brewing; Second edition. Boca Raton, FL: CRC Press. pp. 607–629.

59. Papalexandratou Z, Camu N, Falony G, De Vuyst L (2011) Comparison of the bacterial species diversity of spontaneous cocoa bean fermentations

carried out at selected farms in Ivory Coast and Brazil. Food Microbiol 28: 964–973. doi: 10.1016/j.fm.2011.01.010

60. Papalexandratou Z, Vrancken G, De Bruyne K, Vandamme P, De Vuyst L (2011) Spontaneous organic cocoa bean box fermentations in Brazil are characterized by a restricted species diversity of lactic acid bacteria and acetic acid bacteria. Food Microbiol 28: 1326–1338. doi: 10.1016/j.fm.2011.06.003

61. Medina K, Boido E, Fariña L, Gioia O, Gomez ME, et al. (2013) Increased flavour diversity of Chardonnay wines by spontaneous fermentation and co-fermentation with*Hanseniaspora vineae*. Food Chem 141: 2513–2521. doi: 10.1016/j.foodchem.2013.04.056

62. Joyeux A, Lafon-Lafourcade S, Ribereau-Gayon P (1984) Evolution of acetic Acid bacteria during fermentation and storage of wine. Appl Environ Microbiol 48: 153–156.

63. Ribéreau-Gayon P, Glories Y, Maujean A, Dubourdieu D (2006) Aging red wines in vat and barrel: phenomena occurring during aging. Handbook of Enology: The Chemistry of Wine Stabilization and Treatments, Volume 2, 2nd Edition. pp. 387–428.

64. Millet V, Lonvaud-Funel A (2000) The viable but non-culturable state of wine micro-organisms during storage. Lett Appl Microbiol 30: 136–141. doi: 10.1046/j.1472-765x.2000.00684.x

65. Lagier JC, Armougom F, Million M, Hugon P, Pagnier I, et al. (2012) Microbial culturomics: paradigm shift in the human gut microbiome study. Clin Microbiol Infect 18: 1185–1193. doi: 10.1111/1469-0691.12023

66. Swaffield CH, Scott JA (1995) Existence and development of natural microbial populations in wooden storage vats used for alcoholic cider maturation. J Am Soc Brew Chem 53: 117–120.

67. Swaffield CH, Scott JA, Jarvis B (1997) Observations on the microbial ecology of traditional alcoholic cider storage vats. Food Microbiol 14: 353–361. doi: 10.1006/fmic.1997.0105

Chapter 12

BIOACTIVE COMPOUNDS DERIVED FROM THE YEAST METABOLISM OF AROMATIC AMINO ACIDS DURING ALCOHOLIC FERMENTATION

Albert Mas,[1] Jose Manuel Guillamon,[1,2] Maria Jesus Torija,[1] Gemma Beltran,[1] Ana B. Cerezo,[3] Ana M. Troncoso,[3] and M. Carmen Garcia-Parrilla[3]

[1]Facultad de Enología, Universitat Rovira i Virgili, Marcel·lí Domingo s/n, 43003 Tarragona, Spain

[2]Departamento de Biotecnologia de Alimentos, Instituto de Agroquímica y Tecnología de los Alimentos (CSIC), Agustín Escardino, 7, 46980 Valencia, Spain

[3]Facultad de Farmacia, Universidad de Sevilla, Profesor García González, 2, 41012 Sevilla, Spain

ABSTRACT

Metabolites resulting from nitrogen metabolism in yeast are currently found in some fermented beverages such as wine and beer. Their study has recently attracted the attention of researchers. Some metabolites derived from aromatic amino acids are bioactive compounds that can behave as hormones or even mimic their role in humans and may also act as regulators in yeast. Although the metabolic pathways for their formation are well known, the physiological significance is still far from being understood. The understanding of this relevance will be a key element in managing the production of these compounds under controlled conditions, to offer fermented food with specific enrichment in these compounds or even to use the yeast as nutritional complements.

NITROGEN METABOLISM DURING ALCOHOLIC FERMENTATION

The transformation of grapes into wine is a biotechnological process where

microorganisms, primarily yeast, convert a sugary liquid in a water-alcohol solution of flavour and pleasant aroma. To perform this process, they use the nutrients present in the medium for growth, producing a range of metabolites that yield the complexity of fermented beverage.

The grape must is a very complex food product, with a variety of compounds ranging from mainstream (sugars) to very small but important quantities, from both nutritional (vitamins, minerals, and polyphenols) and organoleptic (flavour and precursors) points of view. However, far from being an optimal culture medium, it is indeed a highly selective medium. This selectivity is due to the high sugar content, present in equimolar concentrations of glucose and fructose between 170 and 280 g/L, low pH (ranging from 2.8 to 3.5), nutrient limitation (especially nitrogen), and some technological practices such the addition of SO_2 (up to 150 mg/L in certain cases) and fermentations with a broad range of temperatures (from 10°C up to 35°C). However, the sugar concentration can reach much higher concentrations in certain cases, such as dehydration and overripening. Overripening could be natural (raisins, attacks from Botrytis and other fungi) or induced during wine making (cooking must, water elimination by reverse osmosis, or using frozen grapes, etc.). Additionally, there is a strong imbalance with the nitrogen fraction, which is in a concentration three orders of magnitude lower (concentrations between 70 and 600 mg/L). This nitrogen component plays a predominant role in the fermentation process. Grape must contains a variety of nitrogen compounds, among which the most important are amino acids, ammonium ion, and small peptides. These nitrogen compounds, excluding proline, constitute what is called yeast assimilable nitrogen. Nitrogen affects yeast cells in two aspects: biomass production during fermentation and the fermentation rate [1]. Therefore, the nitrogen content exerts an action on fermentation by regulating both its rate and its end. In fact, the lack of nitrogen has been pointed as one of the main reasons of stuck or sluggish fermentations [2, 3]. Stuck and sluggish fermentations are detrimental for wine quality as they leave residual sugars that would increase microbial instability and change the organoleptical properties of the final wine. The nitrogen content also affects other pathways in yeast, in particular, through the redox status of the cells, which affects the production of ethanol and other metabolites such as glycerol, acetic acid, and succinic acid [4–6]. Finally, other metabolites very relevant to wine quality are the volatile compounds and Saccharomyces cerevisiae produces different concentrations of those depending on fermentation conditions. Among these conditions, the quality and quantity of the nitrogen sources are critical in the formation of some aromatic molecules [7]. A range of volatile compounds such as acetate and ethyl esters, higher alcohol, volatile fatty acids, and carbonyls, which are the main molecules contributing to secondary or fermenting wine flavour, are mainly synthesised as metabolites derived from the metabolism of nitrogen [7–10]. Thus, nitrogen availability modulates the organoleptic quality and the taste of wine [11].

The presence of nitrogen in any of these chemical forms is highly variable, depending on various factors, including grape variety, degree of ripeness, soil, climate characteristics, and various technological aspects (type of vinification, pressing, etc. [12]). The current context of global warming, which results in overripe grapes, has two very direct effects on the composition of the must: higher sugar concentration and lower nitrogen levels. This combination produces higher fermentation hurdles. Thus, the knowledge of the nitrogen needs of different wine yeast strains used in the wine industry becomes more necessary. The addition of nitrogen to must is a very common practice among winemakers to avoid fermentation problems. Thus, the nitrogen addition should be adjusted to the real needs of each wine strain to prevent excessive concentrations, which would have negative consequences. The most relevant ones are the microbial instability of wines due to the nitrogen availability for the proliferation of other microorganisms or the synthesis of unhealthy substances, such as ethyl carbamate synthesis formed by yeast or biogenic amines, due to lactic acid bacteria during the malolactic fermentation using this residual nitrogen. Therefore, there is a need to optimise the use of the nitrogen by wine yeast, leaving very limited amounts of amino acids. Furthermore, excess nitrogen has also an impact on the organoleptic characteristics, such as the production of ethyl acetate or fruity aromas, depending on the type of nitrogen added (inorganic or organic, [13])

Although nitrogen concentration is a relevant factor, it is also important to underline that not all the nitrogen sources support equally yeast growth. In complex mixtures of amino acids and ammonium, such as grape must, wine yeasts have preference for some nitrogen sources, and the pattern of the preferential uptake of the nitrogen sources is determined by different molecular mechanisms. In S. cerevisiae, the mechanism is known globally as nitrogen catabolite repression (NCR). The NCR allows cells to detect the presence of the best sources of nitrogen by limiting the use of those that do not allow for the best growth. The detection of the rich nitrogen sources triggers a signalling chain that culminates with the activation of genes involved in the transport and metabolism of these rich sources and the suppression of those genes involved in the transport and use of poorer sources. Once the richest sources of nitrogen (ammonium, glutamine, and asparagine) are consumed, yeast metabolism activates the utilisation of the poorer sources of nitrogen (arginine, glutamate, alanine, etc.). Gutiérrez et al. [7] quantified the effect of different nitrogen sources on the three main parameters related to yeast growth (lag phase, generation time, and population size) in four commercial wine yeasts widely used in Spanish wineries, obtaining significant differences in these parameters concerning both the strain and the nitrogen sources. However, this study concluded that the categorisation between "good" sources and "bad" sources was dependent on the carbon backbone resulting from the metabolism of these amino acids. The transamination or deamination of "good" sources, which support rapid cell growth, produces easily assimilable carbon compounds by cell metabolism. This is the case of high growth rate sources such as glutamine, asparagine, glutamate, or alanine, which produce carbonyl derivatives such as α-ketoglutarate or pyruvate, which are readily integrated into the yeast fermentative metabolism. Instead, the amino acids with complex carbon backbones, which need to be detoxified or go through a complex metabolism, support slower growth.

DERIVATIVES OF AROMATIC AMINO ACID METABO-LISM AND ITS PHYSIOLOGIC CONSEQUENCES: THE EHRLICH PATHWAY

Aromatic amino acids are catabolised by the Ehrlich pathway, which starts with the transamination of the amino group and the formation of α-keto acid (Figure 1), such as indole pyruvate, phenyl pyruvate, and 4-hydroxyphenyl pyruvate from tryptophan, phenylalanine, and tyrosine, respectively (Table 1). Subsequently, these keto acids are decarboxylated to the corresponding aldehydes (indole acetaldehyde, phenyl acetaldehyde, and 4-hydroxyphenyl acetaldehyde). Finally, depending on the redox state of the cell, they can be further metabolised to the corresponding aromatic alcohol, indole 3-ethanol (tryptophol), phenyl ethanol, and tyrosol, or are oxidised to the corresponding acids, indole acetic acid, phenyl acetic acid, and 4-hydroxyphenyl acetic acid.

Table 1: Ehrlich pathway intermediates and derivatives of aromatic amino acids

Amino acid	Tryptophan	Tyrosine	Phenyl alanine
α-Keto acid	3-Indole pyruvate	p-Hydroxyphenyl pyruvate	Phenyl pyruvate
Higher aldehyde	3-Indole acetaldehyde	p-Hydroxyphenyl acetaldehyde	2-Phenylacetaldehyde
Higher alcohol	Tryptophol	Tyrosol	2-Phenylethanol
Higher acid	Indole acetic acid	4-Hydroxyphenyl acetic acid	Phenyl acetic acid

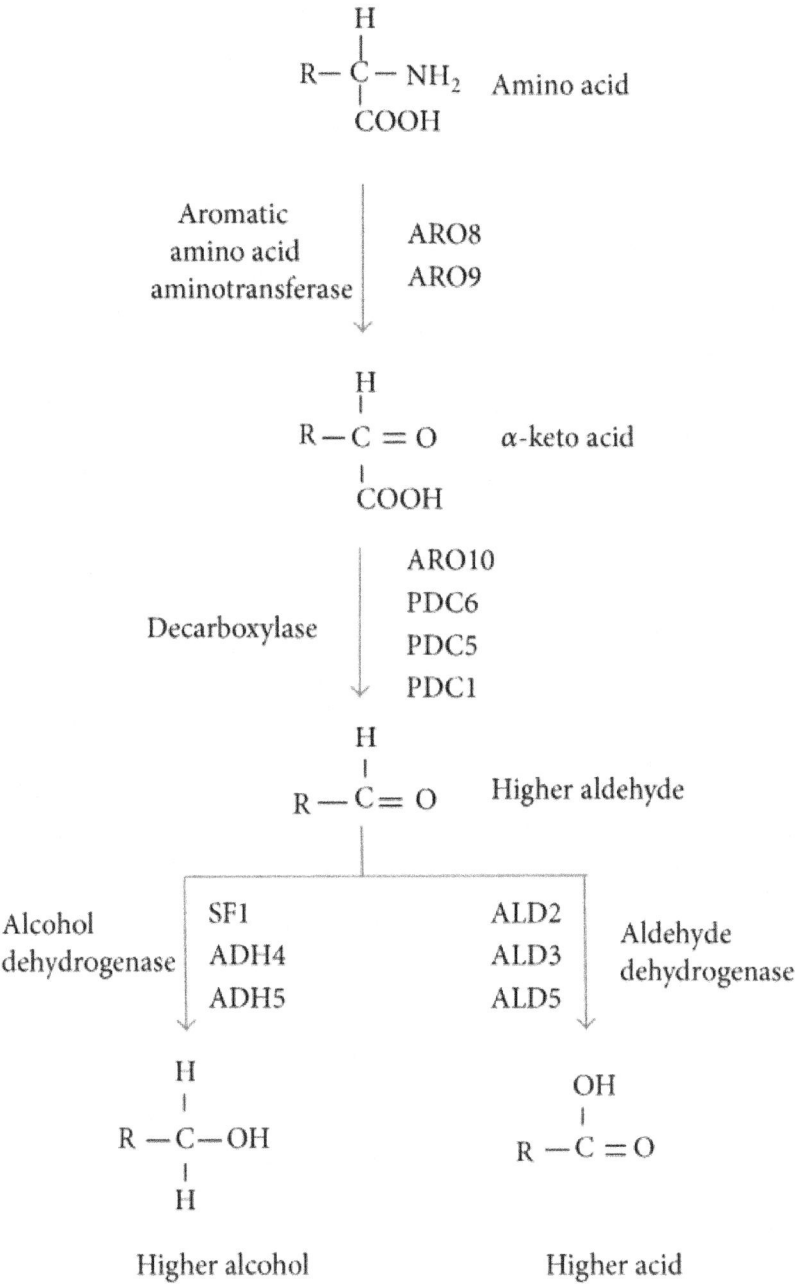

Figure 1: Ehrlich pathway of aromatic amino acids with indication of the enzymes and their coding genes.

As already mentioned, these higher alcohols affect wine aroma, especially 2-phenyl ethanol, which has a nice scent of roses, highly desired in some wines. Because of its industrial importance (it is widely used in cosmetics and as food additive), the production of this compound is well known among the full range of aromatic alcohols produced by yeast (for a review, see Hua and Xu [14]). Less attention has been given to the other types of aromatic alcohols, such as tryptophol or tyrosol, although their concentrations in some cases are also relatively high in wine (reaching up to 50 mg/L [10]). Regardless of the contribution of these higher alcohol levels to wine aroma, they have been recently described as the molecular modulators of some physiological and morphological processes considered involved in cell signalling. Hence, these higher alcohol levels have been linked with the stimulation of pseudohyphal growth in S. cerevisiae, resulting in a decrease in the growth rate [15]. This aspect has been also called as quorum sensing in yeast, which is related to population size and a morphological change from yeast to pseudohypha. The concept of quorum sensing appears in bacterial studies, described as the underlying mechanism that regulates the bacterial population in a variety of situations. In recent years, there have been several studies of quorum sensing in fungi and yeast species [16]. Interestingly, all these cases involve transient morphological changes of filamentous mycelium to yeast state or vice versa. One of the best-studied cases is the dimorphic human pathogenic yeast, Candida albicans. This yeast, depending on the conditions of the culture medium, moves from the nonpathogenic yeast form to the formation of hyphae or pathogenic form. Two alcohols that act antagonistically mediate this change: farnesol (intermediate in the synthesis of sterols) and tyrosol. Farnesol is excreted during cell growth, and when the cell population is high, the synthesis of farnesol increases and inhibits the formation of hyphae [17]. On the contrary, when the culture is diluted to a low cell density, the production of tyrosol promotes the formation of hyphae, and thus the yeast becomes a pathogen [18].

These morphological changes have also been observed in S. cerevisiae and also related to cell density signalling or quorum sensing. Chen and Fink [19] found that this yeast in stationary phase with high cell density and nutritional deficiencies, particularly in nitrogen, underwent pseudohyphal growth. Transcriptionally, this induction was associated with a fivefold increase in the activity of the FLO11 gene, an essential gene for pseudohyphal growth [20]. However, this induction was biochemically linked to two types of aromatic alcohols: phenyl ethanol and tryptophol. The addition of these kinds of alcohol to cultures resulted in very invasive pseudohyphal growth, along with an increased induction of FLO11. The final proof of the involvement of these types of aromatic alcohols in this morphological change was that mutants in ARO9 andARO10 genes, required for the synthesis of phenyl ethanol and tryptophol, dramatically decreased the pseudohyphal growth [19]. In turn, the expression of these genes is induced by tryptophol in a sort of self-stimulatory cycle. Therefore, high population densities produce more types of aromatic alcohols per cell compared with low population densities. These studies point to a direct relationship between the synthesis of these types of aromatic alcohols and the signalling cell density, nutrient deficiency, and entry into stationary phase.

However, the effect of these types of aromatic alcohols on humans is not only related to the flavour effect due to its presence in fermented foods. Tyrosol has been described as an antioxidant in human cell lines [21] and also as a cardioprotective agent [22]. The latter has been related to its presence in wine and attributed to some of the positive actions of moderate wine consumption [23]. On the other hand, tryptophol has been demonstrated to induce sleep in mice [24], although this action could be due to functional analogue or precursor of serotonin or melatonin.

SYNTHESIS OF OTHER BIOACTIVE COMPOUNDS FROM AROMATIC AMINO ACIDS

Moreover, there are other metabolites derived from these aromatic amino acids that are putative bioactive molecules with interesting properties. They have been only very recently described, and their metabolic pathways, regulation, coding genes, and so on are still under research. One of them is melatonin, which has been recently detected in wine, and its presence has been related to the activity of the yeast involved in the fermentation process [25, 26]. Originally, melatonin was seen as a unique product of the pineal gland of vertebrates and was called a neurohormone. However, in the last two decades, it has been identified in a wide range of invertebrates, plants, bacteria, and fungi. Therefore, today it is considered that melatonin is a ubiquitous molecule present in most living organisms [27]. Although little information is available on melatonin biosynthesis in organisms other than vertebrates, in yeast the pathway seems to be similar to the synthetic route and enzymes described in vertebrates [28]. This synthesis route is very simple, with four enzymes involved in the conversion of tryptophan to serotonin and N-acetylserotonin intermediates and finally to melatonin (Figure 2).

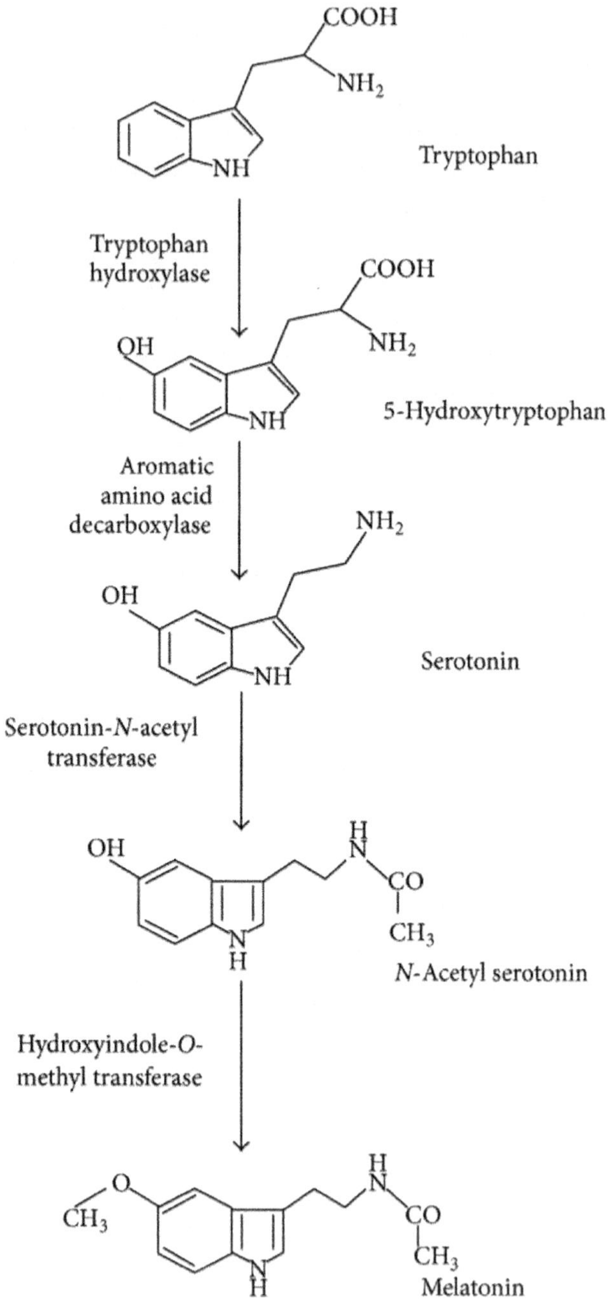

Figure 2: Synthesis of melatonin from tryptophan in yeast.

In humans, melatonin is a hormone that modulates physiological processes, such as circadian rhythms and reproductive functions, and also acts as an antioxidant [29–31]. In animals, melatonin typically occurs in the pineal gland (although subsequently described in other tissues and synthesised in important levels also in the intestine [32]), and its effects are very large, producing a pleiotropic response. Apart from the very well-described role as a regulator of circadian rhythm, melatonin has also been associated with antioxidant effects. These effects are not associated with a typical redox cycle but with a cascade of metabolites that turn into antioxidant activities [33, 34]. These antioxidant effects have also been correlated with an increased longevity [35] and the development of protective mechanisms against mutations [36], which would allow for a radioprotective effect [37]. Melatonin seems also to affect the immune system [38], although the mechanisms of action are poorly established. However, most of its effects suggest a clear neurohormonal activity, which has allowed us to relate their presence to learning and memory processes [39], ageing, and treatment for Alzheimer's disease [40, 41], amyotrophic lateral sclerosis [42], or migraine [43].

Regarding the presence or production of melatonin in yeast, the pioneer study of Sprenger et al. [28] related the presence of S. cerevisiae and the production of melatonin. Later, some reports detected melatonin in wines [44–46] and beer [47]. Recent studies also describe melatonin in grapes and other tissues of the vine, which could indicate that the origin was the substrate [23, 48, 49]. However, all the references that analyse the presence in wines and grape must indicate the production of melatonin during fermentation, being absent in the initial grape must [25, 47]. The description of melatonin in wine has linked its formation with yeast metabolism [25, 26], although the number of references in this case is still rare, indicating the need to pursue further this subject. In addition, all previous studies have focused exclusively on melatonin production bySaccharomyces yeasts, without considering the presence and metabolic activities of non-Saccharomyces wine yeast, significantly present in grapes and at the beginning of alcoholic fermentation. Therefore, the possible relation of the non-Saccharomyces wine yeast with the production of melatonin during alcoholic fermentation needs further evaluation.

Although the functions of melatonin are clear in mammals and animals [50], mainly related to regulatory mechanisms involved in circadian rhythms [51], the role of melatonin in yeast and other microorganisms seems to be even very far from being understood. Indeed, although the presence of circadian rhythms in yeast has been determined [52], this seems to be far from independent daily rhythms and regulated in response to the light produced in the multicellular organisms described. Instead, the response in yeast is induced by temperature changes only after several generations in chemostats and appears to be related to the primary nitrogen metabolism, particularly, to the expression of transporter genes of some nitrogen compounds (MEP2, which is the transporter on ammonium, and GAP1, which is a general amino acid transporter [53]). Thus, although melatonin is a ubiquitous molecule, its function in microorganisms is unknown. However, it has to be emphasised that in the organisms, where it has been studied, it exerts potent regulatory functions.

Melatonin can present up to nine isomers [54], including melatonin itself, because of the different pattern substitutions of the groups (N-acetyl-(2-aminoethyl)) and methoxy in the indolic ring. An isomer of melatonin was detected in wine. Indeed, the MS fragmentation ions of melatonin were different from those of the isomers found in wine [44]. Both melatonin and its isomer are present in different wine varieties, showing that those from the variety Jaen Tinto had the highest amount of melatonin isomer (21.9 ng/mL).

This finding was confirmed later by Gomez et al. [46], who described the isomer in Malbec wine and its formation during the fermentation step. Recently, Kocadağli et al. [55] detected the highest amount of melatonin isomer in yeast-fermented products (red wine, beer, and bread crump). Up to now, there is just one isomer of melatonin (N-acetyl-3-(2-aminoethyl)-6-methoxyindole) that is commercially available. However, when wines were spiked with this standard, the chromatograms showed 3 peaks with identical fragmentation pattern: melatonin, the N-acetyl-3-(2-aminoethyl)-6-methoxyindole, and the isomer not identified yet [44]. An analytical challenge is the position of the methoxy group in the indolic ring by mass spectrometry. Further analysis by NMR is required to elucidate its structure.

Serotonin is present at low levels in many plant-derived food products, including coffee beans, cherries, strawberries, and many others [56, 57]. Its role in the plant kingdom may be related to the regulation of root development probably by acting as a natural auxin inhibitor [58].

So far, there is no evidence for the production of serotonin by S. cerevisiae, although serotonin was some years ago found to be synthesised by yeast in response to UV radiation [59]. In fact, it is assumed that serotonin is an intermediate in the synthesis of melatonin in Saccharomyces cerevisiae, as it is in vertebrates [28, 60].

Serotonin is found in wines at levels ranging from 2 to 23 mg/L, mainly as a result of the malolactic fermentation, and significantly higher serotonin levels were observed when Lactobacillus plantarum was used [61, 62].

CONCLUSIONS

The metabolism of the aromatic amino acids in yeasts can produce a broad array of molecules that could be relevant from different aspects related to both yeast regulation and human health. The activities of these compounds as neurohormones and antioxidants open a new scenario of applications from nutrition supplements to functional foods. However, the role of these compounds in yeast is still far from being completely understood. Thus, it is still beyond our possibilities to modulate their production and their appearance in fermented food. Further research in the field of yeast metabolism related to the presence of aromatic amino acids would provide the theoretical basis for a broad array of applications in modern nutrition.

CONFLICT OF INTERESTS

The authors declare that there is no conflict of interests regarding the publication of this paper.

ACKNOWLEDGMENT

This work was supported by the Ministry of Economy and Competitiveness, Spain (Grant no. AGL2013-47300-C3).

REFERENCES

1.   C. Varela, F. Pizarro, and E. Agosin, "Biomass content governs fermentation rate in nitrogen-deficient wine musts," Applied and Environmental Microbiology, vol. 70, no. 6, pp. 3392–3400, 2004.

2.   L. F. Bisson, "Stuck and sluggish fermentations," The American Journal of Enology and Viticulture, vol. 50, no. 1, pp. 107–119, 1999.

3.   P. Taillandier, F. Ramon Portugal, A. Fuster, and P. Strehaiano, "Effect of ammonium concentration on alcoholic fermentation kinetics by wine yeasts for high sugar content," Food Microbiology, vol. 24, no. 1, pp. 95–100, 2007.

4.   E. Albers, C. Larsson, G. Lidén, C. Niklasson, and L. Gustafsson, "Influence of the nitrogen source onSaccharomyces cerevisiae anaerobic growth and product formation," Applied and Environmental Microbiology, vol. 62, no. 9, pp. 3187–3195, 1996.

5.   F. Radler, "Yeasts—metabolism of organic acids," in Wine Microbiology and Biotechnology, G. H. Fleet, Ed., pp. 165–182, Harwood Academic Publishers, Singapore, 1993.

6.   C. Camarasa, J.-P. Grivet, and S. Dequin, "Investigation by 13C-NMR

and tricarboxylic acid (TCA) deletion mutant analysis of pathways of succinate formation in Saccharomyces cerevisiae during anaerobic fermentation," Microbiology, vol. 149, no. 9, pp. 2669–2678, 2003.

7. A. Gutiérrez, G. Beltran, J. Warringer, and J. M. Guillamon, "Genetic basis of variations in nitrogen source utilization in four wine commercial yeast strains," PLoS ONE, vol. 8, no. 6, Article ID E67166, 2013.

8. A. Rapp and G. Versini, "Influence of nitrogen compounds in grapes on aroma compounds of wines,"Developments in Food Science, vol. 37, pp. 1659–1694, 1995.

9. P. A. Henschke and V. Jiranek, "Yeasts-metabolism of nitrogen compounds," in Wine Microbiology and Biotechnology, G. H. Fleet, Ed., pp. 77–164, Harwood Academic Publishers, Chur, Switzerland, 1993.

10. J. H. Swiegers, E. J. Bartowsky, P. A. Henschke, and I. S. Pretorius, "Yeast and bacterial modulation of wine aroma and flavour," Australian Journal of Grape and Wine Research, vol. 11, no. 2, pp. 139–173, 2005.

11. S.-J. Bell and P. A. Henschke, "Implications of nitrogen nutrition for grapes, fermentation and wine,"Australian Journal of Grape and Wine Research, vol. 11, no. 3, pp. 242–295, 2005.

12. P. Ribéreau-Gayon, D. Dubourdieu, B. Donèche, and A. Lonvaud, Handbook of Enology, John Wiley & Sons, 2nd edition, 2006.

13. D. Torrea, C. Varela, M. Ugliano, C. Ancin-Azpilicueta, I. Leigh Francis, and P. A. Henschke, "Comparison of inorganic and organic nitrogen supplementation of grape juice—effect on volatile composition and aroma profile of a Chardonnay wine fermented with Saccharomyces cerevisiae yeast,"Food Chemistry, vol. 127, no. 3, pp. 1072–1083, 2011.

14. D. Hua and P. Xu, "Recent advances in biotechnological production of 2-phenylethanol," Biotechnology Advances, vol. 29, no. 6, pp. 654–660, 2011.

15. J. R. Dickinson, "Filament formation in Saccharomyces cerevisiae—a review," Folia Microbiologica, vol. 53, no. 1, pp. 3–14, 2008.

16. G. F. Sprague Jr. and S. C. Winans, "Eukaryotes learn how to count: Quorum sensing by yeast," Genes and Development, vol. 20, no. 9, pp. 1045–1049, 2006.

17. J. M. Hornby, E. C. Jensen, A. D. Lisec et al., "Quorum sensing in the dimorphic fungus candida albicans is mediated by farnesol," Applied and Environmental Microbiology, vol. 67, no. 7, pp. 2982–2992, 2001.

18. H. Chen, M. Fujita, Q. Feng, J. Clardy, and G. R. Fink, "Tyrosol is a quorum-sensing molecule in Candida albicans," Proceedings of the

National Academy of Sciences of the United States of America, vol. 101, no. 14, pp. 5048–5052, 2004.

19. H. Chen and G. R. Fink, "Feedback control of morphogenesis in fungi by aromatic alcohols," Genes and Development, vol. 20, no. 9, pp. 1150–1161, 2006.

20. D. van Dyk, I. S. Pretorius, and F. F. Bauer, "Mss11p is a central element of the regulatory network that controls FLO11 expression and invasive growth in Saccharomyces cerevisiae," Genetics, vol. 169, no. 1, pp. 91–106, 2005.

21. C. Giovannini, E. Straface, D. Modesti et al., "Tyrosol, the major olive oil biophenol, protects against oxidized-LDL- induced injury in Caco-2 cells," Journal of Nutrition, vol. 129, no. 7, pp. 1269–1277, 1999.

22. S. M. Samuel, M. Thirunavukkarasu, S. V. Penumathsa, D. Paul, and N. Maulik, "Akt/FOXO3a/SIRT1-mediated cardioprotection by n-tyrosol against ischemic stress in rat in vivo model of myocardial infarction: switching gears toward survival and longevity," Journal of Agricultural and Food Chemistry, vol. 56, no. 20, pp. 9692–9698, 2008.

23. S. Vitalini, C. Gardana, A. Zanzotto et al., "From vineyard to glass: agrochemicals enhance the melatonin and total polyphenol contents and antiradical activity of red wines," Journal of Pineal Research, vol. 51, no. 3, pp. 278–285, 2011.

24. E. M. Cornford, P. D. Crane, L. D. Braun, W. D. Bocash, A. M. Nyerges, and W. H. Oldendorf, "Reduction in brain glucose utilization rate after tryptophol (3-indole ethanol) treatment," Journal of Neurochemistry, vol. 36, no. 5, pp. 1758–1765, 1981.

25. M. I. Rodriguez-Naranjo, M. J. Torija, A. Mas, E. Cantos-Villar, and M. D. C. Garcia-Parrilla, "Production of melatonin by Saccharomyces strains under growth and fermentation conditions," Journal of Pineal Research, vol. 53, no. 3, pp. 219–224, 2012.

26. M. Arevalo-Villena, E. J. Bartowsky, D. Capone, and M. A. Sefton, "Production of indole by wine-associated microorganisms under oenological conditions," Food Microbiology, vol. 27, no. 5, pp. 685–690, 2010.

27. D.-X. Tan, R. Hardeland, L. C. Manchester et al., "Functional roles of melatonin in plants, and perspectives in nutritional and agricultural science," Journal of Experimental Botany, vol. 63, no. 2, pp. 577–597, 2012.

28. J. Sprenger, R. Hardeland, B. Fuhrberg, and S.-Z. Han, "Melatonin and other 5-methoxylated indoles in yeast: presence in high concentrations

and dependence on tryptophan availability," Cytologia, vol. 64, no. 2, pp. 209–213, 1999.

29. E. Serrano, C. Venegas, G. Escames et al., "Antioxidant defence and inflammatory response in professional road cyclists during a 4-day competition," Journal of Sports Sciences, vol. 28, no. 10, pp. 1047–1056, 2010.

30. M. Chahbouni, G. Escames, C. Venegas et al., "Melatonin treatment normalizes plasma pro-inflammatory cytokines and nitrosative/oxidative stress in patients suffering from Duchenne muscular dystrophy," Journal of Pineal Research, vol. 48, no. 3, pp. 282–289, 2010.

31. A. López, J. A. García, G. Escames et al., "Melatonin protects the mitochondria from oxidative damage reducing oxygen consumption, membrane potential, and superoxide anion production," Journal of Pineal Research, vol. 46, no. 2, pp. 188–198, 2009.

32. G. A. Bubenik, "Gastrointestinal melatonin: localization, function, and clinical relevance," Digestive Diseases and Sciences, vol. 47, no. 10, pp. 2336–2348, 2002.

33. D.-X. Tan, L. C. Manchester, R. J. Reiter, W.-B. Qi, M. Karbownik, and J. R. Calvo, "Significance of melatonin in antioxidative defense system: reactions and products," Biological Signals and Receptors, vol. 9, no. 3-4, pp. 137–159, 2000.

34. D.-X. Tan, L. C. Manchester, M. P. Terron, L. J. Flores, and R. J. Reiter, "One molecule, many derivatives: a never-ending interaction of melatonin with reactive oxygen and nitrogen species?" Journal of Pineal Research, vol. 42, no. 1, pp. 28–42, 2007.

35. S. Oaknin-Bendahan, Y. Anis, I. Nir, and N. Zisapel, "Effects of long-term administration of melatonin and a putative antagonist on the ageing rat," NeuroReport, vol. 6, no. 5, pp. 785–788, 1995.

36. M. Karbownik, R. J. Reiter, S. Burkhardt, E. Gitto, D.-X. Tan, and A. Lewiñski, "Melatonin attenuates estradiol-induced oxidative damage to DNA: relevance for cancer prevention," Experimental Biology and Medicine, vol. 226, no. 7, pp. 707–712, 2001.

37. V. Vijayalaxmi, R. J. Reiter, T. S. Herman, and M. L. Meltz, "Melatonin and radioprotection from genetic damage: in vivo/in vitro studies with human volunteers," Mutation Research—Genetic Toxicology, vol. 371, no. 3-4, pp. 221–228, 1996.

38. A. Carrillo-Vico, J. M. Guerrero, P. J. Lardone, and R. J. Reiter, "A review of the multiple actions of melatonin on the immune system," Endocrine, vol. 27, no. 2, pp. 189–200, 2005.

39. J. Larson, R. E. Jessen, T. Uz et al., "Impaired hippocampal long-term potentiation in melatonin MT2 receptor-deficient mice," Neuroscience Letters, vol. 393, no. 1, pp. 23–26, 2006.

40. M. A. Pappolla, M. Sos, R. A. Omar et al., "Melatonin prevents death of neuroblastoma cells exposed to the Alzheimer amyloid peptide," Journal of Neuroscience, vol. 17, no. 5, pp. 1683–1690, 1997.

41. M. Pohanka, "Alzheimer's disease and related neurodegenerative disorders: implication and counteracting of melatonin," Journal of Applied Biomedicine, vol. 9, no. 4, pp. 185–196, 2011.

42. S. Jacob, B. Poeggeler, J. H. Weishaupt et al., "Melatonin as a candidate compound for neuroprotection in amyotrophic lateral sclerosis (ALS): high tolerability of daily oral melatonin administration in ALS patients," Journal of Pineal Research, vol. 33, no. 3, pp. 186–187, 2002.

43. D. W. Dodick and D. J. Capobianco, "Treatment and management of cluster headache," Current Pain and Headache Reports, vol. 5, no. 1, pp. 83–91, 2001.

44. M. I. Rodriguez-Naranjo, A. Gil-Izquierdo, A. M. Troncoso, E. Cantos-Villar, and M. C. Garcia-Parrilla, "Melatonin: a new bioactive compound in wine," Journal of Food Composition and Analysis, vol. 24, no. 4-5, pp. 603–608, 2011.

45. M. I. Rodriguez-Naranjo, A. Gil-Izquierdo, A. M. Troncoso, E. Cantos-Villar, and M. C. Garcia-Parrilla, "Melatonin is synthesised by yeast during alcoholic fermentation in wines," Food Chemistry, vol. 126, no. 4, pp. 1608–1613, 2011.

46. F. J. V. Gomez, J. Raba, S. Cerutti, and M. F. Silva, "Monitoring melatonin and its isomer in Vitis vinifera cv. Malbec by UHPLC-MS/MS from grape to bottle," Journal of Pineal Research, vol. 52, no. 3, pp. 349–355, 2012.

47. M. D. Maldonado, H. Moreno, and J. R. Calvo, "Melatonin present in beer contributes to increase the levels of melatonin and antioxidant capacity of the human serum," Clinical Nutrition, vol. 28, no. 2, pp. 188–191, 2009

48. M. Iriti, M. Rossoni, and F. Faoro, "Melatonin content in grape: myth or panacea?" Journal of the Science of Food and Agriculture, vol. 86, no. 10, pp. 1432–1438, 2006.

49. S. J. Murch, B. A. Hall, C. H. Le, and P. K. Saxena, "Changes in the levels of indoleamine phytochemicals during véraison and ripening of wine grapes," Journal of Pineal Research, vol. 49, no. 1, pp. 95–100, 2010.

50. A. Brzezinski, "Melatonin in humans," The New England Journal of Medicine, vol. 336, no. 3, pp. 186–195, 1997.

51. H. E. Boccalandro, C. V. González, D. A. Wunderlin, and M. F. Silva, "Melatonin levels, determined by LC-ESI-MS/MS, fluctuate during the day/night cycle in Vitis vinifera cv Malbec: evidence of its antioxidant role in fruits," Journal of Pineal Research, vol. 51, no. 2, pp. 226–232, 2011.

52. Z. Eelderink-Chen, G. Mazzotta, M. Sturre, J. Bosman, T. Roenneberg, and M. Merrow, "A circadian clock in Saccharomyces cerevisiae," Proceedings of the National Academy of Sciences of the United States of America, vol. 107, no. 5, pp. 2043–2047, 2010.

53. M. Merrow and M. Raven, "Finding time: a daily clock in yeast," Cell Cycle, vol. 9, no. 9, pp. 1671–1672, 2010.

54. G. Diamantini, G. Tarzia, G. Spadoni, M. D'Alpaos, and P. Traldi, "Metastable ion studies in the characterization of melatonin isomers," Rapid Communications in Mass Spectrometry, vol. 12, no. 20, pp. 1538–1542, 1998.

55. T. Kocadağli, C. Yilmaz, and V. Gökmen, "Determination of melatonin and its isomer in foods by liquid chromatography tandem mass spectrometry," Food Chemistry, vol. 153, pp. 151–156, 2014.

56. A. Ramakrishna, P. Giridhar, K. U. Sankar, and G. A. Ravishankar, "Melatonin and serotonin profiles in beans of Coffea species," Journal of Pineal Research, vol. 52, no. 4, pp. 470–476, 2012.

57. F. A. Badria, "Melatonin, serotonin, and tryptamine in some Egyptian food and medicinal plants,"Journal of Medicinal Food, vol. 5, no. 3, pp. 153–157, 2002.

58. R. Pelagio-Flores, R. Ortíz-Castro, A. Méndez-Bravo, L. Macías-Rodríguez, and J. López-Bucio, "Serotonin, a tryptophan-derived signal conserved in plants and animals, regulates root system architecture probably acting as a natural auxin inhibitor in arabidopsis thaliana," Plant and Cell Physiology, vol. 52, no. 3, pp. 490–508, 2011.

59. M. G. Strakhovskaia, A. M. Serdalina, and G. I. Fraĭkin, "Effect of the photo-induced synthesis of serotonin on the photoreactivation of Saccharomyces cerevisiae yeasts," Nauchnye Doklady Vysshei Shkoly. Biologicheskie Nauki, no. 3, pp. 25–28, 1983.

60. R. Hardeland and B. Poeggeler, "Melatonin beyond its classical funtions," The Open Physiology Journal, vol. 1, no. 1, pp. 1–22, 2008.

61. J. M. Landete, S. Ferrer, and I. Pardo, "Biogenic amine production by

lactic acid bacteria, acetic bacteria and yeast isolated from wine," Food Control, vol. 18, no. 12, pp. 1569–1574, 2007.

62. L. Manfroi, P. H. A. Silva, L. A. Rizzon, P. S. Sabaini, and M. B. A. Glória, "Influence of alcoholic and malolactic starter cultures on bioactive amines in Merlot wines," Food Chemistry, vol. 116, no. 1, pp. 208–213, 2009.

Chapter 13

DEVELOPMENT OF AN INNOVATIVE NUTRACEUTICAL FERMENTED BEVERAGE FROM HERBAL MATE (ILEX PARAGUARIENSIS A.ST.-HIL.) EXTRACT

Isabela Ferrari Pereira Lima [1], Juliano De Dea Lindner [1,2], Vanete Thomaz Soccol [3], José Luiz Parada [3] and Carlos Ricardo Soccol [1],

[1]Bioprocess Engineering and Biotechnology Division, Chemical Engineering Department, Federal University of Paraná, Curitiba, PR 81531-991, Brazil

[2]Research and Development Department, Incorpore Foods, Camboriú, SC 88340-000, Brazil

[3]Post Graduation Program, Positivo University, Curitiba, PR 81280-330, Brazil

ABSTRACT

Herbal mate (*Ilex paraguariensis* A.St.-Hil.) leaves are traditionally used for their stimulant, antioxidant, antimicrobial, and diuretic activity, presenting as principal components polyphenolic compounds. The aim of this work was to develop an innovative, non-dairy, functional, probiotic, fermented beverage using herbal mate extract as a natural ingredient which would also be hypocholesterolemic and hepatoprotective. Among different strains used,*Lactobacillus acidophilus* was selected as the best for fermentation. The addition of honey positively affected the development of *L. acidophilus* and the formulated beverage maintained microbial stability during shelf life. Key ingredients in the extract included xanthines, polyphenols and other antioxidants with potential health benefits for the consumer. Caffeine levels and antioxidant activity were also studied. Acceptable levels of caffeine and large antioxidant capacity were observed for the formulation when compared to other antioxidant beverages. An advantage of this product is the compliance to organic claims, while providing caffeine, other phyto-stimulants and antioxidant compounds without the addition of synthetic components or preservatives in the formulation. Sensorial analysis demonstrated that the beverage had good consumer acceptance in comparison to two other similar commercial beverages. Therefore, this beverage could be used as a new, non-dairy vehicle for probiotic consumption, especially by vegetarians and lactose intolerant consumers. It is expected that such a product will have good market potential in an era of functional foods.

INTRODUCTION

The herbal mate (*Ilex paraguariensis* A.St.-Hil.) is a tree of the Aquifoliaceae family naturally distributed in South America. Aqueous extract of *Ilex paraguariensis* is a typical antioxidant-containing beverage largely consumed in several South American countries. In the USA, herbal mate is listed as GRAS (generally recognized as safe) [1].

Herbal mate extract (HME) has received attention for its health benefits. Some studies have reported that herbal mate is hypocholesterolemic and hepatoprotective [2], a central nervous system stimulant, a diuretic and an antioxidant [1,3–7]. Other studies have suggested that numerous active phytochemicals identified may be responsible for its health benefits as well. The highest concentrations of bioactive compounds are polyphenols (chlorogenic acid) and xanthines (caffeine, theobromine and theophylline) [8], which are responsible for antioxidant [1,9] and central nervous system stimulant effects [6].

Because most strains of lactic acid bacteria (LAB) are considered generally-recognized-as-safe (GRAS), there has been a long history for their widespread use in fermented foods [10]. Members of the genus *Lactobacillus* have been shown to colonize the human gastrointestinal tract. Several species are considered to exhibit beneficial effects, such as antimicrobial activity by acidification and by the production of bacteriocins, which inhibit the growth of food deteriorating and poisoning bacteria [11], anti-carcinogenic activity and their ability to enhance the immune response [12]. Furthermore, Kaizu *et al.* [13] demonstrated that some *Lactobacillus* species possessed antioxidant activity and were able to reduce the accumulation of reactive oxygen species (ROS) during the ingestion of food.

Traditionally, the products proposed for gastrointestinal health have been more popular in Europe (current estimated market at USD 8 billion), Japan and USA, than Latin America, Africa and Asia. Latin American consumers have recently begun to increase the consumption of functional prebiotic and probiotic foods and beverages that, when consumed regularly, could lead to better gastrointestinal health. For example, in 2008, the Brazilian probiotic beverage market was USD 250 million, with an increase of 11%, compared to 2007, which corresponded to 0.6% of the entire food market [14,15]. The Brazilian Association of the Food Industry (ABIA) predicted an increase of 13% for 2010 in the functional foods market, compared to a 6% increase for the entire food market. Brazil is the eighth major market in the world for this type of beverage [15].

Nowadays, there are an increasing number of herbal mate products being developed. The main objective of this study was to develop a functional, innovative, non-dairy probiotic, fermented beverage using HME as a natural ingredient. HME and carbohydrate substrates were fermented by different strains of probiotic LAB, and the final formulation was tested for the presence of bioactive compounds during the extended shelf life. Sensorial comparisons with other commercial beverages containing herbal mate were also examined.

RESULTS AND DISCUSSION

Selection of the Formulation and LAB Strain

The first aspect of this study was to test and to select the best formulation of the herbal mate fermented probiotic beverage. Preliminary tests included changing HME concentrations and decoction conditions (data not shown). Using sensorial and physical criteria, 14 formulations were screened (Table 1) after fermentation using *Lactobacillus acidophilus*, *Lactobacillus sake*, *Lactobacillus casei* subsp. *rhamnosus* and *Lactobacillus casei*. After the analysis of appearance, bacterial development, precipitate formation, taste and odor, unacceptable formulations were eliminated and selected the formulation N (Table 1) fermented by *L. acidophilus*, which contained only HME and honey as the nitrogen and carbohydrate source. This formulation resulted in a sweet taste, moderate acidity, without bottom precipitate and was, thus, demonstrated to be a good substrate for the development of the probiotic strain *L. acidophilus* ATCC 4356.

Table 1: Formulations with different compositions of carbohydrate and nitrogen substrates.

Formulations	Carbohydrate and nitrogen substrates (w/v) a			
	Yeast extract	Honey	Malt extract	Sugar cane molasse
A	0.5	4	-	-
B	0.5	6	-	-
C	0.5	8	-	-
D	0.5	10	-	-
E	0.5	4	-	-
F	1.5	4	-	-
G	-	-	4	-
H	-	-	6	-
I	-	-	8	-
J	-	-	-	4
K	-	-	-	6
L	-	-	-	8
M	-	10	-	-
N	-	14	-	-

[a]percentage; - absent.

Probiotic Fermentation of the HME

Among LAB, *L. acidophilus* has attracted attention for its potential probiotic effects in human health [16–18] and ability to use prebiotic compounds [19]. When selecting a probiotic culture for use as a dietary adjunct in human food, a number of factors should be considered. One of these is the use of a microorganism originating from the human intestinal tract that exhibits host specificity [20]. Additional factors important in selecting a strain that can survive and grow in the intestinal tract include bile and temperature tolerance, adherence capability to the intestinal wall, cholesterol assimilation and antioxidant ability [18,21,22].

L. acidophilus ATTC 4356, a human isolate, is used as a dietary adjunct in various cultured dairy products. In the present work, strain ATCC 4356 was used to ferment a non-dairy formulation, and showed to not only meet the strict criteria of viability but also capable of surviving storage. *L. acidophilus* cultivability was evaluated during fermentation and following shelf-life. Figure 1 showed the growth of *L. acidophilus*, where it was demonstrated to grow exponentially during 10 h of fermentation. During the shelf life period, the microbial count remained similar and was estimated at 10^8 colony-forming units (CFU) mL^{-1}. The presence of sufficient numbers of viable bacterial cells (10^9 CFU) is necessary to provide therapeutic benefits [23]. Therefore, in order to call a product "probiotic", such as in a new beverage, the viability of probiotic bacteria must be maintained. The formulated beverage maintained microbial stability during the 28 days of shelf life at 4 °C. The stable logarithmic probiotic value of 10^8 CFU ml^{-1} found during the shelf life may prevent the development of deteriorative/poisoning microorganisms and deliver a sufficient concentration of probiotics to the consumer.

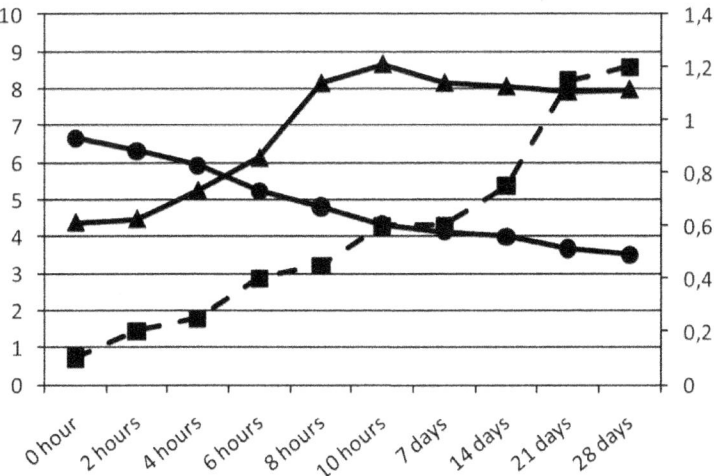

Figure 1: Changes in pH (●), growth trends of *Lactobacillus acidophilus* ATCC 4356 expressed as log colony-forming units (CFU) mL^{-1}(▲) and acidity expressed as mL of 0.1 N NaOH (secondary axis—dashed line ■) during different steps of fermentation and shelf life for the herbal mate beverage.

L. acidophilus was shown to easily ferment a substrate rich in glucose and fructose (honey) and to produce lactic and acetic acid (Figure 2) resistant to acidic conditions after fermentation. The successful use of *L. acidophilus* resulted in a complete fermentation process. The optimum fermentation period of 10 h was determined after longer fermentation tests (data not shown). pH and bacterial growth analyses after 12 h showed a substantial decrease of viability of the cells in an acidic condition at pH 3.67.

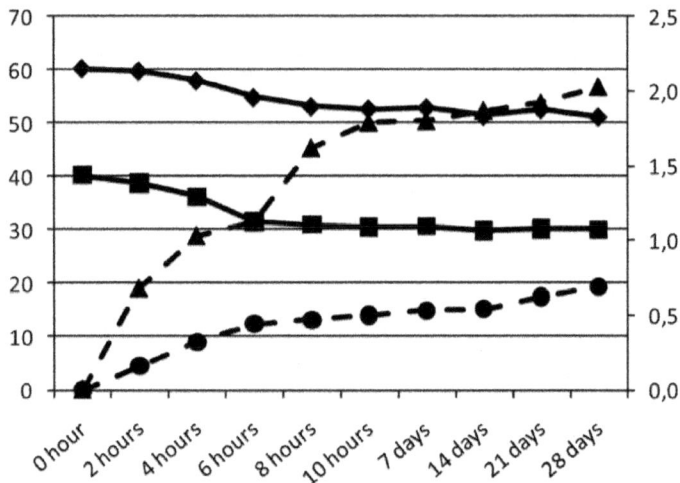

Figure 2: Sugar content (Glucose: solid line ■; Fructose: solid line ◆) and organic acids (secondary axis: Lactic acid—dashed line ▲; Acetic acid—dashed line ●) expressed as g L⁻¹ for formulations and beverage during fermentation and shelf life.

Table 2: pH and mean values obtained after duplicate assays for fermentation parameters, caffeine content and antioxidant activity of formulation during fermentation and beverage during shelf life.

	pH	Acidity a	Bacterial count b	Sugar c Glucose	Sugar c Fructose	Organic acid c Lactic	Organic acid c Acetic	Caffeine d	Antioxidant activity e
0 h	6.65	0.10	4.37 ± 3.12	40.2	60.2	0.00	0.00	nd	nd
2 h	6.30	0.20	4.46 ± 2.15	38.7	59.7	0.68	0.16	nd	nd
4 h	5.92	0.25	5.23 ± 3.94	36.2	57.9	1.03	0.32	nd	nd
6 h	5.23	0.40	6.11 ± 4.18	31.5	54.8	1.13	0.44	nd	nd
8 h	4.80	0.45	8.13 ± 3.24	30.9	53.1	1.62	0.47	nd	nd
10 h	4.30	0.60	8.64 ± 5.14	30.5	52.7	1.79	0.50	6.56	58 ± 2
12 h	3.67	0.70	7.69 ± 4.47	30.1	52.6	1.94	0.53	nd	nd
7 days	4.12	0.60	8.14 ± 4.56	30.6	52.9	1.80	0.53	6.70	56 ± 3
14 days	4.00	0.75	8.04 ± 2.98	29.8	51.4	1.87	0.54	6.82	55 ± 2
21 days	3.67	1.15	7.90 ± 3.35	30.1	52.6	1.92	0.62	6.68	59 ± 2
28 days	3.50	1.20	7.96 ± 4.16	30.0	51.2	2.03	0.69	6.70	58 ± 1

[a]mL 0.1 N NaOH;[b]logarithmic bacterial counts expressed as log CFU mL⁻¹ (±SD);[c]g L⁻¹;[d]mg 100 mL⁻¹;[e]% reduction DPPH (±SD), percentage inhibition defined as [(A517 blank − A517 sample) A517 blank⁻¹] 100⁻¹ (%); nd: not determined.

Virtanen *et al.* [16] and Lin and Chang [24] demonstrated that *L. acidophilus* ATCC 4356 had the ability to scavenge DPPH free radicals. In the present study, antioxidant activities measured directly in the beverage varied from 51 to 62% (Table 2). No positive trend was observed for the confirmation of summation activity from bacterial or lysed cells in the HME.

Caffeine Content

HPLC analysis for caffeine content in the formulation for 10 h and in the beverage during shelf life indicated almost the same quantitative composition. The average values for each sample are presented in Table 2, based on chromatography using a caffeine standard.

The concentration of caffeine in relation to consumer consumption has been found to be approximately 6.7 mg in one regular dose (100 mL) of beverage (Table 2). Comparing coffee and herbal mate tea (approximately 56 and 52 mg per dose, respectively) [25] with this beverage, 6.7 mg of caffeine was detected, a low amount for ingestion of this alkaloid. Caffeine is one of the most studied alkaloids in terms of physiological effects in human beings [26,27]. Some physiological effects associated with caffeine include central nervous system stimulation, acute elevation of blood pressure, increased metabolic rate, and diuresis [28]. The values observed in the present product indicated that consumption of it provided a little intake of caffeine, an amount so small that not affect consumer preference.

Antioxidant Activity

The consumption of herbal mate significantly contributed to the overall antioxidant intake with potentially beneficial biological effects for human health [5]. A number of therapeutic applications have been claimed for herbal mate infusions. *I. paraguariensis* was shown to contain high antioxidant activity which positively correlated with the concentration of caffeoyl-derivatives [1,3]. Newell *et al.* [29] demonstrated that herbal mate tea possessed a much higher antioxidant capacity than green tea. According to Bixby *et al.* [3], herbal mate tea showed greater inhibition against cytotoxicity compared to green tea and or red wine. The ability to quench reactive oxygen species (responsible for cell structure damage after environmental oxidative stress) was examined and correlated to peroxidase-like activity which was related to the polyphenol concentration in herbal mate extract [30]. Higher polyphenol concentrations demonstrated greater peroxidase-like activity [31].

Quenching of the DPPH free radical by the formulations used in this study is shown in Table 2. The formulation during fermentation and beverage during shelf life showed similar amounts of antioxidant activity, approximately 56%. Comparison of the beverage with the study conducted by Ramadan-Hassanien [32] found that the anti-radical performance towards DPPH radicals of the fermented herbal mate beverage was just below those of mango and red grape juice, both rich in phenolic compounds. In other antioxidant beverages, such as hot drinks, especially those rich in caffeine [32], the product demonstrated a large antioxidant capacity, slightly below that of tea with lemon, green tea, black tea, soluble coffee and Turkish coffee.

Sensorial Analysis of the Beverage

The acceptance level for the probiotic and herbal mate Tea beverages is presented in Table 3 and supported by variance analysis (Table 4). Results for the sensorial analysis are expressed as a mean value. The mean value of some attributes presented significant differences and could be observed in the frequency distribution for the values of acceptability factors shown in Figure 3. Sweetness, acidity and astringency attributes for all beverages showed a similar acceptance by the testers. Significant differences for transparency and precipitation were expected when comparing fermented to an unfermented beverage, and these differences did not affect the final product acceptance (AF 71 and 82%, respectively).

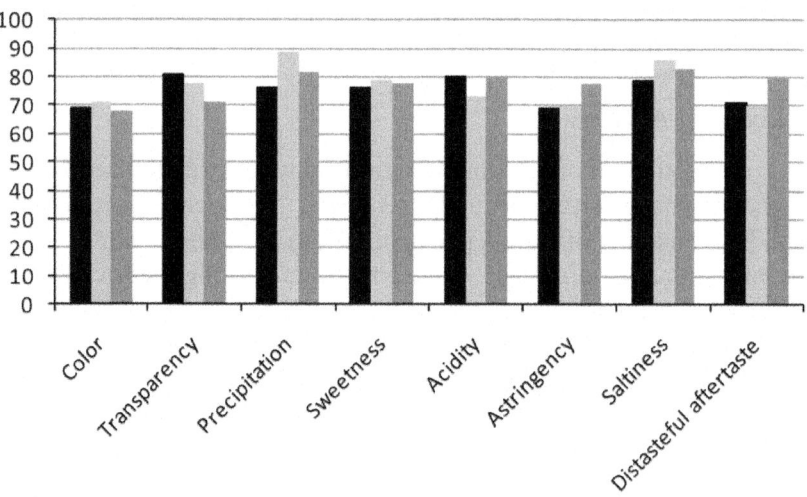

Figure 3: Acceptability factors in percentage (*Y* axis) and attributes (*X* axis) for the herbal mate beverage (black bar) and commercial mate tea A (dark grey bar) and B (pale grey bar).

According to Dutcosky [33], AF ≥ 70% represented good acceptability for the attribute analyzed in a sensorial analysis (Figure 3), and excluding color that presented an AF of 68%, this beverage presented a minimum AF of 71%.

Table 3: Acceptance (average values ± SD) and acceptability factors (AF) for the herbal mate beverage and commercial mate tea A and B.

Attributes (AF)	Commercial Mate Tea A	Commercial Mate Tea B	Probiotic Mate Beverage
Color	6.2 ± 1.4 a (69)	6.4 ± 1.3 a (71)	5.4 ± 1.3 b (68)
Transparency	7.3 ± 1.3 a (81)	7.0 ± 1.2 a (78)	5.0 ± 1.2 b (71)
Precipitation	7.8 ± 1.1 a (76)	8.0 ± 1.1 a (89)	4.9 ± 1.3 b (82)
Sweetness	6.1 ± 1.4 a (76)	6.3 ± 1.3 a (79)	5.5 ± 1.4 a (78)
Acidity	6.4 ± 1.1 a (80)	6.6 ± 1.3 a (73)	5.6 ± 1.0 a (80)
Astringency	6.2 ± 1.7 a (69)	6.3 ± 1.7 a (70)	5.5 ± 1.3 a (78)
Saltiness	7.1 ± 1.1 a (79)	6.9 ± 1.0 ab (86)	5.8 ± 1.0 b (83)
Distasteful aftertaste	6.4 ± 1.3 a (71)	6.3 ± 1.5 b (70)	6.4 ± 1.2 a (80)

a,b Means with the same letter in the same line are not significantly different ($P < 0.05$), according to ANOVA and Tukey's test.

Table 4: Variance analysis (ANOVA) for the sensorial test of beverages.

Source of variation	Degree of freedom	Sum of squares	Mean of squares	F ratio
Color	2	5.6000	2.8000	1.7571 *
Transparency	2	31.2666	15.6333	9.8106 **
Precipitation	2	60.2000	30.1000	18.8890 **
Sweetness	2	3.4666	1.7333	1.0877 ns
Acidity	2	5.6000	2.8000	1.7571 ns
Astringency	2	3.8000	1.9000	1.1923 ns
Saltiness	2	9.8000	4.9000	3.0750 *
Distasteful aftertaste	2	0.0666	0.0333	0.0209 *
Residue	216	344.2000	1.5935	
Total	232	464.0000		

** 1% significance ($P < 0.01$); * 5% significance ($0.01 \leq P < 0.05$); ns not significant ($P \geq 0.05$).

For an overall evaluation, the average values for the sensory analysis and acceptability factors were kept within the acceptance range. No emergent alteration in the values at the final storage period (28 days) was found when compared to the product after fermentation (data not shown). Therefore, product acceptance could be considered as good and comparable to that of commercial beverages available at the local market.

EXPERIMENTAL SECTION

Preparation of the HME

Roasted, milled and sifted leaves and stems of herbal mate (*Ilex paraguariensis* A.St.-Hil.) were obtained commercially from the local markets in Santa

Catarina State (Brazil). HME was prepared by decoction of 15 g in 300 mL of water at 95 °C for 3 min. The extract was filtered under vacuum using a filter paper (Whatman, UK).

HME Formulation and LAB Strains

The formulations were composed by 300 mL of HME and one or two carbohydrate substrates in the proportions described in the Table 1. Honey (Mel de Abelha Superbom, São Paulo, Brazil), malt extract (Acumedia, USA), yeast extract (Acumedia, USA) and sugar cane molasses (local bioethanol producing company) were used with the aim to prepare formulations for the LAB fermentation.

L. acidophilus ATCC 4356, *L. sake* ATCC 15521, *L. casei* ATCC 393 and *L. casei* subsp. *rhamnosus* DEBB H-19 (obtained from the Culture Bank of the Bioprocess Engineering and Biotechnology Division, Federal University of Paraná, Brazil) were used in this work. They were grown overnight in MRS (De Man, Rogosa and Sharpe) broth (Acumedia, USA) at 35 °C. The strains were inoculated at a standard concentration of 10^6 colony-forming units (CFU) mL^{-1} in 300 mL (3%) of the formulations indicated in Table 1.

pH and Acidity

The pH was measured directly using a model HI9321 pH meter (Hanna Instruments, Portugal).

Acidity was measured by titrimetric analysis using the fermented broth diluted 1:50 (v/v) in distilled water. The diluted solution was neutralized with 0.1 M NaOH (Merck, Germany) using an alcoholic solution of phenolphthalein (1% v/v) as an indicator.

Sugars and Organic Acids Concentration

The fermented broths were diluted 1:10 in MilliQ® water and filtered using a 0.22 µm pore filter (Millipore, UK). Sugars (glucose and fructose) and organic acids (lactic and acetic) content were determined by HPLC. Chromatographic analyses were performed using a LC-10AD chromatographic system (Shimadzu, Japan) equipped with a model RID-10A detector (Shimadzu, Japan) and an Aminex HPX-87H cation exchange column (300 mm × 7.8 mm i.d., Bio-Rad Laboratories, CA, USA). Analyses were performed using filtered, degassed 5 mM reagent grade H_2SO_4 (Carlo Erba, Italy) as the mobile phase at a flow rate of 0.6 mL·min^{-1}. Eluates were monitored at 215 nm. Calibration curves were obtained by preparing a standard mix of the sugars and organic acids (Sigma, USA). The resulting peaks area were calculated for duplicate

25 µL injections and plotted against concentration. Each assay was carried out in duplicate and the average values expressed in $g \cdot L^{-1}$ for glucose, fructose, lactate and acetic acid during fermentation were calculated. The estimated error for the chromatographic assays was less than 3%.

LAB Count and Fermentation

Formulations were serially diluted tenfold in $0.05 \ mol \cdot L^{-1}$ sodium citrate (Sigma, USA) buffer, pH 7.5. In order to quantify the cultivable LAB population, MRS agar (Acumedia, USA) was used and incubated at 35 °C for 48 h under anaerobic conditions. Bacterial counts were carried out in triplicate and the standard deviation of mean values was calculated. The estimated error was less than 10%.

After fermentation, the herbal mate probiotic beverage was stored in Scotch flasks at 4 °C during 28 days and the parameters as above were analyzed after 0, 7, 14, 21 and 28 days.

Caffeine Content

The caffeine content, present in the beverage during shelf life, was analyzed using an Agilent 1100 HPLC system (Hewlett Packard, Inc., USA) with a diode array detector (DAD). Chromatographic separation was accomplished using a Zorbax RX Reversed-Phase C18 column, 150 mm × 4.6 mm (Agilent, USA). An isocratic system (MeOH-H_2O 30:70) was used as the mobile phase at a flow rate of $1.0 \ mL \cdot min^{-1}$ at 35 °C. Detection was performed at 272 nm. Broth and beverage were both diluted (5 mL in 250 mL MilliQ water), filtered through 0.22 µm (Millipore, UK) and injected in triplicate (CV < 5%). Peak areas were compared to standard curves. Suitable amounts of caffeine standard were dissolved in MeOH-H_2O (4:6). Standard solutions were injected in triplicate and peak areas were measured. Linearity was evaluated by linear regression analysis, and the precision and accuracy were determined by the coefficient of variation (CV < 4%). The correlation coefficient was $r = 0.988$.

Antioxidant Activity

The antioxidant activity (free radical scavenging capacity) present in the beverage, during the shelf life, was examined by the reduction of the 1, 1-diphenyl-2-picrylhydrazyl radical (DPPH) as described by Ramadan *et al.* [34] with modifications. Five millilitres of a $20 \ mg \cdot mL^{-1}$ DPPH solution in methanol were added to 5 mL of a methanolic solution of the fermented beverage (1:10, v/v). Absorbance was determined spectrophotometrically (UV-1601PC Spectrophotometer, Shimadzu, Japan) at 515 nm after 30 min., and scavenging activity was calculated as percent of radical reduction. The percent of inhibition was defined as $[(A517 \text{ blank} - A517 \text{ sample}) A517 \text{ blank}^{-1}]100^{-1}$ (%). Ascorbic acid was used as a reference compound. The analysis was carried out in triplicate and the standard deviation of mean values was calculated.

Sensorial Analysis

Sensorial evaluation of the beverage was performed on the product after 28 days of shelf-life. Sensorial characteristics of the beverage were compared with two commercial herbal mate lemon flavored teas. The commercial products were purchased from a local market. Mate teas A and B correspond to two different commercial brands.

Sensorial hedonic test of the beverage was carried out by a group of 15 non-trained testers who judged the color, transparency, precipitate formation, sweetness, acidity, astringency, saltiness and distasteful after taste using a hedonic rating scale from 1 to 9 (1: dislike extremely; 2: dislike very much; 3: dislike moderately; 4: dislike slightly; 5: neither like nor dislike; 6: like slightly; 7: like moderately; 8: like very much; 9: like extremely). Sensory tests were performed in individual booths, under white light during the morning shift (9:00 a.m.–11:30 a.m.). Refrigerated (5 °C) samples were served in transparent glass cups. The data obtained were analyzed by ANOVA and Tukey's test according to Monteiro [35] using Assistat version 7.5 software (Assistat, Brazil).

To verify the acceptability of the tested beverages, an acceptability factor (AF) [33] was calculated as the criteria to evaluate each sensorial attribute analyzed:

$$AF = A \cdot 100 \cdot B - 1$$

where A is the mean value obtained for each attribute; B is the maximum value ascribed by the testers for each attribute.

CONCLUSIONS

The interest in herbal mate has risen in recent years because it has shown extraordinary possibilities not only as a consumer beverage but also as a nutraceutical product in the novel food industry. The present work demonstrated a process for preparing a non-dairy, probiotic, fermented beverage from HME. The base formulation ingredients incorporated xanthines, polyphenols and other antioxidant components that have a potential for healthy consumer benefits. The innovative probiotic beverage contained sufficiently high viable counts of *L. acidophilus*. The addition of honey positively affected the development of *L. acidophilus*, and the formulated beverage maintained microbial stability during storage with a stable logarithmic probiotic value of CFU mL^{-1} during shelf life. Acceptable levels of caffeine and large antioxidant capacity were observed for the formulation when compared to other antioxidant beverages. An advantage of this product is the compliance to organic claims, while providing caffeine, other phyto-stimulants and antioxidant compounds without the addition of synthetic components or preservatives in the formulation. Sensorial analysis demonstrated that the beverage had good consumer acceptance in comparison to two other commercial beverages. This beverage can be used as a new, non-dairy vehicle for probiotic consumption, especially by vegetarians and lactose intolerant consumers. The beverage product could have an excellent market potential in the current era of new functional foods.

ACKNOWLEDGMENTS

Authors are thankful to Capes and CNPq for financial support for this work.

REFERENCES

1.   Berté, K.A.; Beux, M.R.; Spada, P.K.; Salvador, M.; Hoffmann-Ribani, R. Chemical composition and antioxidant activity of yerba-mate (*Ilex paraguariensis* A.St.-Hil., Aquifoliaceae) extract as obtained by spray drying. *J. Agric. Food Chem* **2011**, *59*, 5523–5527.

2.   Filip, R.; Lopez, P.; Giberti, G.; Coussio, J.; Ferraro, G. Phenolic compounds in seven South American *Ilex* species. *Fitoterapia* **2003**, *72*, 774–778.

3.   Bixby, M.; Spieler, L.; Menini, T.; Gugliucci, A. *Ilex paraguariensis* extracts are potent inhibitors of nitrosative stress: A comparative study with green tea and wines using nitration model and mammalian cell cytotoxicity. *Life Sci* **2005**, *77*, 345–358.

4.   Lunceford, N.; Gugliucci, A. *Ilex paraguariensis* extracts inhibit AGE formation more efficiently than green tea. *Fitoterapia* **2005**, *76*, 419–427.

5.   Bravo, L.; Goya, L.; Lecumberri, L. LC/MS characterization of phenolic constituents of mate (*Ilex paraguariensis*, St. Hil.) and its antioxidant activity compared to commonly consumed beverages. *Food Res. Int* **2007**, *40*, 393–405.

6.   Heck, C.I.; Demejia, E.G. Yerba Mate Tea (*Ilex paraguariensis*): A comprehensive review on chemistry, health implications, and technological considerations. *J. Food Sci* **2007**, *72*, 138–151.

7.   Lanzetti, M.; Bezerra, F.S.; Romana-Souza, B.; Brando-Lima, A.C.; Koatz, V.L.; Porto, L.C.; Valenca, S.S. Mate tea reduced acute lung

inflammation in mice exposed to cigarette smoke.*Nutrition* **2008**, *24*, 375–381.

8. Athayde, M.; Coelho, G.C.; Schenkel, E.P. Caffeine and theobromine in epicuticular wax of*Ilex paraguariensis*. *Phytochemistry* **2000**, *55*, 853–857.

9. Chandra, S.; DeMejia Gonzalez, E. Polyphenolic compounds, antioxidant capacity, and quinone reductase activity of an aqueous extract of *Ardisia compressa* in comparison to mate (*Ilex paraguariensis*) and green (*Camellia sinensis*) teas. *J. Agric. Food Chem* **2004**, *52*, 3583–3589.

10. Lee, Y.K.; Nomoto, K.; Salminen, S.; Gorbach, S.L. Safety of Probiotic Bacteria. In *Handbook of Probiotics*; John Wiley & Sons: New York, NY, USA, 1999; pp. 17–21.

11. Stiles, M.E. Biopreservation by lactic acid bacteria. *Antonie Van Leeuwenhoek* **1996**, *70*, 331–345.

12. Naaber, P.; Mikelsaar, M. Interactions between *Lactobacilli* and antibiotic-associated diarrhea.*Adv. Appl. Microbiol* **2004**, *54*, 231–260.

13. Kaizu, H.; Sasaki, M.; Nakajima, H.; Suzuki, Y. Effect of antioxidative lactic acid bacteria on rats fed a diet deficient in vitamin E. *J. Dairy Sci* **1993**, *76*, 2493–2499.

14. Associação Brasileira das Indústrias da Alimentação (ABIA). *Mercado Brasileiro dos alimentos industrializados*, 2009. Available online: http://www.abia.org.br/ecopublall17.asp accessed on 3 March 2009.

15. Nielsen Company. Relatórios Executivos de Notícias. *Os Produtos Mais Quentes do Mundo: Informações sobre Categorias de Alimentos & Bebidas*, 2009. Available online: http://br.nielsen.com/reports/reportesejecutivosglobales.shtml accessed on 15 March 2009.

16. Virtanen, T.; Pihlanto, A.; Akkanen, S.; Korhonen, H. Development of antioxidant activity in milk whey during fermentation with lactic acid bacteria. *J. Appl. Microbiol* **2007**, *102*, 106–115.

17. Chen, X.; Xu, J.; Shuai, J.; Chen, J.; Zhang, Z.; Fang, W. The S-layer proteins of *Lactobacillus crispatus* strain ZJ001 is responsible for competitive exclusion against *Escherichia coli* O157:H7 and *Salmonella typhimurium*. *Int. J. Food Microbiol* **2007**, *115*, 307–312.

18. Klaenhammer, T.R.; Altermann, E.; Pfeiler, E.; Buck, B.L.; Goh, Y.J.; O'Flaherty, S.; Barrangou, R.; Duong, T. Functional genomics of probiotic lactobacilli. *J. Clin. Gastroenterol* **2008**, *42*, S160–S162.

19. Barrangou, R.; Altermann, E.; Hutkins, R. Functional and comparative genomic analyses of an operon involved in fructooligosaccharide

utilization by *Lactobacillus acidophilus*. *Proc. Natl. Acad. Sci. USA* **2003**, *100*, 8957–8962.

20. Gilliland, S.E.; Walker, D.K. Factors to consider when selecting a culture of *Lactobacillus acidophilus* as a dietary adjunct to produce a hypocholesterolemic effect in humans. *J. Dairy Sci* **1999**, *73*, 905–911.

21. Percival, M. Choosing a probiotic supplement. *Clin. Nutr. Insights* **1997**, *6*, 1–4.

22. De Dea Lindner, J.; Canchaya, C.; Zhang, Z.; Neviani, E.; Fitzgerlad, G.F.; Sinderen, D.V.; Ventura, M. Exploiting Bifidobacterium genomes: The molecular basis of stress response. *Int. J. Food Microbiol* **2007**, *120*, 13–24.

23. Reid, G.; Beuerman, D.; Heinemann, C.; Bruce, A.W. Probiotic *Lactobacillus* dose required to restore and maintain a normal vaginal flora. *FEMS Immunol. Med. Microbiol* **2001**, *32*, 37–41.

24. Lin, M.; Chang, F. Antioxidative effect of intestinal bacteria *Bifidobacterium longum* ATCC 15708 and *Lactobacillus acidophilus* ATCC 4356. *Dig. Dis. Sci* **2000**, *45*, 1617–1622.

25. Mazzaferra, P. Mate drinking: Caffeine and phenolic acid intake. *Food Chem* **1997**, *60*, 67–71. [

26. Higdon, J.V.; Frei, B. Coffee and health: A review of recent human research. *Crit. Rev. Food Sci. Nutr* **2006**, *46*, 101–123.

27. Satel, S. Is caffeine addictive?—A review of the literature. *Am. J. Drug Alcohol Abuse* **2006**, *32*, 493–502.

28. Carrillo, J.A.; Benitez, J. Clinically significant pharmacokinetic interactions between dietary caffeine and medications. *Clin. Pharmacokinet* **2000**, *39*, 127–153.

29. Newell, A.M.B.; Chandra, S.; Gonzalez de Mejia, E. Ethnic Teas and Their Bioactive Components. In *Hispanic Foods: Chemistry and Flavor*; Tunick, M.H., de Mejia, E.G., Eds.; American Chemistry Society: Washington, DC, USA, 2007; pp. 127–131.

30. Schinella, G.R.; Troiani, G.; Davila, V.; de Buschiazzo, P.M.; Tournier, H.A. Antioxidant effects of an aqueous extract of *Ilex paraguariensis*. *Biochem. Biophys. Res. Commun* **2000**, *269*, 357–360.

31. Anesini, C.; Ferraro, G.; Filip, R. Peroxidase-like activity of *Ilex paraguariensis*. *Food Chem* **2006**,*97*, 459–464.

32. Ramadan-Hassanien, M.F. Total antioxidant potential of juices, beverages and hot drinks consumed in Egypt screened by DPPH *in vitro* assay. *Grasas y Aceites* **2008**, *59*, 254–259.

33. Dutcosky, S.D. *Análise Sensorial de Alimentos*; Champagnat: Curitiba, Brazil, 1996.

34. Ramadan, M.F.; Kroh, L.W.; Moersel, J.T. Radical scavenging activity of black cumin (*Nigella sativa* L.), coriander (*Coriandrum sativum* L.) and niger (*Guizotia abyssinica* Cass.) crude seed oils and oil fractions. *J. Agric. Food Chem* **2003**, *51*, 6961–6969.

35. Monteiro, C.L.B. *Técnicas de Avaliação Sensorial*, 2nd ed; UFPR/CEPPA: Curitiba, Brazil, 1984; p. 101.

CITATION

CHAPTER 1

G Spano, P Russo, A Lonvaud-Funel, P Lucas, H Alexandre, C Grandvalet, E Coton, M Coton, L Barnavon, B Bach, F Rattray, A Bunte, C Magni, V Ladero, M Alvarez, M Fernández, P Lopez, P F de Palencia, A Corbi, H Trip and J S Lolkema, Biogenic amines in fermented foods, doi:10.1038/ejcn.2010.218

CHAPTER 2

Emiliane Andrade Araújo, Ana Clarissa dos Santos Pires, Maximiliano Soares Pinto, Gwénaël Jan and Antônio Fernandes de Carvalho (2012). Probiotics in Dairy Fermented Products, Probiotics, Prof. Everlon Rigobelo (Ed.), ISBN: 978-953-51-0776-7, InTech, DOI: 10.5772/51939.

CHAPTER 3

Takeuchi M, Takino J-i, Furuno S, Shirai H, Kawakami M, Muramatsu M, et al. (2015) Assessment of the Concentrations of Various Advanced Glycation End-Products in Beverages and Foods That Are Commonly Consumed in Japan. PLoS ONE 10(3): e0118652. doi:10.1371/journal.pone.0118652

CHAPTER 4

Tendekayi H. Gadaga , Molupe Lehohla , VictorNtuli, traditional fermented foods of Lesotho, http://www.jmbfs.org/wp-content/uploads/2013/05/jmbfs-0241-gadaga.pdf

CHAPTER 5

Fábio Faria-Oliveira, Raphael H.S. Diniz, Fernanda Godoy-Santos, Fernanda B. Piló, Hygor Mezadri, Ieso M. Castro and Rogelio L. Brandão (2015). The Role of Yeast and Lactic Acid Bacteria in the Production of Fermented Beverages in South America, Food Production and Industry, Prof. Ayman Amer Eissa (Ed.), ISBN: 978-953-51-2191-6, InTech, DOI: 10.5772/60877.

CHAPTER 6

Anne Pihlanto (2013). Lactic Fermentation and Bioactive Peptides, Lactic Acid Bacteria - R & D for Food, Health and Livestock Purposes, Dr. J. Marcelino Kongo (Ed.), ISBN: 978-953-51-0955-6, InTech, DOI: 10.5772/51692.

CHAPTER 7

Ho-Sang ShinEmail author and Eun-Young Yang, Simultaneous determination of methylcarbamate and ethylcarbamate in fermented foods and beverages by derivatization and GC-MS analysis, DOI: 10.1186/1752-153X-6-157

CHAPTER 8

Ingrid Torres-Rodríguez, María Elena Rodríguez-Alegría, Alfonso Miranda-Molina, Martha Giles-Gómez, Rodrigo Conca Morales, Agustín López-Munguía, Francisco Bolívar and Adelfo Escalante, Screening and characterization of extracellular polysaccharides produced by Leuconostoc kimchii isolated from traditional fermented pulque beverage, DOI: 10.1186/2193-1801-3-583

CHAPTER 9

Hideki Okada, Eri Fukushi, Akira Yamamori, Naoki Kawazoe, Shuichi Onodera, Jun Kawabata and Norio Shiomi, structural analysis of three novel trisaccharides isolated from the fermented beverage of plant extracts, DOI: 10.1186/1752-153X-3-8

CHAPTER 10

A. El-Abasy, H. Abou-Gharbia, H. Mousa and M. Youssef, "Mixes of Carrot Juice and Some Fermented Dairy Products: Potentiality as Novel Functional Beverages," Food and Nutrition Sciences, Vol. 3 No. 2, 2012, pp. 233-239. doi: 10.4236/fns.2012.32034.

CHAPTER 11

Spitaels F, Wieme AD, Janssens M, Aerts M, Daniel H-M, Van Landschoot A, et al. (2014) The Microbial Diversity of Traditional Spontaneously Fermented Lambic Beer. PLoS ONE 9(4): e95384. doi:10.1371/journal.pone.0095384

CHAPTER 12

Albert Mas, Jose Manuel Guillamon, Maria Jesus Torija, et al., "Bioactive Compounds Derived from the Yeast Metabolism of Aromatic Amino Acids during Alcoholic Fermentation," BioMed Research International, vol. 2014, Article ID 898045, 7 pages, 2014. doi:10.1155/2014/898045

CHAPTER 13

Lima, Isabela Ferrari Pereira et al. "Development of an Innovative Nutraceutical Fermented Beverage from Herbal Mate (Ilex Paraguariensis A.St.-Hil.) Extract." International Journal of Molecular Sciences 13.1 (2012): 788–800. PMC. Web. 16 Mar. 2016.

INDEX